Evolution and the Big Questions

For my students

Evolution and the Big Questions

Sex, Race, Religion, and Other Matters

David N. Stamos

Blackwell
Publishing

BLACKWELL PUBLISHING
350 Main Street, Malden, MA 02148-5020, USA
9600 Garsington Road, Oxford OX4 2DQ, UK
550 Swanston Street, Carlton, Victoria 3053, Australia

First published 2008 by Blackwell Publishing Ltd

1   2008

Library of Congress Cataloging-in-Publication Data

Stamos, David N., 1957–
  Evolution and the big questions : sex, race, religion, and other matters/David N. Stamos.
    p. cm.
  Includes bibliographical references and index.
  ISBN 978-1-4051-4902-0 (hardcover : alk. paper) — ISBN 978-1-4051-4903-7 (pbk. : alk. paper) 1. Evolution. I. Title.

  B818.S823 2008
  116–dc22
                        2007024791

A catalogue record for this title is available from the British Library.

Set in 10/12.5 pt Classical Garamond
by Prepress Projects Ltd, Perth, Scotland
Printed and bound in Singapore
by Utopia Press Pte Ltd

For further information on
Blackwell Publishing, visit our website at

www.blackwellpublishing.com

Contents

Acknowledgments

Special thanks to Blackwell's Senior Editor Jeff Dean, who suggested I write this book (I had originally proposed an anthology) and who provided extensive and quite perceptive feedback on every chapter; to Prepress Projects' Production Editor Catriona Vernal; to my mentor, friend, and colleague David M. Johnson and my late mentor and friend Robert H. Haynes; to George C. Williams, R. C. von Borstel, Alex Levine, and David Shaner; to my students to whom I taught Mind and Nature, in particular Adriana Iannozzi, Atessa and Mahsa Izadpanah, Robert Curtis, and Andreea Diaconescu; to Sharon Weltman Fixler, who went over the entire manuscript before its final submission; and last but not least to my mate Sandra Javadi. I thank you one and all for your help, advice, and encouragement.

Introduction

There is a debate raging in virtually every college and university in the Western world, and also widely among the public. It is whether evolutionary explanations—Darwinian explanations—can be legitimately extended to the big questions that vitally concern us all, questions that fall outside of biology as normally circumscribed. The big questions concern matters between the sexes, racial issues, religion, and so much more. The debate as a whole is the interdisciplinary question *par excellence*, involving not only biology but philosophy, psychology, anthropology, sociology, feminism, theology, and virtually every other discipline in one way or another. This unique and timely book is devoted entirely to that debate, as a critical introduction. Both its content and its style were written for two main audiences, one the general public, the other students in college and undergraduate university courses in a variety of disciplines. I have also not refrained from developing my own views in every chapter, not only to provoke thought on the part of the reader but to challenge the heavyweights in the various fields. This book operates, then, on a number of levels. But more on all of this below, including chapter summaries.

What this book is *not* devoted to is a defense of the science of evolutionary biology *per se*. That debate is dead among scientists and the intellectual world as a whole. Beginning with Charles Darwin's *On the Origin of Species*, first published in 1859, the evidence for evolution has grown exponentially, such that evolutionary biology long ago became the core and foundation of professional biology. To be sure, there are theories and debates about various aspects of evolution by professional biologists and philosophers of biology (biology would not be a healthy science were it otherwise). But the debates are not about evolution *per se*. Instead, all the debates occur within the framework of evolution as a fact, much the same way debates in modern astronomy occur within the framework of a dynamic rather than a static universe.

The fact is, all professional biological research around the world, in every country that has institutions of professional science—whether that research is on animal behavior, ecology, agriculture, medicine, genetics, or the fossil record—is conducted from an evolutionary point of view. The explanatory power of evolutionary principles is enormous, and that is putting it much too mildly. Evolutionary principles not only sufficiently explain what we find in the biological world, but they alone allow us to predict new findings and to understand the mysteries of life in their manifold diversity. Indeed every new finding by biologists, every new discovery, fits perfectly within the evolutionary framework begun by Darwin. From the changes and transitional forms studied in the fossil record, the geographical distribution of plants and animals, their anatomical relationships, the study of DNA and related mechanisms, to the study of the mutation and evolution of viruses and the evolutionary resistance of bacteria to antibiotics, "Nothing in biology makes sense except in the light of evolution." This is what the renowned geneticist Theodosius Dobzhansky, one of the principal architects of the Modern Synthesis, once wrote in defense of the teaching of evolution in public school science classes (Dobzhansky 1973, 125).[1] At the center of it all is natural selection, Darwin's main mechanism of evolutionary change and his only proposed explanation for the existence of biological adaptations such as beaks and eyes. This mechanism has now been studied and confirmed over and over again in the lab and in the wild, and it remains the core causal explanation of biological adaptation, of complexity and design in organisms.

In short, evolutionary science is one of the greatest and most solid of human achievements, possibly even the greatest of all time. As such, it should be denied to no one. But what is worse, to deny evolution is to deny the very nature and value of evidence itself. Reasoning that is not based on evidence, that ignores it or even fights against it, is reasoning that invites moral condemnation. We would hold a judge or jury in contempt were they to decide court cases on emotions and ideologies rather than on evidence. The offense becomes only worse for the big questions in life. As W. K. Clifford (1879) argued over a century ago, we have both a personal and a social duty to avoid belief unsupported by or opposed to evidence, just as we have both a personal and a social duty to avoid the spread of disease. Disrespect for evidence translates psychologically and socially into a culture of lies and power politics, not a culture that values truth and justice.

The public perception of the status of evolutionary biology, unfortunately, is in many parts of the world quite the opposite of what it should be. One often used to hear of *creation science*, especially of court cases in which special interest groups challenged the scientific status of evolutionary biology or attempted to have creation science taught in public schools as a rival to evolutionary biology. Recently it has reemerged under a new guise known as *intelligent design theory*. Is not evolutionary biology just a competing theory after all? The answer is clearly *no*, for reasons already given and further elucidated in a number of places in this book. Nor is intelligent design theory genuine science. This theory, along with its previous incarnation as creation science, is essentially mythological thinking masquerading in a lab coat. It is the attempt to take a way of thinking common to frightened and ignorant peoples living in pre-scientific societies, a way of thinking possibly rooted deeply in human nature, and to make it intellectually respectable. But no matter how it is dressed, its explanations are not real explanations, it makes no testable predictions (because

---

1   Key terms that are not defined in the text, or that are defined but occur more than once, will find a discussion in the Glossary at the back. It might be a good idea for those not familiar with key terms in biology and philosophy to read the Glossary first, before reading the chapters.

one cannot test the will of an invisible creator or designer), and it opens up no fruitful lines of research. In short, the public has been bamboozled by an enormous propaganda machine driven by the religious right wing. Debate over intelligent design theory does not exist within science itself, and for very good reasons. In all of this the public has been seriously misled. For more on this topic there is the Appendix, which includes plenty of references as well as a discussion on the main misconceptions about evolution. This is needed for removing obstacles in the minds of many readers and thus preparing them for the real debates, the debates that are alive and well and are being fought within one college and university after another.[2]

To return to the purpose of this book, then, it is to accomplish something far more interesting than to argue for the truth of evolution. That is something that has been done over and over again and need not be repeated here. Instead, the purpose of this book is to question whether and to what extent evolutionary biology shines light on the big questions debated in the humanities and social sciences, questions that concern us all. Adding evolution to those questions has the effect of making them controversial in the extreme. The philosopher Daniel Dennett (1995), for example, has argued that Darwinian evolution is what he, approvingly, calls a "universal acid" (63), corroding its way through our cherished beliefs in virtually all areas of life, such as ethics, politics, romance, and religion. Certainly one can choose one's metaphor here, Dobzhansky's light or Dennett's acid or whatever else. Whatever one's metaphor, it remains an extremely interesting question just how far evolutionary explanations can legitimately go.

The main theme that runs throughout the chapters is the debate between evolutionary explanations and what has come to be known as the *Standard Social Science Model* (SSSM). The SSSM is a way of looking at human nature that is commonly found in sociology, behaviorism in psychology, cultural anthropology, Marxism, women's studies, and gay studies. There is a danger in presenting this debate as a dichotomy, since in recent years some of these fields (particularly anthropology and psychology) have in some of their quarters been warming up to evolutionary explanations, while evolutionists on human nature have been paying more attention to the role of the environment. And yet the debate has not changed so much that the distinction between the two competing models no longer holds. Each model is alive and well and competing for allegiance, the social sciences and the humanities still have much of the SSSM in them, and the debates with evolutionists are as hot as ever. While I can only paint in broad strokes in this introduction, we shall see what I mean when we get to the details in the chapters. It will be useful at this point, then, to set the nature of the debate as an opposition between evolutionary models and the SSSM. Understanding each in its pure form will help to recognize and evaluate them when they are mixed.

---

2   There is a simple two-part test that I like to apply to anyone who thinks evolution is just a theory: (i) ask them what books by evolutionary biologists they have read, and (ii) ask them where they think species came from if not from evolution. Invariably the answers are lame. Of course, everyone is entitled to their own opinion, but this must not be confused with the fact that uninformed opinions are a dime a dozen (and that is putting too high of a price on it). The same is true of opinions that are informed to some degree but are motivated primarily by an agenda other than logic and evidence.

One might say the debate between the two models is between an emphasis on evolutionary history and an emphasis on cultural history, but this is not entirely accurate. The debate is *not* nature versus nurture, but rather nature-nurture versus nurture. Biologists routinely argue that a full explanation for a given trait (whether physical or behavioral) requires a genetic and ultimately evolutionary explanation (nature) *and* an environmental explanation (nurture). Take, for example, the height of a particular plant. Reference to the genetics of the plant and the evolutionary history of its species is not enough for a full explanation of its height. For that, reference is also needed to the environmental conditions to which the plant has been exposed, such as the amount of sunshine and water. The same plant, with its genetics and evolutionary history, could have had a different height had it been exposed to different environmental conditions. This is easily seen in the case of clones.

The SSSM, on the other hand, tries its best to play down the role of biology and play up the role of the environment, namely, culture and conditioning. Ultimately it views human nature as enormously *plastic* (moldable), or, to vary the metaphor (following the 17th-century philosopher John Locke), as a *tabula rasa*, a blank slate. For example, Marxists argue that humans are not innately greedy, contrary to the view made popular by the 17th-century philosopher Thomas Hobbes, who thought humans were innately selfish but rational enough to form a social contract. Instead, for Marxists, it is a capitalist system that makes people greedy. Raised in a truly communist system, a system without class distinctions and private property, people would be unselfish and cooperative. Similarly, many feminists and social scientists argue that it is not because of their biology that men are so aggressive and violent toward women. Instead, they claim, it is a patriarchal system that makes men that way. Raised in a truly egalitarian system, men and women, behaviorally, would be basically the same. In gay studies it is similarly claimed that heterosexuality is not the biological norm for humans. Raised in a sexually permissive society, without stereotypes and prejudice, human sexual preferences would be either all the same or an even continuum. Common to each of these three examples, to Marxism, women's studies, and gay studies—indeed it is part of the common denominator of SSSM thought—is the further claim that hierarchy is not innate to the human species but the product of cultural history, in other words a *social construction* (this is the current fashionable phrase). Humans in this view are perfectly capable, in spite of their biology, to live in non-hierarchical social arrangements. It is the environment, past and present, that makes the hierarchies in humans, not genes, so also it is the environment that must be changed to fix the problem.

In the previous century one of the most powerful exponents of SSSM thinking was behaviorism in psychology. For behaviorists, what was true of serial killers and rapists was true of philanthropists and gifted musicians. In each and every case, it was not the person, or their genes, that was responsible, but rather it was the environment that had made them. John Watson, the first great of behaviorism, proclaimed,

> Give me a dozen healthy infants, well-formed, and my own specified world to bring them up in and I'll guarantee to take any one at random and train him to become any type of specialist I might select—doctor, lawyer, artist, merchant-chief, and yes,

even beggar-man and thief, regardless of his talents, penchants, tendencies, abilities, vocations, and race of his ancestors.

(Watson 1924, 104)

Similarly for the last of the great behaviorists, B. F. Skinner, although he recognized the role of genes via instincts more than earlier behaviorists, the due he gave them was superficial. For Skinner as for many others, human nature is still so plastic that he envisioned a utopia in which human society, engineered using behaviorist principles, enjoys previously unknown bliss, a world where "behavior likely to be punished seldom or never occurs," "people live together without quarreling," and people "bear no more children than can be raised decently" (Skinner 1972, 66, 214). (Utopia thinking, indeed, tends to be common among SSSM thinkers.)

What went hand in hand with behaviorism, both ideologically and temporally, was cultural relativism in anthropology. Not only were fashion and art found to be tremendously variable and entirely culturally relative, but so too, as the anthropologist Ruth Benedict (1934) argued, were "mannerisms like the ways of showing anger, or joy, or grief . . . or in major human drives like those of sex . . . in fields such as that of religion or formal marriage arrangements" (59). What is considered normal in one society, she pointed out, might easily be considered abnormal in another. On such a view no culture is right and no culture is wrong, and morality, far from having any innate norm, "differs in every society, and is a convenient term for socially approved habits" (73).

Although genuinely evolutionary explanations of human nature have been around since Darwin, they had difficulty being taken seriously in academic disciplines outside of professional biology until the power and pervasiveness of the SSSM received a number of serious blows. One serious blow (some would say the fatal blow) was delivered by the linguist Noam Chomsky. Beginning in the 1950s, Chomsky argued that language acquisition, contrary to behaviorism, is not simply a matter of basic intelligence and stimulus-response conditioning; indeed, that such a model could not possibly work. He argued, instead, that we humans enter the world with what he called a *universal grammar* hard-wired into our brains, meaning that it is coded for in our DNA (more on this in Chapter 3). Chomsky accomplished a veritable revolution in the science of linguistics, one that is still ongoing but is now widely accepted in its broad outlines. His greater importance, however, lies in the effect his revolution had on the SSSM with its blank slate view of human nature. Not only did Chomsky's revolution open the door to the computer model of the mind-brain (cognitive science), but it also opened the door to evolutionary models of human behavior, namely, sociobiology and evolutionary psychology. *Sociobiology* is the application of evolutionary principles to help explain social behavior in humans and other animals,[3] while *evolutionary psychology* is the application of evolutionary principles to help explain psychological phenomena.[4] These two burgeoning fields have much in common and a lot of overlap.

---

3   Wilson (1975) is the classic text, Alcock (2001) a recent and able defense.
4   Barkow *et al.* (1992) and Crawford and Krebs (1998) are the main anthologies, Badcock (2000) a recent introductory text.

Another serious blow to the SSSM came from anthropology. In 1928, around the time behaviorism was becoming popular in professional psychology, Margaret Mead's *Coming of Age in Samoa*, the *locus classicus* of cultural anthropology, was first published. Mead claimed that in Samoan culture men were not dominant, rape and violence virtually did not exist, nor rivalry and competition, and sexual mores were completely free. A university textbook for many decades, *Coming of Age in Samoa* subsequently proved to be as shoddy as can be, with Samoans turning out to be basically no different than the rest of humanity. Mead, it turned out, got her information not from a careful study of the Samoans but from interviews with 25 teenage Samoan girls in one of the villages, girls who had fun in making up stories. Mead ultimately believed what she wanted to believe, taking most of anthropology with her (Freeman 1983, 1989; see Alcock 2001, 131–134). Although anthropology long remained in denial (Freeman's study of Samoans and of Mead was done in the 1940s and 1960s), many anthropologists in recent decades have increasingly distanced themselves from anthropology's SSSM past, taking seriously evolutionary biology and the search for cultural universals underlying cultural diversity (Brown 1991; Ghiglieri 2000).

The main problem with the SSSM is not that it is completely wrong. Indeed, it is obviously right about many things. For example, it is today plainly clear that the historically widespread view that men have a greater share of reason than women (e.g., Aristotle *Politics* I.12) was a socially constructed myth, as was the view that women are mainly to blame for the woes of humanity (the myth of Eve for the ancient Jews, Pandora for the ancient Greeks). And the SSSM might well turn out to be right about many other matters, for example that human rights are not natural or innate but are merely a social construction.

What is wrong with the SSSM, instead, is that as a way of thinking it produces resistance, even phobia or denial, to the fact that humans are a biological species. We can resist the fact as much as we want, but it remains the fundamental fact of our existence. Our species, *Homo sapiens*, did not pop into existence out of nothing, but instead evolved ever so gradually from an earlier species, which in turn evolved ever so gradually from a yet earlier species, and so on back through evolutionary time. Granted, very few, if any, professional academics who subscribe to the SSSM would wish to deny this (the evidence for evolution is just too great). But what they do wish to deny, instead, are the many *implications* of this fact. And that is where the problems arise. Quite simply, *Homo sapiens* is not just an evolved species but a social species, one that evolved in small hunting-gathering groups. As such, it would be utterly remarkable if this animal species did not evolve special instincts while all the others have. SSSM thinkers, interestingly, have no problem admitting rabbit nature, or wolf nature, or gorilla nature, but when it comes to humans they just do not want to admit that there is such a thing as human nature.

There are many reasons for this resistance, and they are interconnected. One reason simply involves the history of the term "human nature." Deeply rooted in pre-evolutionary doctrines such as those of Plato, Aristotle, and St Augustine, talk of human nature immediately suggests to many today something essentialistic, something that is fixed and eternal, not something that gradually evolves. It also has normative connotations, suggesting not only what is but what ought to be. Indeed, so theory-laden is the term "human nature" that I recently had a prominent evolutionary biologist get angry with me at the very mention of the term. Such anger, however, is misplaced. As our knowledge grows, the meaning of terms naturally change. We still believe in the existence of humans, of course, and that, following Linnaeus in the 18th century, humans are a species. But ever since Darwin we also now recognize that, like other species, we are a constantly evolving species. Moreover, following Darwin but especially following our recent ability to read DNA, we know that variation is the norm for every population or species at the genetic level. But more than that, we also now know that there are *statistical norms* for populations and species at any one slice of time. Given this knowledge, then, there is no reason to abandon the concept of human nature altogether. Instead, our evolved knowledge suggests an evolved meaning, which in biology and related sciences is and should only be that of the *genetically influenced statistical behavioral norm* of the species taken at any one *time slice* in its evolutionary history. There is no essentialism here. There is also nothing in this meaning that need involve evaluative connotations such as that statistical deviations are "deviants," nor need it involve normative connotations such as that the statistical norm is "good" or "best." Instead, the only legitimate meaning of "human nature" in science today is *purely descriptive*. When legitimized as such, hopefully the fear that its folk and pre-Darwinian past evokes in so many will eventually subside and cease to be a barrier to understanding.

Related to this is the fear that talk of genes and human nature invariably involves *biological determinism*. The fear is that once it is granted that genes *influence* human behavior then it must also be granted that they *determine* human behavior, so that the status quo with all its injustice is justified and any hope of progressive change is lost. This fear is understandable, given that there is a history of injustice supported by biological theories, the Nazi doctrines of racial supremacy and inferiority being a striking example among many. Indeed, the fear of what evolutionary biology might mean for human nature has prevented many from taking the time to learn the basic principles of evolutionary biology and genetics. What should become apparent as we go through the coming chapters, however, is that "biological determinism" is a bogey term, one that does not have scientific respectability. To be sure, there are legitimate fears that are involved with the self-knowledge that comes from studying evolutionary biology, but the fear of biological determinism is not one of them.

Finally, there are political reasons for why many find the SSSM appealing and are immediately suspicious of, or will not even listen to, those who provide evolutionary perspectives on human nature.[5] At the core of it all is political correctness, with its goal of a sensitive and fair society, especially with regard to groups that have

---

5   Jumonville (2003) provides a nice discussion on how the debate between the SSSM and evolutionary models relates to politics, in particular the multicultural values of the rising New Left (which embraces group identity values, whether based on race, ethnicity, or gender, and opposes hierarchy) versus the universal values of the Enlightenment. For a sustained critical discussion on how the SSSM embraces the *tabula rasa* model and in effect denies the existence of human nature, see Tooby and Cosmides (1992), Gross and Levitt (1998), Alcock (2001, ch. 7), and Pinker (2002).

suffered and continue to suffer discrimination and oppression. While among the general public there is a lot of division over the value of political correctness, with many thinking it has gone too far (just listen to call-in radio talk shows), in colleges and universities, especially in the humanities and social sciences, it has become quite a dominant force, even to the point of censorship (in many colleges and universities in the United States, for example, racial theorizing is not allowed). While the basic reasons for political correctness are just and laudable, much politically correct thinking is arguably unrealistic and a form of denial. Nowhere is this plainer than in the big questions where biology should be clearly relevant. Indeed, politically correct thinkers routinely give the impression that they could not care less about being *biologically correct*. If there is a conflict between political correctness and biology, then too bad for biology. What we shall see in coming chapters is that political correctness, when it shuts itself off from empirical evidence and argument or flies too easily to the SSSM, easily becomes its own worse enemy. To give a quick example, it can now be argued that the communist experiment failed, in country after country around the world, just as every commune experiment of hippies in the 1960s failed, not necessarily because evil or stupid people were behind the experiments, but because they had the wrong theory of human nature. How many other grand visions of human happiness are destined to fail because they have an erroneous concept of human nature?

The truly interdisciplinary challenge, then, as I see it, and it is the real debate, is to try to figure out as best one can just where the SSSM is right and where it is wrong and to be fearless about it, even if that means throwing political correctness to the wind at times. Biology in general and evolutionary biology in particular need to be taken seriously, both if we want to truly understand the human condition and if we believe knowledge is power and we want to support the most effective ways of bettering the world. Granting that we should take biology seriously, however, is one thing, saying exactly where we should do so and to what degree is another. Indeed it is the hard part, but it is also the most interesting.

As mentioned briefly at the beginning, this book was written for three main audiences and I want to say more about them here. One of the main audiences is the general public, more specifically fairly well-educated people with at least some basic background in high school science. We all know that there is enormous interest out there over evolution versus religion, over whether evolution is true and whether it undermines religion and *vice versa*. But I suspect that there is also enormous interest out there about the wider implications of evolution should it be true. For example, virtually everyone has wondered about whether there is a genetic component to homosexuality or to behavioral differences between men and women. Well if genes are involved then so is evolution. It is that simple. Many also wonder if evolution and religion are not really antipodes but can be combined, and yet few will be aware of the fact (including Catholics) that in 1996 the late Pope John Paul II officially accepted evolution (combining it, of course, with theology) and that this is the official position of the Vatican. Indeed many scholars and intellectuals have combined evolution with theology, including some famous evolutionary biologists such as Theodosius Dobzhansky and Francisco Ayala. But can they do so legitimately? These and many other interesting questions are explored in this book, such that I cannot imagine the general public giving it a cool reception.

An equally important audience is college and undergraduate university students. What they shall learn in this book are not only the basic principles of evolution, but the real debates that are hotly engaging their own professors in department after department and from discipline to discipline. Indeed this book is a brief education in the nature and value of interdisciplinary studies. Whenever I have taught these topics (in the form of a course kit anthology which also contained my critical commentaries), twice in a university and twice in a community college, students became engaged in a way and to a degree that I have never seen elsewhere. This is because most if not all of their exposure in college and university is to SSSM thinking, so they are naturally surprised to find that there is a powerful alternative. It was in fact this situation that ultimately brought me to write this book, all combined with the discovery that there does not seem to be anything out there in the book market that is at all comparable.

Not only are evolutionary perspectives on the big questions normally not studied by undergraduate students in colleges and universities, but students who try to express evolutionary explanations in courses in the humanities and social sciences often find themselves in quite a tempest (as a number of my students have attested when they took courses after my Mind and Nature course). Undoubtedly much of the heat generated—and it is typically heat, not light—comes not from the topics themselves, or from the interdisciplinary approach as a whole, but first and foremost from the professors, from their territoriality which compels them to defend and protect their individual disciplines or subdisciplines from outside explanatory encroachment, and from their egos, which compel them to look down upon outside disciplines with a condescending smile or frown. Indeed there is enormous arrogance in academia, with in-group and out-group mentality. Unfortunately, not only is this egotism and arrogance misplaced, such that these professors do their respective fields a great disservice, but they also infect their students with the air of their high-minded insularity and thereby perpetuate *a priori* a barrier within those students to an increase in knowledge and understanding. In other words, they indoctrinate students into a specialization. Hence, students with an interdisciplinary mindset who attempt to introduce evolutionary explanations in classes outside of biology or philosophy of biology typically experience two walls of resistance: one from the professor, and the other, often before the professor even speaks, from other students in the class, especially those who share the disciplinary perspective of the professor.

Specialization, of course, is extremely important for progress within any field. Nevertheless, if one truly wants to know how the world works, whether the human world or the world of nature as a whole, in other words if one is genuinely imbued with a spirit of inquiry and a thirst for knowledge, then one has little choice but to welcome interdisciplinary studies with open arms. And the reason is simple. Not only is there ultimately only one reality, but examples abound of insights that were gained only by the cooperative efforts of workers in different fields. A prime example is the emerging explanatory paradigm of mass extinction #5 (which included the dinosaurs), a paradigm that involves the fields of geology, paleontology, evolutionary biology, chemistry, geophysics, astronomy, and astrophysics (Glen 1994). Another example (though far less one of cooperation) is the species problem, the problem of determining the nature of biological species, a problem that has enjoyed an enormous amount of input and benefit not only from biologists but also from historians and philosophers (Stamos 2003, 2007). The present book, it is hoped, will awaken students to many other possibilities.

Although the expository level of this book is designed primarily for the general public with at least a high school education as well as college students and lower-level university undergraduates, it would be more greatly enjoyed by upper-level undergraduates, students with some background in disciplines such as philosophy, biology, anthropology, psychology, sociology, women's studies, gay studies, or theology. With students from a variety of backgrounds and the present book, both the course director and the students should experience (as I and my students have) a course they shall never forget!

As for specific courses, one obvious choice is philosophy of biology. However, it should be kept in mind that this book does not focus on some of the traditional topics in philosophy of biology (such as the species problem, the levels of selection problem, the problem of reductionism, or the problem of whether there are laws in biology). Instead, it casts a much wider net, a net that not only makes it more attractive than the standard philosophy of biology fare but that makes it attractive to many courses in the humanities and social sciences, especially courses that focus in one way or another on the topic of human nature.

The final audience for this book is professional scholars. While much of the content of this book is devoted to the clearest possible exposition of various views, concepts, and theories involved (again, it must be as clear as possible for students and the general public), I often critique the views of others and argue my own position on the various questions. In this way I hope to engage professional scholars and present a serious challenge to their views. This book, then, operates at more than one level and fluctuates between them, the one being clarity of exposition for students and the general public, the other being critical arguments designed for student research and professional scholars.

What this further means is that the style in each of the following chapters is not linear but mainly dialectical, in that each chapter moves forward typically by working through opposing views. This style is not only more engaging for the reader, I find, but it also provides the materials and perspectives necessary for responsible and informed conclusions on the various questions. For this purpose there are plenty of references to guide further research.

In all of this, whatever conclusions one comes to on evolution and the big questions after reading this book, they should always be tentative, and one need not even go that far. Instead, one may simply choose to suspend judgment on the various questions, pending further research and reflection. In either case, one cannot help but come away with a much more informed perspective on whether, where, and to what degree evolutionary explanations legitimately extend beyond biology as traditionally circumscribed. It is an exploration, indeed, that should not only enrich one's life but that should continue for the rest of one's days.

I should also say that my primary concern is to do justice to each of the big questions and to not hold back on what I think and why I think it. While an approach is needed that is sensitive to people's feelings, since many of the questions and arguments in this book have the potential to disturb and even offend, my primary concern is to stimulate thought and to be sensitive to logic and evidence, which involves the value of being biologically correct. Throughout it all my attitude is that, to paraphrase the Christian existentialist Søren Kierkegaard, there is no thought I am afraid to think whole. Nor do I think should you be afraid.

Now on to the chapter summaries.

Chapter 1 deals with epistemology, the study of knowledge, and we shall examine two questions that involve evolution. First, centering around the evolutionary epistemology of Karl Popper, we shall examine arguments pro and con on whether evolution evolved in us an ability to find truth, or whether that ability is a byproduct of other abilities that evolution evolved in us. We shall then examine a more narrow question, but nonetheless epistemological, namely, whether the science of evolutionary biology is capable of giving us any knowledge about human nature. After a brief discussion on postmodernism, the focus shall then be on the views of three critics, namely, the biologists Stephen Jay Gould and Richard Lewontin, and the anthropologist Marvin Harris. Much of this discussion paves the way for what is to come in the rest of the book.

Chapter 2 is devoted to evolution and consciousness. The problem of consciousness is arguably the central problem in philosophy of mind. The problem is whether this remarkable thing called consciousness, which seems so utterly different than matter, can receive an adequate explanation from evolutionary principles, or whether one needs to go outside of them. For some, such as Richard Swinburne, the existence of consciousness is literally a miracle, while for others, such as John Searle, Horace Barlow, and Gerald Edelman, it is perfectly natural.

Chapter 3 is devoted to evolution and language, a problem closely related to that of the previous chapter since many theorists connect consciousness irrevocably with language use, in the full-bodied sense of sentence formation (on this view, dogs and human babies, for example, are not conscious). The focus in this chapter is on Noam Chomsky's claim that humans are born with what he calls a universal grammar (UG), a language organ distinctly human (animals do not have it),[6] hard-wired into our DNA, in a sense the common denominator underlying natural languages such as English and Chinese (which for Chomsky are superficial surface phenomena, humans really have only one language). We shall examine the basic ideas in Chomsky's theory of the UG as well as problems raised against the UG from an evolutionary perspective. Accepting the UG (in some form), we shall then examine Chomsky's argument for why he thinks the UG is not the product of evolution by natural selection, as well as two very different attempts to fit Chomsky's UG into the evolutionary picture, attempts by the evolutionary linguists Steven Pinker and Derek Bickerton.

---

6   Although from a biological point of view humans *are* animals, throughout this book I shall follow linguistic convention and use the word "animal" to refer only to non-human animals. This is much more convenient than repeatedly using the phrase "non-human animals," which quickly becomes tiring.

In Chapter 4 we turn to matters of sex. Four questions shall be examined here. First, we shall look at the argument by the evolutionary psychologist David Buss that evolution evolved different mating strategies in men and women. Second, we shall focus on the question of why men rape, specifically the argument by the anthropologist Michael Ghiglieri for the evolution of a rape instinct in men, which makes rape at bottom a matter of sex and reproduction, as well as critical arguments that place the cause solely in culture, as a matter of male dominance and hatred toward women. Third, we shall look at the question of homosexuality, examining two very different evolutionary theories that attempt to explain why there is homosexuality, the kin selection theory of the sociobiologist E. O. Wilson and the X chromosome theory of the behavior geneticist Dean Hamer. We shall also examine some critical arguments, one of which is that heterosexuality, homosexuality, and bisexuality are not natural human kinds but social constructions. Fourth and finally, we shall look at the incest taboo, which is universal or near universal in human culture, specifically the argument by E. O. Wilson for the evolution of an instinct for incest avoidance and the argument by Marvin Harris that it is totally based on culture.

In Chapter 5 we deal with questions raised by feminism. One question is why women's studies courses are characterized by what has been termed *biophobia*. Another question is whether sexual selection theory in biology legitimately applies to humans. Related to this is the question of gender roles. Feminists often argue that gender roles are social constructions. Certainly at least some are, but we shall examine arguments for whether on the whole they have an evolutionary basis. We shall also examine an argument, by the well-known feminist biologist Anne Fausto-Sterling, that not only gender roles but the very dichotomy of male-female is not biologically justified but is another social construction, so that what we really have is a continuum or single sex. Finally, we shall take a look at the question of whether the science of evolutionary biology is sexist and if so whether scientific knowledge must always be biased.

In Chapter 6 we deal with questions concerning human races. We shall begin by looking at some popular misconceptions about race that are undermined by a basic understanding of evolutionary biology. That will take us to the interesting debate over whether human races really exist. On the one hand we shall examine the now standard arguments for why many biologists say we should not name human races, focusing on a paper by Stephen Jay Gould. We shall then examine a number of arguments, much more recent, that attempt to reintroduce the concept of race, one using ecology, the other using cladistic taxonomy. We shall then examine the question of race and IQ, assuming that there are human races for the sake of argument. Finally, we shall examine the question of whether racism, which is a human universal, is acquired by culture and environment or has a deeper cause, one rooted in our evolutionary past.

In Chapter 7 we examine many of the questions that evolutionary biology raises for ethics. We shall begin with an examination of Social Darwinism, the older kind of evolutionary ethics which holds that we should apply the principle of natural selection to the human species. This will be followed by a discussion on the distinction (all-too-often confused even by professional philosophers) between the is-ought and naturalistic fallacies. We shall then examine the modern version of evolutionary ethics, the sociobiological approach, which argues that a variety of moral instincts evolved in the human species. We shall focus in particular on conscience and altruism. We shall also examine the arguments of some of the critics of the sociobiological approach, who focus on moral reasoning and who view ethics as an autonomous discipline. As a counterbalance, we shall then examine the argument by Peter Singer, the don of practical ethics, for why politically left thinkers (they tend to dominate in colleges and universities and are typically SSSM thinkers) should adopt a Darwinian view of human nature. Along the way we shall examine some arguments on the issue of moral values, whether they can have an objective existence given evolution, or whether we project them onto nature. In connection with this, we shall examine the question of whether natural, innate, universal human rights are objective or a social construction.

In Chapter 8 we examine the implications of evolution for religion. The human world teems with religion, and this cries out for explanation. Three questions shall be examined here. The first is whether the ubiquity of religious beliefs and practices can be adequately explained by *memetics*, which is the application of evolutionary principles to the spread of *memes* (the units of cultural evolution, the analog of genes in genetics). The second question is whether religion requires a deeper explanation rooted in the evolution of human nature, in other words, whether there evolved in humans a religion instinct, such that the world's many religions are simply surface phenomena. Finally, we shall examine whether theology and evolutionary biology can be legitimately combined, as the late Pope John Paul II and Stephen Jay Gould and many others have attempted to do. I shall examine arguments pro and con and argue that they cannot be legitimately combined, a large part of the reason being the implications of evolution for the problem of evil, another being the ultimately incompatible approaches to knowledge and truth of science and religion.

In Chapter 9, the final chapter, we take a look at what it's all about, at whether evolution adds to the meaning of life, is neutral or irrelevant, or dissolves it away completely. We shall also examine whether evolutionary biology is compatible with a major movement in philosophy and literature known as *existentialism*.

And so there it is. A common theme running throughout this book is that, as with other species, one should expect evolution by natural selection to have evolved in the human species a variety of instincts and that, moreover, these instincts play a continuing and significant role in human behavior, from language, sexual mating, racism, morality, religion, and even the search for truth and the meaning of life. Probably the most famous quotation in philosophy on the subject of instincts is from the metaphysician F. H. Bradley, who in the Preface to the first edition of his magisterial *Appearance and Reality* (1893) wrote, "Metaphysics is the finding of bad reasons for what we believe upon instinct" (x). What is usually overlooked with this quotation is the clause that completes the sentence: "but to find these reasons is no less an instinct." Certainly food for thought in a book like this, and probably more to the skeptic's liking. But much more important than the quotation itself is the context in which it is found. It is none other than an expression of the true university mind. Not only is it the recognition that "to gain an education a man must study in more than one school" (viii), but it is also the recognition that the greatest enemy to just reasoning is the "dogmatic frame of mind" (ix). Thus along with Bradley hopefully we can all agree, no matter what our views on the big questions examined in this book, that "I would rather keep my natural place as a learner among learners" (ix).

1

Evolution and Knowledge

Epistemology is the study of knowledge, whether there is such a thing and, if it exists, how we can acquire it. In modern philosophy, whether we are talking about rationalists (those who focus on reason) or empiricists (those who focus on the senses), epistemology has been dominated by the approach of René Descartes, who lived in the early 1600s. In order to find out if anything can be certain, Descartes in his *Meditations* took systematic doubt to its limits, finding not only that he could doubt authorities and experts, but even his senses, so that he could doubt the existence of an external world, a world outside of himself, even the existence of his body, even the existence of God and other minds, even that one plus one equals two. But there was one thing, he claimed, that he could not possibly doubt no matter how hard he tried, namely, the existence of his mind while he was doubting (or more generally thinking). This is his famous *Cogito ergo sum* (I think, therefore I am). But the *Cogito* is a position of *solipsism* (the view that only my mind and my ideas exist). To get out of the *Cogito*, and ultimately back to the external world and other minds, in other words to provide epistemology (including science) with a foundation, Descartes found it necessary to prove the existence of God and that God is not a deceiver, because only then can reason trust its senses and intellectual powers. It is commonly acknowledged in philosophy today that Descartes' three proofs of God's existence do not work. It is even possible that Descartes himself would have secretly agreed (Stamos 1997). At any rate, arguments that attempt to establish the existence of God have an interesting but ultimately bad history, involving flaws of one sort or another (Martin 1990). Partly because of this, modern epistemology has been plagued ever since with the problem of creating and justifying an epistemology from the inside out, an epistemology that starts with the inner self—one's mind and the ideas found inside it including sense data—and works its way out to the external world (e.g., Russell 1940).

Because of this history of what is apparently a dead end, a number of philosophers more recently have argued that the only way out is to begin from an altogether different starting point, namely, modern science. Hence the program known as *naturalized epistemology* (Quine 1969; Kornblith 1994). Given evolution, one obvious candidate is to place knowledge on an evolutionary foundation, hence *evolutionary epistemology*, truly an interesting and growing field (Radnitzky and Bartley 1987; Maienschein and Creath 1999).

Evolutionary epistemology is the subject of this chapter, and two major questions shall first be examined, both following the useful division of evolutionary epistemology outlined by Michael Bradie. Bradie (1994) distinguishes between the *evolutionary epistemology of mind*, which attempts to "account for the cognitive mechanisms in animals and humans by a straightforward extension of the biological theory of evolution," and the *evolutionary epistemology of theories*, which "attempts to account for the evolution of ideas, scientific theories and culture in general by using models and metaphors drawn from evolutionary biology" (454).

There are many examples of the former view, such as W. V. O. Quine's (1969) suggestion that "Creatures inveterately wrong in their inductions have a pathetic but praiseworthy tendency to die before reproducing their kind" (126), or Susan Blackmore's (1999) claim that "We are designed by natural selection to be truth-seeking creatures" (202). The don of philosophy of biology, Michael Ruse (1986, ch. 5), has attempted to be more specific, arguing that the basic tools and methods of science—inductive and deductive reasoning, analogical reasoning, mathematics, appeal to laws of nature (regularities), observation, experiment, etc.—are not only "rooted in our biology" but "have their being and only justification in their Darwinian value, that is in their adaptive worth to us humans—or, at least, to our proto-human ancestors" (155). Ruse's book is aptly titled *Taking Darwin Seriously*, and it is interesting too that Darwin provided an evolutionary epistemology of the first kind. Noting that throughout human evolutionary history humans existed in social groups or tribes, and that tribes would compete with and supplant other tribes, Darwin (1871 I) argued that the main though not exclusive cause of success would be "art" (by which he meant technology, both for subsistence and for battle), that tribes with superior art would increase in number and supplant other tribes, and that since art is the product of the intellect it is "highly probable that with mankind the intellectual faculties have been gradually perfected through natural selection" (160).

Examples of the second kind of evolutionary epistemology are common too, going right back again to the time of Darwin. For example, the biologist T. H. Huxley, famously known as "Darwin's bulldog" for his public defense of Darwin's theory of evolution, wrote in a book review of Darwin's *Origin* that

> Mr. Darwin's hypothesis of the origin of species will take its place among the established theories of science, be its consequences whatever they may. If, on the other hand, Mr. Darwin has erred, either in fact or in reasoning, his fellow-workers will soon find out the weak points in his doctrines, and their extinction by some nearer approximation to the truth will exemplify his own principle of natural selection.
>
> (Huxley 1859, 148)

Similarly, the Harvard botanist Asa Gray, Darwin's main advocate in the United States, wrote in a review of Darwin's *Origin* that

> A spirited conflict among opinions of every grade must ensue, which,—to borrow an illustration from the doctrine of the book before us—may be likened to the conflict in

nature among races in the struggle for life, which Mr. Darwin describes; through which the views most favored by facts will be developed and tested by "Natural Selection," the weaker ones be [sic] destroyed in the process, and the strongest in the long run alone survive.

(Gray 1860, 154)

Much more recently the psychologist Donald Campbell (1974) advocated a "selective retention paradigm [i.e., natural selection] to *all* knowledge processes" (56), not just to scientific knowledge.

Our focus in this first part of the chapter shall be on the evolutionary epistemology devised by the philosopher of science Karl Popper, hailed by Campbell (1974) and many others as "the modern founder and leading advocate of a natural-selection epistemology" (89). Popper is no longer today the biggest name in evolutionary epistemology, but he is nevertheless an excellent focus for our purposes, first because his is still indeed a big name in the field, second because he provides us with both kinds of evolutionary epistemology as described by Bradie above, and third because some of his key ideas recur in later chapters in this book.

Following our examination of the questions raised by Bradie's two divisions, the final question we shall examine is whether the science of evolutionary biology is capable of telling us anything about human nature. For this question we shall begin by taking a brief look at the loose collection of ideas and agendas known as *postmodernism*, not only because it is good to know what it is about, but because, again like Popper, some of its ideas and agendas recur in a number of places in later chapters. We shall then focus on some critics who, unlike postmodernists, accept the evolution of species as a fact but who nevertheless take a negative position on our final question, namely, the biologists Stephen Jay Gould and Richard Lewontin, and the anthropologist Marvin Harris.

All of this, questions raised by Bradie's two divisions and the question of whether evolutionary biology is capable of shedding light on human nature, helps set the stage for the remaining chapters in this book, making it important to keep the debates in mind as we go through chapter after chapter.

Turning to Popper, then, who gives us an epistemology of mind and of theories, in his article titled "Natural Selection and the Emergence of Mind" (1978) he begins the final section with a standard view in science, which is that life gradually "evolved or emerged" from non-living matter and that mind gradually evolved or emerged from living things. This is a genuine view of *emergence*, one that rejects the philosophical flights of fancy known as *hylozoism* (the theory that all matter possesses life to some degree) and *panpsychism* (the theory that all matter possesses mind to some degree). Popper is not terribly disturbed by the question of the origin of life, thinking (like many others) that it will eventually be solved by science, but the emergence of mind from life seems to him a much more difficult problem, for which he offers some "speculative conjectures."

Following the evolutionary biologist Ernst Mayr, Popper supposes that natural selection first evolved in some organisms what Mayr calls "closed behavioral programs," behavioral programs that are very detailed and rigid and are built into the DNA of the organisms, such as web-weaving in spiders, honeycomb-making in bees, and nest-building in birds. From these closed behavioral programs, natural selection then evolved what Mayr calls "open behavioral programs," behavioral programs that are not rigid but leave open different possibilities, such as imprinting in goslings (Mayr's example). Imprinting is a genetic program, such that the newborn gosling permanently takes to be its parent the first moving object it sees which makes the appropriate sounds, but it is an *open* genetic program in that the adopted parent need not be either of its actual parents (though it invariably is in the wild) and may instead be something else—such as a clever biologist doing studies on imprinting. Indeed Mayr (1976) claims that "On the whole, and certainly among the higher vertebrates, there has been a tendency to replace rigidly closed programs by open ones or, as the student of animal behavior would say, to replace rigidly instinctive behavior by learned behavior" (24).

Open behavioral programs are still instincts, behavioral programs that are coded in the organism's DNA. How does one get from these to genuine learning behavior, indeed to what we readily recognize as knowledge? Popper supposes that there are four stages, all of them evolutionary stages produced by natural selection (and here we get some overlap with our next two chapters). The first stage is the evolution of pain and fear, the selective advantage of which ought to be obvious, as they increase survival to adulthood and consequently reproduction. Popper sees this stage as involving *"real* trial-and-error behavior" (151), and it is not difficult to see why. Organisms that experience pain or fear, or pleasure we might add, are attempting to adapt to their environment, by experimentation and feedback. But because they are being direct about it, these organisms are also directly putting their lives on the line. Hence Popper's second stage, where natural selection evolved *"imagined* or vicarious trial-and-error behavior," evolved by natural selection since this ability confers a selective advantage over mere (real) trial-and-error behavior. The advantage here is that the organism can play out trial-and-error scenarios in its mind, using memory based on its past experience, such as remembering to avoid certain kinds of animals or plants. From the second stage evolved the third stage, "the evolution of more or less conscious aims, or ends: of purposive animal actions, such as hunting" (152). Clearly Popper thinks that not only consciousness but teleology (having goals or ends that effect present action) is not confined to humans but is shared in various degrees by many animal species. I would agree with Popper, but it is a question for our next chapter. From the third stage, natural selection evolved the fourth stage, found thus far only in humans, namely, the evolution of language. It is only with language, in the full sense of sentences (words combined with grammatical rules), that we *"formulate* our hypotheses, and criticize them," so that "Let our conjectures, or theories, die in our stead!" (152). While the evolution of language is a question we shall examine in Chapter 3, it is Popper's claim that trial-and-error elimination is the essence of knowledge that shall concern us here. It is a profound evolutionary epistemology, which links humans with the rest of the animal world, and which has implications for many areas of philosophy and human life.

At the core and heart of Popper's evolutionary epistemology—and indeed of his philosophy of science as a whole, in which genuine science is demarcated from both pseudo- and non-science—is the law of logic known as *modus tollens*. The law looks like this:

If A, then B
Not B
Therefore, not A

No matter what one plugs in for the place-holders A and B, the argument is always valid. If it is raining, then my Jeep is wet. My Jeep is not wet. Therefore, it is not raining.[1] It does not matter whether one applies this law to the expectations in animals or to theories of science, the process of knowledge for Popper is always the same. A scientist begins with a theory. An animal begins with an expectation. If the expectation is not met, in that either nothing happens or something other than what was expected happens, then the animal changes its expectation. Indeed Popper thinks that all animals come into the world with inborn expectations and that these are the source of its first problems. From these first problems knowledge grows by the *modus tollens* of trial-and-error elimination (see Popper 1979, 258–259).

Some comments are necessary before we proceed further. First, the idea that animals can have knowledge will immediately be rejected by many. A common view in epistemology, around which revolves much critical discussion, is that knowledge is "justified true belief." For some (e.g., Johnson 1988), animals cannot possibly have knowledge because they lack beliefs, and they lack beliefs because they lack language. We commonly say beliefs are true or false, but properly speaking truth or falsity are only properties of sentences, as in "The cat is on the mat" is true (if in fact the cat is on the mat), so beliefs do not exist except in sentences. Since animals are incapable of making sentences (a question we shall examine in Chapter 3), it follows that they are incapable of having beliefs. And even if this is mistaken and they are capable of having beliefs, others would deny that they could justify them (because the act of justification requires language), so that they could not have knowledge.

The problem here is too narrow a meaning of both belief and truth. Surely a dog, say, can have a belief that the person coming down the stairs is its friend and caregiver, a belief which may turn out to be true or false. And surely it can have knowledge of something more about this person (like the person's voice), so that it makes no mistake when it is in that person's presence. Similarly, surely a squirrel can have knowledge of where to find the kind of food it likes. What should be required of a theory of knowledge is that we begin with examples of many different kinds of knowledge (assuming we allow knowledge in the first place) and then ask how such knowledge is possible. To begin with uniquely *human* knowledge is to prejudice the matter. To begin with only propositional knowledge (knowledge expressed in sentences) and a corresponding theory of belief that precludes animal belief—and hence animal knowledge—is to repeat the same prejudice.

Consider this from another perspective. A common distinction in philosophy is the distinction between knowing *that* and knowing *how*. One might say that a spider knows how to make a web, even though its knowledge is innate (in its DNA). Similarly,

---

1   Briefly, in logic a *valid* argument is one where *if* the premises are true (they do not need to be) then the conclusion must be true. A *sound* argument is a valid argument with *in fact* true premises. The distinction is a useful one, since sometimes one cannot know whether the premises are true, but one can still assess the argument in terms of validity.

a beaver knows how to make a dam, a bird a nest. In Chapter 3 we shall find Chomsky claim that humans have innate, unconscious knowledge of language, where behavior is merely evidence of knowledge and knowledge is a mental structure in the brain (e.g., Chomsky 1980, 47–49, 93–103). Whether making a dam or making a sentence, and whether innate or learned, both humans and animals know how to do things, the latter in many cases better than we do, and we in many cases better than they. But is their knowledge confined to only knowing how?[2] The philosopher of language Gilbert Ryle (1949) called the attempt to reduce all knowing *how* to knowing *that* the "intellectualist legend" (29). For Ryle, the one cannot be reduced to the other for the simple reason that people often perform actions without being able to put into sentences how they should be performed. Ryle had pathetically little to say about animals, but presumably he would have allowed that animals have knowing *how*. But if we grant that animals have knowing *how*, it seems yet another intellectualist legend to deny animals the capability of knowing *that*. Just as knowing *how* does not require sentences, does not require the ability to recite rules or criteria or propositional knowledge before the performance, neither does knowing *that* require sentences. A dog knows *that* the person in front of it is its friend and caregiver. A squirrel knows *that* this is the kind of food it likes. Sentences are not needed for such knowledge. To think otherwise is yet another intellectualist legend, in this case also an intellectualist prejudice.

Returning to Popper, at the heart and core of knowledge in his view is the *modus tollens* of trial-and-error elimination, of "conjectures and refutations," of "learning from our mistakes." Even amoeba can have knowledge of a sort and grow in knowledge. "From the amoeba to Einstein," he says, "the growth of knowledge is always the same . . . a process of elimination" (Popper 1979, 261). Again, the main difference between scientific knowledge and knowledge in simple forms of life such as amoeba is that we do not die with our theories. A further difference, in some senses a more important one, is that although the evolution of life produces an ever-expanding tree, scientific knowledge evolves in an opposite direction, toward unified theories, toward an all-inclusive understanding, in a way an upside-down tree (262).

There are further features of Popper's evolutionary epistemology that yet need to be discussed before we enter into critical analysis. First, truth is taken in the standard way to mean *correspondence to reality*. A statement, or an expectation, is true for Popper if it corresponds to the way the outside world really is. By having our expectations or theories bump up against the world, we either keep them or find that we have to modify or even reject and replace them. In this way we get "nearer to the truth" (264), a concept Popper in various places calls *verisimilitude*. As our rejected theories are replaced with more inclusive ones that pass the tests their forebears failed, our theories get progressively closer to the truth. But for Popper although we naturally want our theories to be true, in everyday life no less than in science, we

---

2 Granted, Chomsky thought that knowledge of language is not so much a matter of knowing *how* as of knowing *that*, as in knowing whether this or that is a well-formed sentence, but we can afford to ignore his claim here.

can never know when they are in fact true. All we can know is that so far they have survived refutation. Hence the *modus tollens* path toward knowledge is a negative one. We can know that theory x is false when it has failed its predicted observations or tests. But we cannot know that it is true when it has passed them. In the latter case it is *corroborated* only, it is not confirmed or verified. Theory x might not simply be nearer to the truth but in fact be true. But we can never know this. We can only know when theories are false. This is a salutary check on dogmatism, something Popper always stressed.

Equally important for Popper is the rejection of inductive reasoning and so-called inductive knowledge. An inductive argument is an argument where the conclusion goes beyond the premises. For example, using as a premise the claim that all crows observed thus far have been black, an inductive conclusion would be that all crows are black, or that the next crow observed will be black. This is simple induction by enumeration. We naturally think that the more black crows we observe, the more probable the truth of the conclusions above, just so long as no non-black crow is observed. But for Popper the probability is always zero. No matter how many crows we observe, the number is always finite, whereas the conclusion that all crows are black (which by the way is implicit behind the other inductive conclusion, the one about the next crow) is about an unlimited number of crows (past, present, and future). A finite number, no matter how large, when divided by infinity always results in zero. Hence an inductive conclusion, contrary to our intuitions, must always have zero support (Popper 1963, 281).

This is a devastating critique of inductive reasoning, but there is more to it. The Scottish philosopher David Hume (e.g., 1748, 24–45) provided what many consider to be the first devastating analysis of inductive reasoning. Hume noted that all claims about matters of fact, such as that this is a crow or that this piece of bread will nourish me, are based on causal reasoning, which in turn is based on inductive reasoning. But, he asked, what is the foundation of inductive reasoning? Since our reasoning about the future goes beyond past and present experience, it is natural to conclude that it can only be probable. But for Hume it cannot even be probable. This is because all reasoning from the past to the future involves as a major premise what is often called "the principle of the uniformity of nature." This idea is sometimes stated as "the future will resemble the past," sometimes as "like causes will produce like effects," but the latter amounts to basically the same idea as the former. The problem is that the principle of the uniformity of nature cannot be deductively demonstrated to be true, since there is no contradiction in thinking that the future will not resemble the past. In other words, it is logically possible that the secret powers behind the superficial qualities we perceive as things could radically change at some time. For example, there is no contradiction in asserting that this piece of bread will kill me, that the sun will not rise tomorrow, that it could snow in Jamaica in July, that ice will melt my skin off, and so on. Moreover, the principle of the uniformity of nature cannot be concluded to be probable, since it is based only on past experience, not on future experience. Therefore, since inductive arguments use as a premise the principle of the uniformity of nature, they are *circular* or *beg the question* (they take for granted precisely what is in dispute). They argue in a circle since they attempt to establish that the future will resemble the past (e.g., that this piece of bread will nourish me) by *assuming* that the future will resemble the past.

Hume's conclusion, then, is that it is not by argument or reason that we conduct ourselves in everyday life, but by "natural instincts," including induction and "principles of association," which, he adds, "no reasoning or process of the thought and understanding is able, either to produce, or to prevent" (39). Nature has made us in such a way that we are creatures primarily of instinct, not of reason, indeed that reason is only a "slave" of the former, to use Hume's label (1739, 266).

Popper accepts Hume's critique of induction but takes it a step further. For Popper, we do not even reason inductively! Instead, we reason deductively, by the *modus tollens* of trial-and-error elimination. We do not arrive inductively at theories or conclusions, but rather we *begin* with them and then test them against the world. One problem with Popper's evolutionary epistemology here, however, is that induction may well be compatible with evolutionary epistemology after all. This is because the major premise in inductive reasoning, the premise that the future will resemble the past, might well be an instinctual belief evolved in us and other animals because it confers a selective advantage over those animals that did not have it, or had it to a lesser degree, those that operated more or mainly by straightforward trial-and-error elimination. If so, then inductive reasoning might not really be invalid, since the major premise itself, contra Hume, is not arrived at by induction (hence no circularity) but instead by eons of trial-and-error elimination. The major premise might in fact be false, but inductive arguments could still be valid, just not sound if the major or any other premise is false. Hence, one might argue that evolutionary epistemology, contra Popper, gives us both induction and possibly even genuine epistemology. If this is so, then Hume (even though he came before Darwin) was closer to the truth than Popper. Interestingly, Hume (1748) adds that if reasoning about matters of fact were left to our intellect, it would be maladaptive, as the operations of reason are slow and if left to themselves tend to skepticism. Instead, since inductive reasoning is "so essential to the subsistence of all human creatures," it is probable, he says, that nature "implanted in us an instinct, which carries forward the thought in a correspondent course to that which she has established among external objects" (45)—i.e., an instinct for induction. The fact that inductive reasoning seems so intuitive to us and that we are shocked to hear that we never reason inductively swings the arrow, I would say, toward a Darwinized Hume rather than a Darwinian Popper.[3]

---

3   Indeed, Popper was not much of a Darwinian when it really came down to it. For a start, until he had a change of mind in 1977 he long thought of natural selection as not testable because it is "almost tautological" (Stamos 1996), a tautology being a sentence that is true simply because of the meaning of its words. The fact of the matter, however, is that natural selection is not simply a matter of words but is a well-documented *process* occurring in nature (Endler 1986). Moreover, contra Popper (1979), Darwin did not show that "in principle" natural selection can "simulate the action of a Creator" or that "*in principle* any particular teleological explanation may, one day, be reduced to, or further explained by, a causal explanation" (267). Darwin argued for and accomplished much more. Employing what an older contemporary of his, William Whewell, called a "consilience of inductions," Darwin brought together evidence from a wide variety of fields to argue *for* the fact of evolution by natural selection and *against* creationism, a method today known as *inference to the best explanation*. (More on this in the Appendix and in Chapter 8.)

Another interesting philosopher to compare Popper's evolutionary epistemology with is the German philosopher Immanuel Kant. Like Hume, Kant looms large in the history of philosophy. Interestingly, Kant (e.g., 1783) was struck by Hume's skeptical conclusions and attempted to put knowledge on a solid foundation, one on which we can claim certainty. For Kant, we do not simply receive impressions from the world, as Hume and others thought. Instead, our mind has *categories* built into it, categories *a priori* (prior to experience), which process what we receive from the world in specific ways. Some of the categories are space, time, causality, and subject-attribute. What the world in itself is really like we can never know, because we can never get beyond these categories which necessarily frame and shape all our experience of the world. Hence knowledge is the product of two contributors, one the world in itself, the other the categories in our mind. One has to wonder if this is an epistemology that really gives us objectively valid knowledge, as Kant thought it does. At any rate, for Popper, Kant was right in supposing that theories (categories) are *a priori*, but he was wrong in thinking them valid and unchanging (see Popper 1979, 24, 68 n, 92–93, 328–329; 1963, 47–48, 117). Kant was an idealist, whereas Popper considered himself a realist. Trial-and-error elimination allows our minds (via our theories) to hook up better and better to the real world.

But has Popper really Darwinized Kant? And does it matter? Donald Campbell (1974) argues *yes* in answer to both questions. The problem is that Popper never thought of natural selection as evolving in us finely tuned categories of the mind. Instead, he only talks about *theories* or *expectations*. Granted, a major theme in Popper's writings is the theory-dependency of observation, that there are no such things as neutral observations, that every observation presupposes a theoretical point of view. But this is not the same as Kantian categories. The most Popper says (1963, 47–48) is that the "instinctive" expectation of finding regularities "corresponds very closely" to Kant's *a priori* category of causality. The problem is that theory dependency is a far cry from processing information from the world via categories. Not only does Popper believe that our theories change as they clash with experience, whereas Kant believed that experience itself is shaped and formed by these categories and is not possible without them, but those who attempt to Darwinize Kant, like Campbell himself, have more in mind a permanent hard-wiring of categories in our brains by evolution, something not found in Popper's theorizing. It is important to notice this because, arguably, a Darwinized Kant is really a contradiction in terms. Not only, from a Darwinian point of view, would Kantian categories be expected to evolve differently in different species (or at least in some of them), but Darwinism is populational, entailing that variation in a population is the norm, not uniformity (heritable variation is what natural selection operates on). Arguably, then, Darwinism precludes Kantian categories of the mind.

Problems return for Popper with his Darwinizing of scientific progress. Popper (1979) calls it "*the natural selection of hypotheses,*" by which he means "our knowledge consists, at every moment, of those hypotheses which have shown their (comparative) fitness by surviving so far in their struggle for existence; a comparative struggle which eliminates those hypotheses which are unfit" (261).[4] For Popper, the demarcation of science from pseudo- or non-science is the *modus tollens* of "falsifiability" (the potential for falsification). Real scientific theories entail test predictions. Indeed for Popper the empirical content of a scientific theory is not really what it says about the world but what it prohibits, what it says does not or should not happen, what he calls the "potential falsifiers" (Popper 1959, 86, 113). Theories that entail potential falsifiers are genuine scientific theories, theories that do not, that are immune to possible falsification, are not genuine scientific theories. (This has the interesting consequence that falsified theories remain genuine scientific theories.) Hence for Popper astrology, theology, Marxism, Freudianism, and many other bodies of theory are not genuine sciences because their theories do not entail falsifiable predictions.[5]

Problems arise, however, with Popper's theory of scientific progress as *verisimilitude*. Popper thinks that theories which explain what their falsified forebears explained are closer to the truth. This may be, but how can Popper know this without already knowing the truth? Moreover, the entire concept of *verisimilitude* looks like an example of induction, a kind of reasoning we have seen that Popper rejects. Indeed one might take a Darwinian view of the evolution of scientific theories, of a succession of theories related by descent, but not hold that they evolve toward truth at all, only that they evolve in an adaptive sense to the changing needs of scientists. This is the rather skeptical and relativistic view that the historian and philosopher of science Thomas Kuhn (1970) provides at the end of his book, one in which progress in theories is not toward the truth but consists simply of better "instruments for puzzle-solving" (206). For Kuhn, scientists are locked in what he calls *paradigms* (a major body of theory in which normal science is done, such as the creationist paradigm before Darwin, the Darwinian paradigm, the Newtonian paradigm, the Einsteinian paradigm, etc.), scientists are parts of social groups and are subject to social pressures, their different paradigms are like different languages or even different worlds, conversion in a scientist (if it does happen) is rather quick

---

4   Also often quoted (e.g., in Campbell 1974, 49) are passages from Popper's first book, *The Logic of Scientific Discovery* (1959), first published in German in 1934, in which he wrote, "what characterizes the empirical method is its manner of exposing to falsification, in every conceivable way, the system to be tested. Its aim is not to save the lives of untenable systems but, on the contrary, to select the one which is by comparison the fittest, by exposing them all to the fiercest struggle for survival" (108). Again, "We choose the theory which best holds its own in competition with other theories; the one which, by natural selection, proves itself the fittest to survive . . . A theory is a tool which we test by applying it, and which we judge as to its fitness by the results of its applications" (108). The problem with taking these early passages as examples of an evolutionary *epistemology* of theories is that Popper at that time was still arguably an *instrumentalist* (believing theories are either better or worse for prediction, but not objectively true or false). He did not begin to advocate a correspondence theory of truth until shortly after the publication of his book, when he first became enamored with the correspondence theory of truth enunciated by the logician Alfred Tarski, according to whom "Snow is white" if and only if snow is white. It was only as a consequence of accepting Tarski's theory that Popper went on to develop his own theory of *verisimilitude*.

5   For example, no astrologer ever says, "If this were found to be true it would completely undermine the science of astrology." Similarly Freud, in having an explanation for each and any possible behavior we might make, provided according to Popper no scientific explanation at all, since he made no falsifiable predictions. As for theology, we shall deal with the topic in Chapter 8.

and is akin to religious conversion (in that it occurs for mainly psychological and social reasons), and a scientific revolution is the displacement of one paradigm by another, it does not build on the previous paradigm. Needless to say, Kuhn has many detractors in science as well as in philosophy of science (e.g., Lakatos and Musgrave 1970; Chalmers 1999). As a scientific realist myself, I think the Kuhnian approach fails to see the nature of science as *inference to the best explanation*, which arguably is the heart and core of science (see Chapter 8 and the Appendix), or that later theories are in an important sense more inclusive than their predecessors. At any rate, the Kuhnian approach does show that a Darwinian view of theory evolution need not necessarily be an epistemological one.[6]

A further problem for Popper's natural selection theory of theories concerns the generation of scientific hypotheses or theories. Every biologist agrees that in biological evolution the variations upon which natural selection feeds are random with respect to the environment. There is even now strong evidence, contrary to what Darwin thought, that much of the variation is the product of pure, genuine chance, in the form of point mutations caused by quantum chance events at the subatomic level (Stamos 2001). At any rate, it does not at all seem to be the case that scientists generally construct their theories randomly, let alone by chance. Popper attempted to circumvent this problem by claiming that theory production is a matter for biography or psychology, not for philosophy of science (Popper 1959, 31–32). Campbell (1974) torturously tried to fit scientific theory construction into the mold of "blind" variation by claiming that scientific theories by their very nature go beyond what is known and "In going beyond what is already known, one cannot help but go blindly" (57). All of this is smoke and mirrors. As Ruse (1986) points out, it is to confuse "something being unknown and something being 'haphazard'" (59). Scientific theories undoubtedly have an aspect of the unknown to them, but this does not mean that they are created haphazardly. For Ruse, "Scientific evolution is not Darwinian" (57). Using the example of Darwin, Ruse points out that the development of his evolutionary theory was hardly haphazard and random. Instead, "one sees a dedicated alert mind, probing into the unknown in a systematic way . . . Even the mistakes are directed" (60).

And yet, there is perhaps here a mistaken notion of natural selection. For Darwin himself (1859), "This preservation of favourable variations and the rejection of injurious variations, I call Natural Selection" (81). For the don of 20th-century evolutionary biology, Ernst Mayr (1982), "natural selection is a two-step process. At the first step, the production of genetic variability, accident, indeed, reigns supreme. However, the ordering of genetic variability by selection at the second step is anything but a chance process" (519–20). But arguably Mayr is deviating from Darwin here. If one says that the process of the *ordering* of the variations is natural selection (as Darwin seems to say), then the *production* of the variations upon which natural selection operates is not *part* of the process of natural selection. (The production

---

6   Smithurst (1995) provides an interesting attempt to develop and defend what he thinks the later Popper would have replied to the critics who use evolution to support epistemological skepticism.

would be one of the processes in the *evolutionary* process, but it would be a separate process from the *natural selection* process.) My point is that if one takes this latter view of natural selection that I have suggested and which seems to have been Darwin's view, then it would be acceptable to speak of a natural selection process applying to scientific theories. The nature of the origin of the variations, of the theories, would be irrelevant.

A further problem still remains, however, for the natural selection of theories theory, namely, as Ruse (1986) points out, "science is progressive. It moves toward an understanding of reality" (57). And of course this process is conscious, it is the goal of scientists. In nature, however, species evolution does not move toward any final goal. There is no ultimate goal (a topic we shall return to in the final chapter), only changing adaptation to a changing environment. But again, granting that this is so, it still may be acceptable to call theory competition and change in scientific theories a natural selection process. It only means that the *evolution* process in science is different than in nature, not that a natural selection process does not operate there.

Stephen Downes (2000) adds some further interesting arguments against evolutionary epistemology that I would like to address. For a start, Downes appeals to David Hull's (1988) close study of theory competition in professional taxonomy, in which personality and social factors are made much of, examples of selection pressures being the desire for credit and status, competition for research grants, competition for tenure-track university positions, professional allegiances, and so on. And yet what must not be overlooked in Hull's work—in what arguably is not a typical example of science to begin with—is that like most philosophers of science Hull does not see science as merely a social construction. Instead, he argues that the social structure of science largely affects the content and success of scientific theories only "locally," over the short run but decreasingly over the long run, so that over the long run science does indeed progress more and more toward the truth.

Being skeptical about the social nature of science, Downes provides further criticism of the evolutionary epistemology of theories that lead to criticism of the evolutionary epistemology of mind. Here matters get more interesting. Noting that the standard theory of truth in evolutionary epistemology, for both science and everyday life, is the correspondence theory of truth, he provides some quibbling philosophical problems with that theory, none of which should concern us here except his claim that the correspondence theory

> is of no help to those wishing to establish a connection between truth and selection. If we focus on the relation between abstract objects, such as propositions, and the world they portray, there hardly seems any room for having an explanatory role.
>
> (440 n. 5)

Unfortunately Downes does not tell us why! Granted, philosophers have had a tendency to view propositions as abstract Platonic entities, existing outside of space and time. But one does not have to think of them that way. Rather, one may take propositions to be linguistic expressions of beliefs, beliefs which are in the head (such that animals have them) and which have, in the language of Paul and Patricia Churchland (1990), a "calibration content" (306). For the Churchlands, an organism that has a belief has a representation of some aspect of its environment, and that representation is a particular pattern of neuronal activity in its brain. Just as the waggle dance of bees is calibrated by natural selection, so are beliefs such as "This is a kind of animal that eats kinds of animals like me." With animals such as humans the matter is only more complicated, but not really different. Their bet is that the story "will build on the more basic story of calibration semantics for simpler organisms, following the steps of evolution itself" (308). Simply quoting, as Downes does, Wittgenstein's (1922) naive statement that "Darwin's theory has no more to do with philosophy than any other hypothesis in natural science" (25) simply won't do, given Wittgenstein's very narrow concept of philosophy (which was that its sole concern is to uncover and eliminate conceptual confusions in language). Interestingly, Downes does admit that "if it turns out that there is a complex of genes for a set of mechanisms that usually lead to the formation of true beliefs that are important in guiding behavior, then I will have to take back what I am arguing in this paper" (441 n. 7). We could well reply that vision would be one such mechanism, also the sorts of abilities that Ruse (1986) claimed, such as inductive, deductive, and analogical reasoning.

But Downes' cannot be dismissed so easily. What is truly important is his focus on human evolution as occurring in "bands," namely, small nomadic hunting-gathering groups. This is an extremely important part of human evolution, something evolutionary epistemologists cannot afford to ignore. Many do, however, happen to ignore it (Darwin, as we have seen, was not one of them). At any rate, Downes characterizes the "standard story" as imagining two neighboring bands of hunter-gatherers in, say, a savanna with similar plants and animals. The band with "true beliefs about the plants will survive to reproduce" (433). Again and again, Downes characterizes the standard story as having it that "the truth" (435) is selected for, "the band with true beliefs" as just quoted. In so doing, however, he is preparing to refute by setting up a caricature of the opposing position, a fallacy known as the *straw man* (or straw person) fallacy. The reality is that most if not all evolutionary epistemologists, Popper and the Churchlands for example, recognize truth to be a matter of degree, so that if a higher degree of something confers an evolutionary advantage and it is hereditary, it can be selected for. The same is true for other traits such as beaks and vision, all of which is standard in evolutionary theory. Adaptive traits are not a matter of have it or not but of degree. Evolution is gradual and cumulative.

Having caricatured the standard story, Downes proceeds to claim that human bands would have had a fluctuating hierarchical structure and division of labor, making the selection for true beliefs less likely. Noticing further the present variety of ways that we use the concept of truth (e.g., not lying, discovering the intentions of others, discovering how many beers are in the fridge, discovering the nature of subatomic particles, etc.), he suggests that truth may be something too general to have been selected for, just like the human hand with its various current uses, very few of which were selected for by natural selection. For Downes, instead, what evolved as adaptations were a relatively small number of "cognitive mechanisms" which were later "co-opted to solve the vast array of problem solving tasks we face now" (440), and in this he clearly includes "doing science" (439). In other words, his suggestion seems to be that truth finding is an *exaptation*, that with the advent of civilization a small handful of cognitive adaptations were co-opted into the variety of activities we recognize today as truth finding. *Exaptation* is the term coined by Gould and Vrba (1982) to refer to a trait that *looks* like it was evolved for its present function, where in reality it was evolved for some other function and later co-opted for the present function (e.g., feathers in some small dinosaurs, which were originally evolved for temperature regulation and were later co-opted for flight).

This is an intriguing possibility, which arguably fails if proto-humans faced truth-finding situations largely similar to the ones we all face today (and if science, as many claim, is simply a refined and polished version of truth-finding in everyday life). Where Downes certainly did not err is in his focus on the fact that humans evolved in small hunting-gathering groups. It is this fact that creates some serious problems for the optimists in evolutionary epistemology, but not in the way that Downes imagined.

I have in mind here the late philosopher Howard Kahane, who took very seriously the fact that we evolved in small hunting-gathering groups and who was also a first-rate expert in the field of critical reasoning (informal logic), which includes the study of fallacies. This combination of interests produced in Kahane some very profound insights, to which I shall add some thoughts of my own here and there.

For a start, Kahane (Kahane and Cavender 2006, ch. 6) begins by noting that human beings routinely self-deceive, but he adds that self-deception often has survival benefits so that we should not expect natural selection to have weeded out instinctive self-deception. For example, when faced with a sudden and great danger the self-deception that the danger is not all that great allows us to think fast and react accordingly, where otherwise we might be paralyzed with fear and inaction. In the past it would be dangers such as predators, today it is dangers such as driving and hitting black ice (thankfully I'm still here to write about it). Similarly, the self-deception common to soldiers that they're not going to be the ones who die allows them to go into battle and fight with courage and vigor, whereas someone not deceived would calculate the odds and in many or most cases run away. Indeed war is pervasive in human history and a cultural universal that goes very far back into our past, arguably back to the proto-human stage, given the evidence from our nearest cousins the chimpanzees (more on this in Chapter 6).

Self-deception about death itself, in the many forms of denial—denial that death is always potentially only a few seconds away, denial that it is utter and final annihilation of the self, denial that deceased loved ones do not exist and so do not wait for us on the other side, and so on—greatly reduces anxiety and stress and the paralyzing effect of fear, allowing us to go about our daily routines. Indeed medical science has discovered the intimate connection between stress and the immune system, such that prolonged stress can be a major factor in the onset of illness. Stress, anxiety, and depression are also known to increase the chance of miscarriage in pregnant women. We should not be surprised, then, that wishful thinking in the form of religious beliefs—belief in divine providence, including belief in life after death and communication with the dead—has such wide appeal cross-culturally, both throughout the present and the past. The propensity is quite possibly an evolved adaptation (more on this in Chapter 8).

In connection with death, and also with suffering, we should add self-deception with regard to chance. On the surface in the very least, it should be obvious that chance rules at every level of life, from conception to death, and yet relatively few people seem to have the hardihood to admit this unsettling fact and to face it. It is so much easier to think that everything happens for a reason, that there are purposive agents or forces guiding our lives. To face the fact of chance on a daily basis can result in depression or worse. Hence the widespread belief, in defiance of all logic and evidence, in astrology and psychics and the power of prayer.[7] To all of this we might add the benefits of positive thinking, another kind of self-deception (since it is often not realistic thinking) that would confer survival value, simply because it confers a positive state of mind in the face of adversity.

What is rarely thought of is the survival value of stereotyping. In our politically correct times, "stereotyping" has become a contemptuous term. But stereotyping is not all bad. For example, those in our evolutionary past who failed to stereotype tigers as dangerous carnivores tended to end up their next meal. Stereotyping from a single instance, in the case of a tiger, would be a very beneficial instinct to have. But what is adaptive in one environment can be maladaptive in another. In the world of human affairs, when dealing with members of our own species rather than with members of other species, the quickness and ease of the instinct to stereotype can easily go astray, as can readily be seen in the scourge of racism. Indeed it takes practice, a real and concentrated effort, to resist the urge to stereotype.

Kahane makes much of the herd instinct in humans, the instinct to feel a part of something bigger than ourselves (we see this today in political parties, religions, cults, mass movements, gangs, and so forth). Herd instinct is not only the instinct to be part of a group and to follow but to let the group do our thinking for us. Truth belongs to the group, we find foolish what the group finds foolish, we dichotomize people into in-groups and out-groups, we desire to gain status in the group to which we belong, we are puffed up by praise and feel diminished by blame, and so on. In all of this, herd mentality could not work unless we also had an underlying gullibility and a propensity to be indoctrinated. As E. O. Wilson, the don of sociobiology, put it (1975), "Human beings are absurdly easy to indoctrinate—they *seek* it" (286), such that "Men would rather believe than know" (285). This is not always true, of course, but it should not take an enormous amount of observation to see that

---

7   In the case of astrology, believed in by slightly over half of all Americans, not only does it flatly contradict the basics of modern astronomy and physics, but logically also it makes no sense, for it entails that everyone born at the same time and place should have the same fortune (since they are exposed at birth to the same alleged celestial influences). But this, of course, does not happen. In the case of prayer, it should be obvious that it cannot possibly work when people with conflicting interests are praying for different outcomes. Moreover as Downes (2000, 441 n. 9) and Hume (1779, 186–187) and so many others have pointed out, since all religions flatly contradict each other on a great many matters, so that *at least* most of them cannot be true, it follows that, since most people guide their lives by their religious beliefs to some degree, most people guide their lives by false beliefs.

it is statistically true. (Arguably, a liberal education is the best counterbalance.) In the realm of religion alone, with its many thousands of conflicting divisions and doctrines, people easily become true believers. The same is hardly less true in politics (see Hoffer 1951). What is remarkable about true believers is that each thinks they belong to the group that has the truth, and they repeat the arguments taught to them with the fullest sincerity, conviction, and sense of righteousness, all the while never seeming to realize how much they share in common with true believers of different groups. In all of this there is no evidence of a spirit of truth. Indeed, as the 19th-century German philosopher Friedrich Nietzsche wrote in his book *Human, All Too Human* (1878, §483), "Convictions are more dangerous enemies of truth than lies" (234). A big part of the problem is that herd mentality makes us feel loyal to the group, so that disloyalty is considered harmful to the group, a betrayal, and consequently a sin. And, of course, when the fortune of the group goes sour, scapegoats are always sought. Indeed, much of fallacious thinking can be traced back to the herd instinct, fallacies such as false cause, mob appeal, false dichotomy, double standard, abusive *ad hominem*, appeal to fear, appeal to pity, appeal to the authority of the one (the leader), appeal to the authority of the many (the group), appeal to tradition, and so on.

In the business world, herd mentality is known as *groupthink*. Groupthink is a big problem in organizations as it as a symptom of excessive integration and cohesion. Businesses want team players, which is understandable, but groupthink stifles critical thinking. Thoughts contrary to the group are either self-censored or rationalized away. Warning signs and negative feedback are ignored. Criticism from within is considered a sin—a concept that has groupthink written all over it (whether in the business world, in politics, or in religion). Whistleblowing from within is the ultimate sin, the unforgivable sin, which results in ostracism, while critical outsiders are a less vile form of enemy. Indeed, one should always be wary when an organization likes to call itself a "family," and worse when the members call themselves "brother" and "sister," for this is the language of groupthink at its strongest. Labels like these make one feel good when received, but the fact remains that they are a major technique of manipulation, for both conformity and exploitation. (George Orwell's "Big Brother" in his *1984* is the classic warning.)

Interestingly, conformity studies in psychology lend weight to the argument that we have an instinct for herd mentality and groupthink. For example, in a famous series of experiments conducted by Solomon Asch (1955), groups of college students were shown two sets of lines, one line on the left and three lines on the right, and were asked to pick the line on the right that was most similar in length to the line on the left. Only one student in each group was the test subject. All the others were plants, instructed to pick the same but wrong line. What Asch found is that when the test subjects in discussion with the group learned their answers were different from the rest of the group, almost all went against the evidence of their senses and changed their answers in conformity with the group. In a subsequent study conducted by Bogdanoff *et al.* (1961), it was found that those few subjects who maintained their answers against the group suffered more anxiety than those subjects who switched their answers in conformity with the group, such that one subject was even dripping with sweat. What needs to be kept in mind in all of this is that the test subjects did not experience the pressures of conformity commonly found in the business world, or the worlds of religion and politics. Instead, the pressure came solely from within, suggesting the triggering of an innate instinct.

Groupthink has the potential to do enormous moral damage, evident not only in the business world with scandals such as Enron, but also in society as a whole. At the extreme end of the scale, every example of genocide, such as the Nazi holocaust, is an example of groupthink and herd instinct at its strongest. Remarkable are the arguments given to justify these actions, arguments believed by so many as to contain the truth, arguments given with full sincerity and the most intense conviction. Not only are convictions greater enemies of truth than lies, but they are also often the greatest obstacle to moral progress.

In his book *The Gay Science* (1882, §121), Nietzche argued that "Life is no argument. The conditions of life might include error" (177). Even though Nietzche had an excellent grasp of the herd instinct in humans and was fond of denigrating it, he displayed a poor understanding of Darwinian evolution. But his insight remains. As we have seen, adaptive instincts need not be truth-giving ones. In later chapters, we shall see arguments for various evolved instincts, such as a language instinct, a variety of mating instincts, moral instincts, religious instincts, and even an instinct for the meaning of life. On the other hand, humans (and not just humans) did arguably evolve epistemological features along with non- and anti-epistemological ones, the latter of which the scientific method (really a family of methods) evolved against as a system of inquiry combined with checks and balances best adapted to overcome the human, all-too-human part of us.

This brings us to the remaining question in this chapter, which is whether the science of evolutionary biology has the power to shed light on human nature, or whether we must rest content with entirely social-environmental explanations, the SSSM. The approach shall be to take a critical look at some prominent challenges to an evolutionary point of view.

One challenge comes from the growing trend in academia known as *postmodernism*. This is a philosophical orientation that developed in Continental Europe and that has spilled over into North America, becoming popular especially among graduate students in the social sciences and humanities. Although there are many different currents of thought that are covered by this umbrella term, there are a number of main elements (Berlocher 2001). For postmodernists, the distinction between fact and fiction, truth and falsity, objectivity and subjectivity, is an illusion. Instead, everything is a matter of perspective and interpretation, and there are no privileged perspectives or interpretations in an epistemological sense. Michel Foucault, for example, whose writings are very popular not only in Continental philosophy but also in sociology, argued that the only privileged perspectives come from positions of power. Foucault saw everything as a matter of "will to power," following Nietzsche's claim that will to power, or more specifically exploitation, is the essence of life. For Foucault, modern science has succeeded not because it is good at getting at the truth but instead because of political forces, forces elevating one group over another for the purpose of control and exploitation. The word "knowledge" is simply part of the language of power. Aside from its political power, science has no more claim to truth than the stories of the ancient gods. Each is a social construction, one had its day, the other is currently having its day. For Jacques Derrida, another leading figure in postmodernism, language cannot adequately capture reality, texts are a matter of interpretation, even nature is a kind of text, and language in a sense *creates* reality.

Postmodernism prides itself on taking the moral high ground. It embraces political correctness and multiculturalism and calls for an equal voice for marginalized and oppressed groups. Rather paradoxically, postmodernists do not extend the same friendliness to Western institutions and achievements, which are their favorite targets for criticism. Bashing the USA and bashing science go naturally hand in hand for them. Like the youths of ancient Athens who enjoyed following Socrates around in his public debates exposing the pretensions and false wisdom of people in high standing, postmodernism tends to attract young intellectuals.

A further attraction is that one really does not have to know what one is talking about in order to offer a postmodernist critique of this or that. Postmodernist critics of science, for example, show an appallingly poor grasp of the science they focus on. All that one really has to learn is the lingo and the basic ideas of relativism and skepticism. The language of postmodernism, which includes terms such as "deconstruction," "hegemony," and "hermeneutics," is typically thick and obscure, giving the illusion of depth and profundity. However, it is anything but. The lie of postmodernism was strikingly illustrated by the American physicist Alan Sokal. Tired of the postmodernist critiques of science that were becoming more and more prevalent in left-wing academia, Sokal decided to write a Benedict Arnold article, his own postmodernist critique of physics, which he sent for publication to a leading postmodernist journal, *Social Text*. Sokal's article (Sokal 1996) was quickly accepted and published, so happy were the editors to find a real scientist giving credence to their side. But the article was a hoax, a parody exposed by Sokal himself shortly after publication. Clothed in the style of serious scholarship, it was simply full of the bafflegab and rhetoric and thick nonsense common to postmodernist literature. A little later, Sokal co-authored a book expanding on the topic, aptly titled *Fashionable Nonsense* (Sokal and Bricmont 1998; see also Gross and Levitt 1998; Koertge 1998; *Lingua Franca* 2000).

Relativism and skepticism are hardly new. The ancient Greeks, for example, had the Sophists, whose teachings Plato battled in his written dialogues such as *Meno* and *Gorgias*. For Plato, skepticism and relativism are not so much intellectual positions as they are matters of personal character, namely, lack of integrity (saying what you do not really believe), laziness (the pursuit of truth requires a lot of work), and pessimism (negativity, negativity, negativity). But postmodernism brings in new twists and slants to an old philosophical tradition. I discussed it briefly above because, even though postmodernists do not share the basic premise of this book (that evolution is a fact not a theory, or more generally that there is such a thing as knowledge), postmodernist arguments recur in various places in this book, such as in the chapters on sex, feminism, and race. It is important to recognize these kinds of arguments and what is wrong with them. For a start, if one dismisses logic and evidence the way postmodernists do, and focuses instead on power politics and motives and such, then no amount of logic and evidence could possibly dislodge one from a postmodernist position, so that it is best to simply walk away from a debate with a postmodernist.

But more importantly, one needs to think of arguments as autonomous entities, such that arguments are good or bad irrespective of the nature of the person who gives them or their motives. To reject an argument because of alleged defects in the nature of the person who gives it, whether the defects are real or not, is to commit the *abusive ad hominem* fallacy, to reject an argument because of an alleged motive behind it, the *circumstantial ad hominem* fallacy. It is entirely possible that a defective person could give a perfectly good argument, likewise a person with an ulterior motive. Nietzsche might be right that arguments should be viewed as *symptoms* (I happen to think that he is largely right), but *why* a person gives a particular argument needs to be separated from *whether* the argument is good or not, whether the evidence and reasoning provided *in* the argument should command the assent of a reasonable person.

What remains in this chapter is to examine a number of arguments against the application of evolution, when taken seriously, to matters of human nature, arguments by some prominent scientists who accept that species arose by evolution but who advocate social rather than evolutionary arguments for human nature. These are preliminary matters that set the stage for much of the rest of this book.

We begin with a very popular article co-authored by the late paleontologist Stephen Jay Gould (famous for his popular essays and books on evolution) and the geneticist Richard Lewontin (famous for his work in the 1970s on genetic variation). In their joint article (Gould and Lewontin 1978), they provide a sustained critique of what they call "the adaptationist program," the tendency they perceive among evolutionary biologists "to focus exclusively on immediate adaptation to local conditions" (75) and to hold "the near omnipotence of natural selection in forging organic design and fashioning the best among possible worlds," such that adaptation by natural selection becomes "the primary cause of nearly all organic form, function, and behavior" (76). Often, they point out, the only criterion for an adaptationist story is "*consistency* with natural selection," so that adaptive storytelling often passes "without proper confirmation" (79). For Gould and Lewontin, instead, not only does natural selection often not produce optimization, such that there can be selection without adaptation, but there are many *major* causes and explanatory factors for organism features besides natural selection, such as (see the Glossary) genetic drift, pleiotropy, allometry, exaptation (the term had not yet been coined), phyletic constraints, developmental constraints, and architectural constraints (the latter three involving the chance entrenchment of basic body plans, or *Baupläne*).[8]

Perhaps the most controversial part of their critique is that there can be "adaptation without selection" (84). The Oxford zoologist Richard Dawkins (1986), for example, has famously argued that natural selection is the "only" explanation for "adaptive complexity" (288). I do not wish to get into this controversy here, except to say that what concerns us for the rest of this book is the question of recognizing a behavior or trait as an adaptation produced by natural selection. (Gould and Lewontin do

---

8   A point usually overlooked is that, in spite of their analysis, Gould and Lewontin still do regard natural selection as "the most important of evolutionary mechanisms" (81).

recognize that many adaptations were produced by natural selection.) The problem is that what looks like an adaptation produced by natural selection might not be an adaptation after all. Gould and Lewontin were not the first to point this out. G. C. Williams, for example, in his now classic *Adaptation and Natural Selection* (1966), states that much of his book is devoted to combating "unwarranted uses of the concept of adaptation," such that the concept of adaptation by natural selection "should be used only as a last resort" (11). (One of Williams's famous examples is that of flying fish—their return to water is not an adaptation but simply a matter of gravity.) Although not the first to point out problems with adaptationist thinking, Gould and Lewontin's article had the salutary effect of making biologists and others think more carefully than ever before about selectionist explanations. In his widely used textbook on evolution, Douglas Futuyma (1998, 356–360), for example, provides four criteria for recognizing genuine adaptations by natural selection: (i) complexity, (ii) correspondence with design an engineer might use to accomplish a goal, (iii) experiments which show that a feature enhances survival *or* reproduction, and (iv) the comparative method (the independent evolution of a trait in different species indicates convergent evolution, the same adaptive solution to the same adaptive problem, such as Bergmann's Rule: larger body size in colder climates as the solution to regulating body temperature).[9]

Of equal importance for the rest of this book are the critical analyses provided individually by Gould and Lewontin against the adaptationist program when applied to human nature, known as *human sociobiology* and more recently *evolutionary psychology*. We begin with two articles by Gould. Turning to the first, Gould (1980a) characterizes much adaptationist thinking in sociobiology as "just-so" stories, following the label of the novelist Rudyard Kipling, who introduced fanciful and humorous stories for how the leopard, for example, got its spots, or the camel its hump. For Gould, sociobiologists are telling just-so stories since they use "mere consistency with natural selection as a criterion of acceptance" (256). Again, "Virtuosity in invention replaces testability as the criterion of acceptance" (254). This denigration of mere consistency with natural selection and the call for testability are important points, and we shall have to keep them in mind when we turn in later chapters to claims for evolved instincts in humans. We shall also need to keep in mind Gould's claim that sociobiology "rests on firmer methodological ground when it seeks broad correlations across taxonomic lines" (257). This is basically the same idea as what Futuyma above calls the comparative method.

Gould's point about testability, however, although an important one, should not be taken to be the be-all and end-all of a scientific theory. Instead, as we have already seen above with Ruse and shall see in various other places in this book, there is a growing view in philosophy of science that testability is but one of a number of *epistemic values* that characterize modern science, some others being Ockham's

9 See Dennett (1995, 238–251) for an extended philosophical reply to Gould and Lewontin (1978), which includes an interesting discussion of the highly unorthodox theory that modern humans evolved from aquatic apes, a theory used to explain features such as hairlessness and the diving reflex.

Razor (that we should favor the simpler theory among competitors, all other things being equal), fertility (the opening up of new research programs), and consistency with what we already know. Hence, a theory in sociobiology might not really be testable but might yet be scientifically a good one for other reasons.

There are some further problems with Gould's (1980a) paper. Arthur Caplan (1982), for example, argues that Gould has mischaracterized sociobiological theories. For a start, he says they are not merely "consistent" with evolutionary theory but rather they are "directly *derived*" from it (268). This is a subtle but important point. Think of it this way. Darwin's theory of evolution is consistent with Newton's theory of gravity, but it would be entirely mistaken to say that Darwin's theory is derived from Newton's. Caplan (268–269) adds that sociobiological theories are theories of history and as such they share a number of criteria that distinguish them from just-so stories and myths, such as internal consistency, an avowed intention to be factual, and a willingness to test them against publicly available evidence. Genuine just-so stories share none of these features. G. C. Williams (1996, 23–27), interestingly, says there might be some analogy with Kipling's just-so stories, but adaptationist stories, he says, share far more in common with detective stories, which are guided not simply by rationalizing known facts but by predicting and seeking new facts, such that one theory is abandoned in favor of another if it provides a better explanation. In all of this, the detective wants to get at the truth. More recently still, John Alcock, in his *The Triumph of Sociobiology* (2001), devotes his fourth chapter to Gould's claim that sociobiologists place little value on testability, arguing that not only are sociobiological theories in general testable, but many have been tested and have been either confirmed or rejected.

Interestingly, and rather ironically given what we have just seen, Gould in an earlier article (Gould 1974a) says, "the new biological determinism rests upon no recent fund of information and can cite in its behalf not a single unambiguous fact" (238). What comes out clearly from this and other writings of Gould is that he clearly favors the SSSM, even though by profession he was a paleontologist (i.e., an evolutionary biologist).

Four points need to be made about Gould's claim above. First, the recurring derogatory slogan of "biological determinism," commonly used by critics of sociobiology and evolutionary psychology, contains a serious confusion. For a start, the word "determinism" in the history of philosophy and science means (if it means anything) the conjunction of two theories: first, that every event or thing has a cause and, second, that the same or similar causes necessarily have the same or similar effects. Since biologists routinely recognize that biology is statistical, and that biological traits are usually the result of the *combined* influence of genetic and environmental factors, it follows that no biologist is a *determinist*, let alone a *biological* determinist. But there is more. In his discussion of Rose, Kamin, and Lewontin's *Not in Our Genes* (1984), a title which aptly indicates its content, Richard Dawkins (1989) does a quick but good job of bringing to light the background ideology and fallacies common to critics of an evolutionary view of human nature. The background ideology is left-wing antireductionism, the denial that sociology and psychology can be reduced in any way to biology. The main fallacy concerns the bogey phrase "biological determinism," which Rose *et al.* and others like them continue to use in spite of responses like those of Dawkins. Rose *et al.* charge that sociobiologists such as E. O. Wilson and Dawkins contradict themselves in holding on the one hand that human behavior is biologically determined by our genes and on the other that we have the free will to go against our genes. They even charge that this forces Wilson and Dawkins into a position of "unabashed Cartesianism" (283), i.e., mind-brain dualism. In doing so they commit the *straw man* fallacy. As Dawkins puts it, "it is perfectly possible to hold that genes exert a statistical influence on human behaviour while at the same time believing that this influence can be modified, overridden or reversed by other influences" (331). Dawkins uses the example of sexual desire, which presumably Rose *et al.* would agree evolved by natural selection. As we all know, we can curb or resist the desire entirely if we really want to. Moreover, every time we put on a condom or use a birth control pill we are, says Dawkins, going against our genetic disposition to reproduce. In all of this there is no supernatural ghost in the machine. As well, says Dawkins, even though our genes built our brains, "We, that is our brains, are separate and independent enough from our genes to rebel against them" (332).

Second, the reader should keep in mind that the quotation from Gould above was first published in 1974. As such, it needs to be examined in the light of research before and since then, some of which is to be found in the present book. A greater source, although not involving much that is in the present book, is Alcock's *Triumph* (2001). I will leave it at that. Readers will need to decide for themselves whether it is true that no new evidence has been brought forward on the biological or evolutionary origin of human nature (if that indeed is what Gould is claiming).

Third, it should be noted that new theories in science, including what turn out to be true theories, need not be the result of any new findings but rather may be the result of looking at a difficult and seemingly intractable problem in a new way. Even if it were true that evolutionary theory from Darwin's time to the present can cite "not a single unambiguous fact"—it is not true, unless one defines "unambiguous" in the most contentious way—it remains true that the basic theory of evolution is today considered a fact because of the innumerable pieces of evidence which *together* point in one and only one direction, namely, that evolution really happened and is continuing to happen. Indeed, Darwin's *Origin of Species* (1859), although it employed some of Darwin's own research and findings (e.g., from the Galapagos islands), could have been written solely on the basis of evidence that was known to all of his fellow naturalists, the vast majority of whom were creationists. What Darwin did was look at the evidence in an entirely new way, a way, it turned out, that made much more sense of the evidence than any other way. In short, *inference to the best explanation* need not require any new evidence or facts.

Fourth and finally, in defense of his claim about no new facts, Gould (1974a) raises a problem for the distinction between *homologies* and *analogies* (between similarity due to common evolutionary descent and similarity due to independent adaptation, also known as *convergent evolution*). Gould says this difference is hard enough to make out when it comes to the physical features of organisms, so that "How much harder it is to tell when similar features are only the outward motions of behavior!" (241). Since sociobiology and evolutionary psychology are mainly about behaviors, this raises a problem for the comparative method. Nevertheless, it does not raise an insuperable problem. Instead, it is more of the nature of what the 18th-century Irish bishop and philosopher George Berkeley called "we have first raised a dust and then complain we cannot see." If we take evolution seriously, we have to accept that there are going to be homologies and analogies. The logic of the former for the comparative method was aptly stated by Alcock (2001):

> If two species are very closely related, they have a very recent common ancestor [ancestral species], that is, one that lived a few million years previously, from which they will have inherited a large number of genes. Some of the ancestral genes that both lineages have received are likely to remain unchanged over a relatively short period of time, geologically speaking, and therefore could be responsible for some of the shared attributes between the species. Detailed similarities between two very closely related species could therefore be the product of shared ancestry and need not have evolved *independently* from different genetic backgrounds. If so, we can use the similarities between these carefully selected species to infer what traits were present in their shared ancestor, one step back in these species' history.
>
> (75–76)

In short, some similar behaviors between us and our closest cousins the apes, for example, might indeed be shared from our common evolutionary past (such as our hair standing on end when alarmed—in us, goose bumps, in them, raised hair), while other behaviors in our species might well be independent adaptations to the same kinds of problems faced independently by other species. In both cases, we cannot know for sure by stepping into a time machine and observing evolutionary history. Instead, inferences have to be made. But that is the nature of science. The point is that some inferences are going to be stronger than others, but that does not mean we should forego such inferences. In the case of figuring out which species are more closely related to us than others, we can now be confident not simply by outward appearance but by the amount of shared DNA. In the case of figuring out human nature, then, behavioral homologies with closely related species are going to be a powerful form of evidence—indeed Darwin was the pioneer here yet again, this time in his work on emotions (Darwin 1872)—while analogies too can also be powerful when they are supplemented by other kinds of evidence, such as evidence indicating that the analogical behavior increases reproductive success. I fail to imagine why anyone would want to disparage this kind of scientific reasoning, unless of course they are motivated by an ideology with contrary commitments.

We turn now to a book by Richard Lewontin, titled *Biology as Ideology* (1991). What makes Lewontin such an interesting critic of evolutionary approaches to human nature is that Lewontin is a highly respected geneticist, famous for pioneering a technique in the 1970s for reading genes in chromosomes (gel electrophoresis) and for revealing greater genetic variability in the genomes of species than previously thought. Lewontin begins his chapter titled "A Story in Textbooks" with a discussion on political theories and the theories of human nature upon which they are based. That the different political theories of Plato, Hobbes, Locke, and Marx, for example, were based on different theories of human nature is obvious to anyone who takes the time to study political theory. It should also be obvious that not all of these different theories can be correct. What is interesting is that Lewontin himself is an ardent Marxist (as was Gould to a lesser degree) and Marxism itself presupposes that there really is no such thing as human nature (or that if there is, it is fundamentally plastic), that the way humans are is because of the social systems in which they were raised and live. To repeat the example I gave in the Introduction, according to Marxists we should not say that humans are inherently greedy, but rather that the capitalist system, by reason of private property and competition, makes them so; in a truly communist system, humans would not be greedy. It is therefore interesting that Lewontin, although himself a pre-eminent geneticist, is one of the most vociferous critics of human sociobiology. As an ardent Marxist, he cannot help but see human sociobiology (and by implication evolutionary psychology) as "the ruling justifying theory for the permanence of society as we know it," "the latest and most mystified attempt to convince people that human life is pretty much what it has to be and perhaps even ought to be" (63).

But is that what human sociobiology and evolutionary psychology are really about? Are they really social constructionist tools for maintaining the status quo? Like it or not, a very real possibility is that sociobiologists and evolutionary psychologists are trying to best understand the way we are, using the best scientific evidence and theories available, such that rather than trying to maintain the status quo they are trying to use that understanding to improve our lot in life. For example, the psychologist Mihaly Csikszentmihalyi (1993, 43–44) does not use an evolutionary explanation for why dieting is so hard because he wants to defend obesity and see it remain as a defining element of American culture. Rather, he uses it so that we can understand why we so easily become addicted to foods high in sugar, salt, and fat and why fast-food restaurants are so pervasive (and why, for example, the "Happy Meal" *really* has the word "happy" in it).[10] Armed with that understanding, we can avoid useless diets and be better able to change our eating and exercise habits so as to live healthier lives. Similarly, turning to some of the topics in Chapter 4, E. O. Wilson (1978) and Dean Hamer and Peter Copeland (1998) are not out to increase homophobia, but rather an understanding of, and a tolerance for, homosexuality, while Michael Ghiglieri (2000) and Thornhill and Palmer (2000) are not out to justify rape and give lawyers the ability to get their rapist clients off by reason of their genes, but rather first and foremost to understand the phenomenon of rape itself, coolly and clinically, and then to determine how best to fight it. (In medicine this goes on all the time: before treatment or a possible cure, one needs the correct diagnosis.) Lewontin apparently cannot see any of this as a possibility. At any rate, rejecting arguments because of a supposed ulterior motive is to commit the *circumstantial ad hominem* fallacy (even with an ulterior motive, one might provide a good argument). While the kind of

---

10   The answer, by the way, for why obesity is such a big problem and dieting is so hard is that our predilection for foods high in sugar, salt, and fat evolved in our prehistoric ancestors as an adaptation to environments where they often had to go without these essential nutrients for days. Since many of us now live in the land of plenty, these instincts for us are now maladaptive and cause us various problems, such as heart disease, diabetes, and joint problems.

argument a person gives might tell us something about the kind of person that person is, an argument has to be evaluated based on the merits of the argument itself.

So strong is Lewontin's conviction that it leads him to make one radical claim after another. We find, for example, that not only is human sociobiology an *"ideology of biological determinism"* (23)—Lewontin pays lip service to the reply that sociobiology is a statistical theory (28)—but that there are not enough genes in the human genome to code for all the variety of human social circumstances (72). This, of course, completely misses the point, since sociobiology is a theory of evolved instincts, and the human genome, as with other species, clearly has plenty of genes for that. This should be clear not only from the fact that the human brain has plenty of genes for its numerous bodily structures, but from the fact that the brains of animals with far smaller brains than our own have numerous suites of instincts coded in their DNA—instincts such as hunger, thirst, suckling, playing, mating, herding, migration, fight-or-flight, attraction to certain kinds of food, fear and avoidance of certain kinds of predator, predatorial strategies, web making, honeycomb making, nest making, dam making, and so on. Even so, evidence for instincts in humans— such as male dominance, xenophobia (fear or hatred of foreigners or strangers), religiosity, competition, aggression, introversion, extroversion, eroticism, depression, conservativism and liberalism—are, says Lewontin, "totally absent" (70). Organisms do not even adapt to their environments. Instead, he says, "They create them" (83). Lewontin also rejects the comparative method. Analogy (similarity in different species not due to common evolutionary ancestry) "is in the eye of the observer" (69). Kin selection theory and reciprocal altruism theory (to be discussed in Chapters 4 and 7) are also rejected as being essentially just-so storytelling, since they can be used to explain "the natural selection advantage of any trait imaginable" and plausible stories are not necessarily true stories (74). Wilson's kin selection theory of homosexuality (examined in Chapter 4) is specifically rejected for a variety of reasons, one of which is that there is a continuum for human sexuality, another is variation between cultures, while another is that "there is absolutely no evidence that there are any genetic differences between individuals of different sexual preferences" (76). Lewontin even vehemently rejects Wilson's claim that "Man would rather believe than know," saying it is "more in the line of barroom wisdom" (65). Every possible reason for skepticism is used, including developmental biology (the study of the production of variation in an organism due to developmental processes), such that *"developmental noise,"* the "chance element in development" (27), creates so much distortion or variation in an organism that it ruins the genetic program model of DNA. Never mind the enormous similarity between twins and clones! Indeed, Lewontin rejects studies on identical twins (same DNA) raised in different environments, raising problems such as small sample sizes and biased selection of evidence (32). To think otherwise, for Lewontin, is to buy storytelling. In our brains alone, he says, "There may be large random differences in the growth of our central nervous systems" (27), except for "only the most general outlines of social behavior" (72). Gone is the validity of the human genome project, the mapping of the entire DNA of a human individual, which he says is motivated mainly by big business and Nobel prizes (46–51). In short, "At the surface of this theory of human nature [human sociobiology] is the obvious ideological commitment to modern entrepreneurial competitive hierarchical society . . . the priority of the individual over the collective" (67).

In a book review, Lewontin (1999) writes, "The real question about human social and individual behaviour is not why men belong to some type, but why they are so extraordinarily variable in time and space" (729).[11] With this statement Lewontin *begs the question*, since he assumes that humans are so variable. Certainly he should not be taken as an authority here, even though he is a famous geneticist, for the simple reason that the topic of human nature is one where the authorities, the experts, some of whom are geneticists like Lewontin himself, disagree. But views like Lewontin's should be taken as a counterbalance to an all-too-easy acceptance of evolutionary theories of human mind and behavior.

Finally, the cultural anthropologist Marvin Harris (1999), one of the most influential anthropologists alive today, provides a number of reasons for concluding that, though nature versus nurture is an "empirical question," "the great majority of cultural traits are overwhelmingly shaped by socially mediated learning" (19), in other words, "The overwhelming majority of cultural innovations . . . do not get selected for or against as a result of their contribution to the reproductive success of the individuals who adopt the innovation" (107). Unlike some anthropologists, Harris defines "culture" as a matter of both thought and behavior (19–21). This is a good definition, because in later chapters we shall see arguments for evolved instincts in humans that involve not only behavior but also thoughts. Harris's definition of "culture" would not be a good one, however, if it means that cultural traits cannot be genetic traits as a matter of definition, for then Harris is taking an avowedly empirical matter and attempting to settle it with a definition. At any rate, Harris (100–109) provides three reasons for rejecting sociobiological and evolutionary psychological theories of human nature, to which I should like to provide some brief replies. Critically examining his objections is necessary because they represent ideas that are very widely held in the academic world.

First, he states that "cultural selection often does not favor behavioral and ideational innovations that increase reproductive success," which he thinks it should

---

11   Gould (1981) attempts an explanation for "flexibility as the hallmark of human behavior" (330) based on two human features: First, our vast intelligence (compared with other species) housed in an "oversized brain," such that "I think it probable that natural selection acted to maximize the flexibility of our behavior" (331). Second, *neoteny*, the process where "rates of development slow down and juvenile stages of ancestors become the adult features of descendants" (333). Because of neoteny, human adults resemble among their closest cousins the chimpanzees the juveniles rather than the adults (the juveniles are the chimps we typically see on TV and in movies). The two explanations go hand in hand. As Gould puts it, "In other mammals, exploration, play, and flexibility of behavior are qualities of juveniles, only rarely of adults . . . The idea that natural selection should have worked for flexibility in human evolution is not an ad hoc notion born in hope, but an implication of neoteny as a fundamental process in our evolution. Humans are learning animals" (333). In short, natural selection evolved high intelligence and flexible behavior in us by making us "permanent children." The problem is that even if all of this is true (neoteny in human evolution is generally accepted, that our brain is a general-purpose learning device is not), it does not negate the possibility that evolution also evolved in us a variety of instincts, some of which are dismissed as cultural artifacts. We shall have to decide the matter not by grand strokes as Gould does, but on a case by case basis, a program pursued in the following chapters.

if culture is a matter of evolutionary biology. In support of his claim, he cites as facts that poor people tend to out-reproduce wealthy people, in both the short and long runs, that many classes and castes practice female infanticide, and that adoption does not support the view that humans have an instinct for parenting given the widespread practice of child abuse. The facts Harris cites, however, granting that they are all facts, do little to show that human sociobiology and evolutionary psychology fail to help explain human nature. Unless, of course, one takes the latter views as theories of biological determinism. But as we have already seen, that charge is a straw man. People may in general have an instinct not only for sex but for parenting (keep in mind that biology is statistical, so that some just won't have it), but that does not mean that the instinct cannot be reduced, thwarted, or redirected (such as toward pets). Genes are not the only influence on our brains. Moreover, Harris's supporting evidence does nothing to refute claims for instincts (many of which would fall under his definition of "culture" if it is not a question-begging definition) such as differences in mating strategies between men and women, rape, homosexuality, morality, and religion.

Second, Harris claims that "reproductive success is almost impossible to measure in human populations," that it can seldom be directly measured in humans in terms of alternate behaviors. If true, one would think that this would negate any confidence in saying the poor out-reproduce the rich. But aside from that, genetic factors that increase reproductive success, if not directly measurable, can be inferred indirectly, especially if a number of indirect lines of evidence converge (such as evidence from reproductive success in closely related species, evidence that shows an increase in sex drive, and so forth). And after all, science is not simply about making direct observations. Often it involves inferences about unobservables, such as the evolutionary tree itself, or the origin of life, or (to go outside of biology) subatomic structure, or the expanding universe. In short, science is about *inference to the best explanation.*

Third and finally, Harris claims that the theory he advocates, which he calls "cultural materialism," is "more parsimonious" than Darwinian theories since it does not require any data about reproductive success. Cultural materialism is a theory of "cultural selection" where the factors that place selection pressures on cultural evolution having nothing do to with selection at the genetic level but instead have everything to do with the material basis of human life, calculated in terms of health and well-being. As Harris puts it, "In the presence of groups with conflicting interests, selection for or against innovations depends on the relative power that each group can exert on behalf of its own interests" (143). Hence, to use one of his examples (146), slavery is a matter of cultural selection not because we have genes for slavery but because the interests of slave-owning populations are served even though the interests of slave populations are not. Other cultural innovations might benefit all groups concerned (a point, he says, that distinguishes cultural materialism from Marxism). At the other end, cultural innovations where no group gains but everyone loses, such as suicide cults, can arise on this theory but they are going to be short-lived. In any case, there is no need, according to Harris, to resort to genes. Genetic evolution is generally very slow, whereas cultural innovations are often rapid. The reason this can happen is because cultural innovations are not coded in genes but rather in the software of brains, allowing for rapid spread from brain to brain and consequently for rapid evolution. Indeed, says Harris, "no matter how culturally divergent any two human societies may become, they can always exchange cultural features (or the information necessary to construct such features)" (108). This is very unlike biological evolution. For all of the above reasons, then, Harris sees explanation of human nature by cultural evolution as being more parsimonious than Darwinian evolution, and hence more desirable epistemologically.

Harris has an important point here. If two theories are of roughly equal value in terms of explanatory power, it is an epistemic virtue to favor the simpler theory, the theory that involves less explanatory entities (Ockham's Razor). Perhaps the most common reason given to support this idea is the belief that the simpler theory is less likely to be false. At any rate, because of Ockham's Razor, and the obvious fact that many features of human behavior and thought are purely cultural (such as the miniskirt and the game of chess), the temptation might arise to affirm the SSSM, including Harris' theory of cultural materialism, as the default position when compared with Darwinian models.

The SSSM, however, is not as simple as it at first appears. In fact, its simplicity is highly deceptive. This is because if one grants, as the vast majority of SSSM theorists do, that humans are an evolved species, then one needs a theory for why the very same processes that produced rabbits with their rabbit nature and wolves with their wolf nature produced humans with an enormously plastic human nature (which is what the SSSM holds). What the SSSM requires, then, is a biological/evolutionary theory of human nature. And so all of a sudden the SSSM is not so simple.

If we move from the global level to the level of particular cases, even with a plasticity theory the SSSM does not automatically become the simpler model. This is because it might turn out to be the case that what appear to be merely cultural constructs have not only an underlying universality but also a past reaching back through human prehistory. In such cases, Darwinian models would seem the simpler, with the burden of argument resting much more heavily on the shoulders of the SSSM.

In short, then, when it comes to evolution and the big questions one should make an inference to the best explanation on a case by case basis, employing as much evidence and established background theory as necessary, rather than getting caught up on what should be the default position. In the chapters that follow, we shall see more clearly how this and other epistemological themes play themselves out when applied to specific questions and cases.

2

Evolution and Consciousness

Arguably the biggest problem in philosophy of mind, in trying to figure out what mind is, is the problem of consciousness (Nagel 1979, ch. 12; Kim 1996, ch. 7). Part of the problem is that consciousness is intangible and seems so utterly unlike matter that its very nature seems to defy any naturalistic or scientific explanation. Consciousness is private, open only to introspection. We can open up and study the brain all we want, but we will never see consciousness. And yet nothing could be more obvious when I am conscious than that I am conscious. But how can I know that you are conscious, or you know that I am? The best we can really do is only make inferences, we cannot directly know. (This creates some interesting questions in philosophy, questions sometimes explored in science fiction, but I shall ignore them here.) Nevertheless, we do seem to *know* when not only ourselves but others are conscious, when we are fooled and when we are not, and the words "conscious" and "consciousness" are common enough in everyday language.

So just what are we talking about when we are talking about consciousness? And can biological evolution, if not help us to solve the problem of consciousness, at least help us to understand it a little better?

In this chapter we are not going to solve the problem of consciousness. Instead, what we shall see is how consciousness might fit into an evolutionary picture. Some solutions to the problem of consciousness, however, can be quickly dismissed. One is the approach made famous by the Austrian philosopher of language Ludwig Wittgenstein, which became very popular in the mid 20th century for solving problems of meaning. For Wittgenstein (1953), "the meaning of a word is its use in the language" (20). There is no point trying to figure out what a word should mean, or might mean. To know its meaning, simply confine oneself to how it is used in a language community. Granted, this approach is useful for certain words, such as "or"—if you look at how we use this word, we sometimes use it to mean "either of two alternatives but not both" (the exclusive sense) or "either of two alternatives and possibly both" (the inclusive sense). But the Wittgensteinian approach works horribly for many other words, such as key terms in science. Granted, Wittgenstein added to the above quotation the qualifier "For a *large* class of cases," but my point still stands.

When it comes to scientific terms the Wittgensteinian approach is superficial at best. One is never going to figure out what a biological species is, or a law of nature, simply by examining how language users use these terms, even if we confine ourselves to a language community of experts. (There is more to Wittgenstein's approach to the meaning of consciousness, and I shall turn to it later in this chapter.)

Another linguistic approach that we can quickly dismiss is what is known as the *causal theory of meaning*. Beginning with proper names, the philosophers Saul Kripke (1980) and Hilary Putnam (1975) noticed that we can in fact refer to something, like the historical person Aristotle—this is what we *mean*—even if we are completely wrong about everything we believe about Aristotle, just so long as our use of the name "Aristotle" is part of a causal-historical chain going back to the initial use of the word (the "baptism," as they like to call it). Of course, we not only use words to refer to things such as individuals, but also to kinds, events, and processes. Kripke and Putnam extended their theory of meaning to these as well. But even so, we should still want to know, to return to the word "consciousness," just what we are referring to when we use that word (which may turn out to be more than one kind of thing or process). Even if our use of the word "consciousness" is indeed part of a causal-historical chain going back to the first use of the word, we are still in the dark as to what consciousness is.

In order to know what we are talking about when we talk about consciousness, then, we need to scratch beneath the surface of language use, whether the synchronic (horizontal) approach of Wittgenstein or the diachronic (vertical) approach of Kripke and Putnam. And for that we need to get interdisciplinary.

What we shall see when we bring in evolution is that there are basically three approaches to an evolutionary understanding of consciousness. One is that consciousness is a direct adaptation, like beaks or eyes, that it evolved because it increased reproductive success in a field of competition. Another is that it is not a direct adaptation, that it is not something directly produced by natural selection, but that it is a byproduct of something else that was evolved directly by natural selection, something like feathers were originally evolved for temperature regulation but proved useful for gliding.[1] The remaining approach is to look at consciousness, granting biological evolution, as so utterly different from matter and material processes that it can only be explained, quite literally, as a miracle. Since the last of these will appeal to the majority of readers, let us begin with that and then work our way through the second and first approaches.

Probably the most famous approach to consciousness as literally a miracle, all the while granting biological evolution, is the argument provided by Richard Swinburne, Professor of Philosophy of Religion at Oxford University. What I like about Swinburne is that, although a theologian, he takes science seriously, including

---

1 This is a recurring theme throughout this book. For example, in later chapters we shall see it argued that language is merely a byproduct of the evolution of bigger brains or of greater intelligence, that rape is merely a byproduct of the evolution of promiscuity, and that the projection of gods or God onto nature is merely a byproduct of the evolution of agent-detection systems in our brains. Indeed, how to distinguish an adaptation from an adaptive byproduct is no simple matter, but we shall see attempts at solutions when we get to those problems.

evolution. Indeed he is a theistic evolutionist, believing that God guides evolution in some way (more on that approach in Chapter 8). But does Swinburne take science, in particular evolution, seriously enough? That is the important question, and I shall argue that he does not. Seeing why his kind of argument fails will also help pave the way for explanations of consciousness that take evolution more seriously.

Swinburne (1987) begins by making a distinction between physical properties or events and mental properties or events (hereafter I shall simply refer to events). Physical events are public, mental events are private (only I have privileged access to the mental events inside my head). Among mental events are conscious events, such as perceptions, sensations, purposings, and even dreams and hallucinations. Thus inside my head right now there are two utterly different kinds of events going on, brain events and mental events. While brain events can cause other brain events, and mental events can cause other mental events (as when doing arithmetic), Swinburne allows that brain events can cause mental events (as with pain) and mental events can cause brain events (as when I choose to do something). This latter possibility is known as *downward causation*, and Swinburne has no problem with it whatsoever. Indeed throughout it all, whether brain events are causing mental events or vice versa, Swinburne maintains it as "evident" that "there are two distinct goings-on" (214). Swinburne also allows that higher animals have consciousness and that at certain points in the evolution of life different facets of consciousness first appeared, although he will not speculate on when or where. In all of this, Swinburne is careful not to commit himself to the view known as *mind-body dualism*, where mind is considered a distinct and potentially separate kind of stuff. He says quite clearly that he is "not assuming here that minds are substances" (215). Traditionally in philosophy, the concept of substance was not necessarily that of a physical substance but merely anything that can exist independently and has properties. Swinburne is not against the view that mind can exist without body; instead, his argument does not require it.

So why then does Swinburne not believe that consciousness can be explained scientifically? The problem is that science would have to explain how the evolution of physical systems gave rise to conscious events. Beginning with physics and chemistry, and then looking at evolutionary biology, Swinburne argues that science is not capable of doing this.

Physics and chemistry, according to Swinburne, would have to catalog all the correlations between a given kind of mental event and the corresponding kind (or kinds) of brain events, and do this for every kind of mental event. If that could ever be done, and of course he doubts that it could, he then makes the further claim that physics and chemistry "cannot possibly explain" (216) why a particular kind of brain state is correlated with, or gives rise to, a particular kind of mental state.

One reason Swinburne gives is that a scientific explanation requires an appeal to at least one law of nature. If we want to explain why a ball came back down to the ground after we threw it up into the air, for example, we need to refer to the law of gravity. If we want to explain why the water in the pot came to a boil, to give another example, we need to refer to the properties of water. Laws state physical or natural necessity for natural kinds. Thus to give a scientific explanation is to state the initial conditions and the relevant laws. Mere correlation, on the other hand, "does not explain, and because it does not explain, you never know when your correlations will cease to hold" (218).

Moreover, Swinburne trades on a distinction I discussed at the start of this chapter. Physics and chemistry can only study public events. When it comes to consciousness, the relevant public events are brain events. But "Brain events are such different things qualitatively from pains, smells, and tastes, thoughts and purposings, that a natural connection between them seems almost impossible" (218).[2] The history of science, says Swinburne, has plenty of examples of successful reduction, such as the reduction of temperature to the movement of molecules, but the supposed reduction of mind to brain, he says, is quite different, since mental events are radically different kinds than physical events. This kind of reduction he does not think can ever happen.

Swinburne then turns to evolutionary biology to show why it in particular fails, even if the physics and chemistry part above could be solved. Here it is interesting to note what Swinburne concedes. He admits the possibility that it might be easier for evolution using DNA to make an organism with sensations than one that has the same repertoire of behavior but without sensations. He admits the same for conscious awareness compared with an organism without conscious awareness. He even admits that it might be impossible for evolution and DNA to build a brain with complicated abilities to react to the environment without also having mental (conscious) states. Moreover, he admits that organisms with true beliefs (in the sense of correspondence to reality) are more likely to survive and reproduce than animals with false beliefs, hence the former will be favored by natural selection. And yet in spite of all of these admissions, Swinburne says that "It is not easy to see what the selective advantage of having a mental life is" (221). With sensation it is not clear to him what it adds to mere stimulus-response. With conscious awareness it is not clear to him what it adds to unconscious disposition. This is the crux of his argument with regard to evolution. If it is not clear what the selective advantage of consciousness is, then evolutionary biology cannot possibly argue that consciousness evolved by natural selection.

To aid his argument, Swinburne turns to a conundrum popular in philosophy of mind known as the *problem of inverted spectrum qualia* ("qualia" refers to the inner or qualitative nature of our mental states). For example, when looking at this tomato, I see what I call "red" but you might see what I call "blue," and yet in spite of our different qualia our mental states perform the same function. The problem of inverted spectrum qualia is a problem for the theory of mind known as *functionalism*, according to which *all* mental states are functional states.[3] Clearly if we can have

---

2 When following an argument, whether in science or philosophy or anywhere else, one always has to pay close attention to the use of words, much like a lawyer. In this case it is not clear whether Swinburne really means "almost" or something stronger. Near the end of his article he asserts that brain states and mental states "have no natural connection" (224), the implication being that they can only have a *supernatural* connection. Either Swinburne contradicts himself, then, with the "almost" above, or near the end of his article he has only become more confident in his argument.

3 On this view, mind is simply the functioning of the brain. If two different brain states, or a brain state and a computer state, are performing the same function, then they have the same mental state. In this sense, mind is not reducible to the physical, even though mind cannot exist without the physical.

different qualia, then not all mental states are functional states. Swinburne, however, uses the problem against an evolutionary explanation of consciousness, since it seems clear to him that evolution cannot explain why we would have different qualia.

If science, including evolutionary biology, cannot explain the fact of consciousness, then what can? For Swinburne, there are basically two kinds of explanation. One is the scientific one, involving laws of nature. The other kind is personal explanation. For example, we can explain scientifically why this water boiled by stating that the water was at sea level and was raised in temperature to 100° centigrade and all water at sea level boils at 100° centigrade (law of nature). We can also explain why the water boiled by stating that Joe put the pot of water on the hot stove. Indeed, appeal to a purposive agent is a kind of explanation we accept all the time in everyday life. Thus, for Swinburne, given the failure of science to explain mind in general and consciousness in particular, it seems only right to appeal to God, the ultimate purposive agent. God would explain the mysterious correlations between brain states and mental states since God is the one who instituted those correlations in the first place, not natural necessity. And why would God do that? The answer, for Swinburne, is "to allow them [humans] to share in the creative work of God himself" (224).

There are numerous problems with Swinburne's solution to the origin of consciousness. One has to do with the nature of theological explanation itself. Right from the start, it cannot help but raise further problems, such as proving the existence of God and solving the problem of evil (why there is so much evil in the world if God exists). None of these problems, of course, are inherited by purely scientific explanations. But perhaps the most interesting problem with theological explanations involves simplicity itself. In science it is a guiding principle to find the simplest, the most economical, the most parsimonious explanation of diverse phenomena (Ockham's Razor). It is often claimed by Swinburne and others that "It is simpler to postulate one unseen God" (Davies 1992, 190). But is it really? In his recent book, *The God Delusion* (2006, ch. 4), the Oxford biologist Richard Dawkins argues that the concept of God is not at all a simple one but instead is extraordinarily complex. This is because God cannot be simple if he is to do all the things that he is claimed to do, such as creating and maintaining the universe and answering prayers. The more complex an entity is, however, says Dawkins, the more improbable it is. That does not mean that God does not exist, of course, but it does mean that he is the most improbable being of all! Hence, an explanation in terms of God is not really a simple explanation, first because it is an appeal to the most complex being of all, and second because, as Dawkins puts it, "His existence is going to need a mammoth explanation in its own right" (149).

A further problem for Swinburne is that it is no longer the consensus in philosophy of science that a scientific explanation requires an appeal to one or more laws of nature. This view of scientific explanation, known as the *deductive-nomological* model of scientific explanation (or *covering law* model), made famous by Carl Hempel (1965), was popular in the mid 20th century, mainly because philosophy of science at that time focused almost exclusively on physics, which was considered by many to be the paradigm science. With an increased focus on developments in biology, however, philosophy of science underwent a profound change. One change was that the deductive-nomological model of scientific explanation became old-fashioned, since it applies little if at all to explanation in biology (Knowles 1990; Beatty 1995). The particularities of the standard genetic code, for example, with its synonymy of codons (Glossary), along with the non-standard variations in certain organisms and organelles such as yeast and mitochondria, are not explained by laws of chemistry or physics, but rather by the contingencies of evolutionary history (Stegmann 2004).

This brings us to Swinburne's claim about two distinct kinds of events, namely, brain events and mental events, and his claim that they can have no natural connection. This is simply a matter of the fallacy known as *begging the question*, the fallacy of taking for granted precisely what is in dispute.[4] Introspection, of course, seems to provide strong support for Swinburne's claim. But introspection can easily be wrong. It is wrong when it tells us that pain is bad (from an evolutionary point of view pain evolved because it increases survival and ultimately reproduction). It is probably wrong when it tells us that mental events precede brain events in acts of willing (soon below we shall see strong evidence that the opposite is true). It may well be wrong when it tells us that we have an enduring inner self. Certainly introspection biases our thinking toward mind, inclining us to underestimate matter and to overestimate mind. Science, on the other hand, points toward the opposite view. Perhaps, then, introspection is just plain wrong about the nature of mental events. Certainly downward causation is not an established fact. If anything, it is a very troublesome concept, along with the concept of mental causation itself (Heil and Mele 1993).

As for Swinburne's correlations, a correlation does not necessarily mean that there is not a natural causal connection. A prime example of correlation is the genetic code, namely, the various correlations between the codons (the 64 triplets of RNA letters) and the 20 amino acids that make up the proteins that make up our bodies. The codon AUA, for example, codes for isoleucine in most animals but codes for methionine in mitochondria. In neither case is there natural necessity (law), and yet both are cases of physical causation.

Swinburne, of course, would reply that the correlation here is between physical kinds. And so it is. But again we must not assume that mental kinds are radically different kinds from physical kinds. Nor must we assume that denying they are radically different kinds commits us to some kind of reductionism. Indeed Swinburne's argument against reductionism seems somewhat of a red herring. John Searle, for example, as we shall see later in this chapter, is an antireductionist (mind for Searle is a macro-level property of brains, much like liquidity is a macro-level property of H_2O), and yet he does not think this means that mind is non-physical and cannot be explained scientifically. Moreover, the history of successful reduction should make us think twice about assuming that mental kinds are radically different kinds than physical kinds. For a long time *life* was thought to be a radically different kind than non-living kinds of things, something added, something substantial, some sort of life force or stuff. But life has been successfully reduced to chemistry (including DNA). Science sees clearly now that life is not a force or a stuff. Life *simply is* cellular processes

---

4   The famous brain scientist Sir John Eccles (1987, 1989) is guilty of the same fallacy on this topic.

and nothing more. Perhaps mind in general and consciousness in particular will suffer the same fate, as eliminative materialists hold that it ultimately will (Churchland 1988). Or perhaps mind will be reduced explanatorily but not ontologically, the way information is explained in the biological domain. A gene, for example, is not the physical piece of DNA or RNA that biologists commonly refer to (contrary to their linguistic usage), but the information (something non-physical) it carries, much of which evolved by natural selection (Williams 1992, 10–13). This is not the same as Searle's approach to consciousness, but it is something more like it.

Arguably the biggest problem for Swinburne is not his antireductionism but rather that he just does not seem to get it when it comes to the basic principles of evolution. Quite simply, any trait (whether morphological, mental, or behavioral) that exhibits heritable variation and nonrandom differential fitness is subject to evolution by natural selection. It really is that simple, and consciousness does seem to exhibit these characteristics. Missing this, Swinburne then fails to grasp the evolutionary advantages that consciousness might indeed confer, his concessions notwithstanding. This is where the real debate is, and his focus on qualia does nothing but make him miss the real action. From an evolutionary point of view, different qualia might be a variable that natural selection is indifferent to. What matters, instead, is that my qualia are internally consistent and yours are internally consistent. Inconsistent, fluctuating qualia in a single individual under the same conditions would be maladaptive and selected against. But not consistent qualia. Indeed, they aid in discriminating things from other things, to a better degree than color blindness, for example, and hence would have an evolutionary function.

Consciousness might be a miracle after all. For Swinburne and others it literally is. But it will take an argument much stronger than Swinburne's to show that this is so. Moreover, there are much better evolutionary explanations than Swinburne lets on, and we need to examine them.

Perhaps before asking what consciousness is we should begin by asking what it does. This is the kind of question that would immediately occur to an evolutionist. Maybe consciousness is an adaptation. But immediately we are faced with a problem, namely, evidence which suggests that consciousness does nothing but is a mere byproduct of something else.

One powerful piece of evidence suggesting that consciousness does nothing, that it is a mere *epiphenomenon* (a non-causal byproduct) of the brain, comes from the work of the neuroscientist Benjamin Libet, conducted from the late 1970s to early 1990s (Libet 1993). In a series of experiments, Libet and his co-workers attempted to determine the time difference between a subjective, conscious event and the neuronal activity correlated with it. For example, in one experiment they asked their subjects to watch a digital clock (which showed seconds and milliseconds) and to record the exact time they willed a particular action, such as flexing a finger. Libet had electrodes attached to their brains. Invariably the neural activity correlated with the activity of the will occurred roughly half to a full second *before* the time reported by the subject. There is much more to Libet's experiments and there are a number of possible interpretations of his results (see Dennett 1991, 154–167; Penrose 1994, 386), but that consciousness is an epiphenomenon of the brain is one of them.

Another kind of evidence for the byproduct thesis comes from the study of language. Perhaps the simplest evidence is that our earliest childhood memories stretch back typically to around the time we started making genuine sentences. This might just be a coincidence, dependent on any of a number of developmental factors, or it might be the result of a causal link with language. Indeed, many thinkers from a variety of backgrounds have argued that consciousness is invariably tied in one way or another with language. Language here is taken in the full-bodied sense of the term, that of sentence formation (semantics and syntax), not simply symbol meaning (semantics). The difference is enormous, as we shall see in the next chapter. What is remarkable about this view of consciousness is that it means that animals and pre-linguistic children, as well as adults who have been deprived of language acquisition or have lost their linguistic ability because of brain damage, are not conscious! I do not want to give the impression that this is the consensus view in philosophy of mind, but it does have a long history and it is still fairly widespread. I want to take a brief look at some representatives of this view before I get to some more substantial evidence which supports it.

First on the list is the 17th-century philosopher René Descartes. In Part V of his *Discourse on Method* (Cottingham *et al.* 1985), Descartes gives what he calls "two very certain tests" (139–140) for whether something has mind. The first is the ability to put words together into a meaningful sentence, the second is the ability to genuinely solve problems (as opposed to instinctual or habitual behavior). By the first test, parrots, the best speakers besides humans, utterly fail the test. They can repeat not only words but whole sentences, and yet one never finds that they make new sentences with new meanings. Thus, for Descartes, all animals are mindless automata, machines without minds, in spite of their behavior and the noises they make which seem to suggest otherwise, while only humans are machines with minds. Or rather minds with machines. Descartes was a substance dualist, thinking that humans are made of two substances, mind and body, the former continuing on while the latter is destroyed. Very few thinkers in the area of philosophy of mind (neuroscientists included) believe that sort of thing anymore. There is too much evidence strongly suggesting that mind is either the functioning of the brain, or is in some other way the product of the brain, or is simply just the brain, namely, the variety of evidence from localized brain damage, from split-brain research, and from PET and MRI scans (see Crick 1994, ch. 12). Certainly evolution provides no basis whatsoever for a belief in substance dualism. And yet it is still easy to find the Cartesian view on the absence of consciousness in animals, as expressed for instance by the Descartes scholar Zeno Vendler (1972), in which he wrote, "we must agree with Descartes that animals, strictly speaking, cannot be conscious of their sensations and other experiences" (162).

Another example of the linguistic approach to consciousness comes from Ludwig Wittgenstein. Although his writings are typically cryptic at best, so that trying to figure out what he meant is more often than not a game among scholars, he has written some things that have suggested to many that consciousness is embedded in language, particularly given what he wrote on pain and private language in his posthumously published book *Philosophical Investigations* (1953). Advocating a behaviorist approach to meaning, such that mental concepts do not refer to inner private states but rather to public behavior, Wittgenstein argued (§§241–317) that there is no such thing as private language. Instead, Wittgenstein emphasized the role of convention and social practice in language, such that although he allowed for the possibility that animals might indeed have conscious experiences such as pain, he added that they and we have no established convention enabling us to say we know that they are in pain. We can only do this with other humans in our language community, where shared use of words is a matter of public agreement. Hence, for Wittgenstein, we have no good reason to suppose that animals have conscious experience. From this it is a small step indeed to the rather common view among language philosophers in the tradition of Wittgenstein, the view that "there can be no thoughts and no thinking if there is no language" (Hartnack 1972, 551).

The most radical (and to my mind fascinating) connection made between language and consciousness was made by the late Princeton psychologist Julian Jaynes (1976), and it is interesting to see how he works in evolution here. Not only are all animals not conscious, according to Jaynes, but humans only began to be conscious roughly 3,000 years ago! For Jaynes, consciousness is not perception, for while driving a car we can daydream (we perceive and respond accordingly, but we are not conscious of what we are doing). Instead, for Jaynes, consciousness is a kind of mind-space generated by language. Just as seeing requires an eye that does the seeing, consciousness requires an analog "I" that does the seeing in mind space. Look, he says, at the language we use to describe consciousness: we see theoretical problems from a *point of view*, we can be *narrow*-minded or *broad*-minded, we can get something *out of* our mind, or *off* our mind, or keep it *in* mind. Moreover, the analog "I" that introspects is not simply recognizing one's face in a mirror (chimpanzees and other animals can do that), but an "I" that has a consciousness of itself over time. On this account, one has to wonder what humans were like before they became conscious. For Jaynes, when language evolved in humans an "offshoot" (216) of this development was what he calls the *bicameral mind*, minds divided into basically two components (neither of which were conscious), one the commanding side located typically in the right hemisphere of the brain, the other the obeying side located typically in the left hemisphere. The commanding side was the voice of the gods (not real gods, of course), which literally spoke to people, as indicated in ancient religious writings, drawings, and sculptures. People were unconscious automatons, obeying the voices inside their heads. The bicameral mind began to break down, however, roughly 3,000 years ago, probably beginning in Mesopotamia, where a number of crucial events occurred roughly at the same time. One of these was the invention of writing, which "was gradually eroding the auditory authority of the bicameral mind" (208). Another was an enormous volcanic eruption or series of eruptions on the island of Santorini, roughly 60 miles from Crete. This natural calamity resulted in widespread climate change and enormous human migrations and invasions of refugees, failing to disrupt Egypt but succeeding in Assyria, the biggest center of commerce, where social organization became chaotic with the resulting overpopulation. In this place and under these circumstances, the authority of the bicameral mind broke down and consciousness was born. Although Jaynes stresses that the origin of consciousness and its spread throughout the rest of the world was mainly a cultural phenomenon, not requiring any change in the brain, he does add that natural selection would have augmented this process, since those humans who had genes that predisposed them to a breakdown of their bicameral minds would have survived and reproduced over those who had genes that made them resist the breakdown of their bicameral minds (220–221). One by one, over a relatively rapid period, bicameral minds broke down, the voices of the gods becoming silent in one human head after another, the only remaining vestiges today being found in the verbal hallucinations of schizophrenics and in the phenomenon of hypnosis (indeed, say what one will about Jaynes's thesis, he has a surprisingly powerful explanation for why hypnosis works).

Although the views of the various authors surveyed above vary in details and perspectives, they have in common the view that consciousness is inextricably mixed with language, such that language is a sign of consciousness, or language is a necessary part of consciousness, or consciousness is an emergent property of language, perhaps with some causal force of its own, or consciousness is an epiphenomenon of language, a noncausal byproduct. For those of these theorists who accept biological evolution (Descartes is the obvious exception), their view would have to be that consciousness is a byproduct of evolution, either an exaptation or not, so that if language is an adaptation then consciousness is a first-order byproduct, and if language is itself a byproduct of something else (a possibility we shall examine in Chapter 3) then consciousness is a second-order byproduct, a byproduct of a byproduct. In any case, many readers will think that all of this is crazy, since it means that their sweet little cats and dogs and pre-linguistic children are not conscious beings at all but automatons, their sparkling eyes windows to nothing but unconscious blanks.

One should not, however, let what one thinks *ought* to be the case cloud one's judgment about what *is* the case. And in fact there is some evidence to support the linguistic view, the most notable of which is the Wada test. In the Wada test, sodium amytal is injected into the left carotid artery, the artery that feeds blood to the left hemisphere of the brain. This causes the cortex in that hemisphere, which contains in most people their language and speech centers, to be anesthetized along with the rest of the hemisphere. The result is that the anesthetized subject cannot talk. So far, the Wada test only supports the lateralization of language. But there is more. The left hemisphere controls the right side of the body, the right hemisphere the left side. If one places, say, a spoon in the left hand of a person whose left hemisphere has been anesthetized, the spoon is both seen and felt only by the right cortex. Now if one removes the spoon before the anesthetic wears off, and after it wears off asks the subject what was in the hand, the subject of course is now able to speak but claims to remember nothing of the spoon and accordingly denies that a spoon was put in the left hand. Apparently the subject was not conscious of the spoon when it was placed in the left hand, otherwise the subject would have reported it when asked. The Wada test suggests to some, such as the evolutionary linguist Derek Bickerton (1990, 212), that consciousness requires the functioning language center of the brain. If that center is not functioning, then there is no consciousness.

There are problems with this conclusion, however, not the least of which is that at best it proves only that consciousness and language reside together in the cortex of the left hemisphere of the brain, not that consciousness is dependent on language.

The speech center in the left side of the brain is complicated, involving Broca's area near the front, Wernicke's area near the back, and the arcuate fasciculus, a bundle of nerve fibers beneath the cortex connecting the above two areas and which has been implicated in storing linguistic information and acting as a language generator (Springer and Deutsch 1993, 152–154). To get the conclusion that Bickerton and others want, one would have to anesthetize only those specific areas of the left hemisphere, being careful to anesthetize nothing more, and then see if the same results happen. To my knowledge this has never been done.

Perhaps the biggest mistake is to think that consciousness is a matter of yes or no, either you have it or you do not. Again, we cannot possibly observe consciousness in others, only in ourselves, but even given this limitation we should be able to note some aspects of consciousness in ourselves and then seek evidence for those aspects in others, particularly in animals. This might lead us to an evolutionary explanation of consciousness as a direct adaptation.

For a start, we can see that consciousness is a matter of degree and has different levels or kinds. I, for example, am never fully conscious when I wake up. I am not a "morning person." But this does not mean I was unconscious in my sleep. While dreaming I also seem to have a kind of consciousness, totally internal and usually illusory. (I say "usually" because I once had a dream so bad that I said to myself in my dream, "Don't worry, Dave, it's just a dream.") Even if after waking I fail to remember what I was dreaming I still arguably had a kind of consciousness. Of course, when I am fully awake I am far more conscious than I am in even the most lucid dream. In either case, whenever I am conscious it seems true that I am always conscious *of* something, as Searle (1992, 84) and others before him have maintained. Consciousness always has a content, whether it is of something external to itself, such as someone talking to us, or not external to itself, such as feeling a pain.

Moreover, consciousness does not always seem to involve language. Even though we might rightly be called, as E. O. Wilson (1998, 132) calls us, "the babbling ape"—cell phones more than anything make me think Wilson is right—it remains that much of our conscious life seems to be language free. This is apparently true even in cases of the deepest scientific thinking. Albert Einstein (1954), for example, claimed that scientific theorizing, for himself at any rate, involved mainly "more or less clear images" (25), while "Conventional words or other signs have to be sought for laboriously only in a secondary stage" (26).[5] Even in the first stage, the stage of images, we get different kinds of consciousness. Visual consciousness of, say, a spoon is a different kind of consciousness than the tactile kind. This should be obvious to every one of us, but it is especially obvious in experiments on animals for what is called *cross-modal transfer*. Monkeys, for example, can be conditioned to respond to the visual image of a particular thing such as a spoon, but they fail to repeat the response when only feeling the spoon (Ettlinger and Blakemore 1969).

---

5   Bickerton (1990) says Einstein "failed to comment on whether it [thinking without words] is done without syntax" (270), but this does not seem to me a good reply, first because it misses Einstein's follow-up point, and second because it begs the question on consciousness and syntax, which leads Bickerton to assume (38, 199–200) that language can fully capture what we consciously think and perceive.

Sometimes it is said that the only real kind of consciousness is consciousness of self, either simple self-recognition or consciousness of a self with a past, present, and future, as Jaynes (1976, 62–63, 460) seems to require for the concept of person (which as we have seen he bases on language). While self-consciousness is arguably a higher level of consciousness, it by no means always accompanies the lesser kinds of consciousness, even in ourselves. I am not always conscious of myself when I am looking around in this room or outdoors. Indeed, most of my conscious states do not involve self-consciousness at all. Why then should only self-consciousness be granted the honorific term "real"?

One sometimes hears that self-consciousness separates humans from animals, or in Jaynes's case modern humans from bicameral humans and animals. Jaynes (1976, 459–460) dismisses the experiments where chimpanzees apparently recognize themselves in mirrors, since the same sort of experiment is successful with pigeons and he doubts that pigeons have a self-concept. Bicameral humans he puts in the same category with chimpanzees and pigeons. The cognitive ethologist Donald Griffin (1992, 134, 249–250), on the other hand, treats such experiments as demonstrating self-consciousness in both chimpanzees and pigeons. No doubt most who have pets such as cats and dogs and who have spent a considerable amount of time playing with them will side with Griffin here. If I may use my own experience as an example—and I know that this sort of evidence is anecdotal and unscientific, but that does not necessarily make it worthless—my late cat Prince was a stray I took in when he was very young, maybe a couple months. For some mysterious reason, he had an insane hatred for other cats. I say "insane" without fear of exaggeration, as in "temporary insanity," not only because he would chase cats like a dog, but because he would do so quite viciously. Indeed, his distemper would be so great that he would rip into my hand if I tried to touch him. It would take maybe fifteen to twenty minutes after the cat was gone for him to settle down and be his normal good-natured self again. Now here's the point. I'll never forget the first time he saw himself in a mirror. He viciously attacked his image and bloodied his nose. But that was the only time. Ever since that first encounter with his mirror-image, whenever he would again see himself in a mirror he would simply stand there for minutes on end, apparently admiring his good looks. I find it hard to believe that Prince did not have a self-concept. Once he recognized himself in the mirror, he never attacked his image again. This could not have been because he felt intimidated. He was placid, without distemper, completely unlike whenever he would see another cat (and he was not the kind of cat to back down from any cat). Nor could his behavior have been the product of stimulus-response conditioning. This was a one-shot affair. The best explanation, it seems to me, is that he quickly realized that the cat in the mirror was not another cat but the image of himself. Did he also have the concept of himself as an extended self, with a past and a future? Did he remember himself in happy times and visualize himself outside every night while he meowed for me to take him outside? I do not think we can ever know in such cases, but neither do I think we should abruptly shut the door to the possibility.

Connected with the problem of self-consciousness is the problem of free will, specifically the *feeling* of free will. Here we run into similar issues. Although I often have the feeling of free will, far from all of my states of consciousness involve this feeling. Why then should we privilege this one kind of consciousness over others? I shall have more to say about the feeling of free will a little later below, but for now it should suffice to say that having the feeling of free will is not necessary for having consciousness.

Given the degrees and kinds of consciousness that we can see in ourselves, there remains no good reason to deny consciousness to animals. Important here is the work of Griffin (1992). Rejecting the view of Descartes and others that consciousness depends in some way on language (syntax and semantics), Griffin draws on a wide variety of examples of the versatility of animal behavior and argues that animals in many species act not simply from instinct but from consciousness and subjectivity and that they communicate thoughts and feelings in various ways. Peter Singer (1975) extends this kind of analysis to ethical issues. For Singer, states of pain are far more primitive than language and have nothing to do with language, and the question of whether animals feel pain is settled for him, as with many others, by *argument by analogy*. Humans have complicated brains and are wired for feeling pain, animals have similarly complicated brains and are similarly wired for feeling pain; humans display pain behavior, animals display similar pain behavior; humans consciously feel pain; it is therefore highly probably that animals consciously feel pain as well (and of course by the same token, it is extremely unlikely that plants feel pain).

Granting that consciousness is not a matter of yes or no but instead a matter of degree and of kind, as well that humans and animals are parts of the very same evolutionary tree, united by common ancestry, it remains to ask what if any evolutionary advantages might be conferred by consciousness. In other words, it remains to ask whether consciousness is an adaptation evolved by natural selection, which boils down to whether it evolved because it increases survival and reproduction.

The focus on degrees and kinds is extremely important here. Eyes are a perfect example of an adaptation. They are as intricate as one could wish for and accordingly they give the illusion of intelligent design. Given evolution, they had to start somewhere, and, as Darwin noted (1859, 186–189), one can see various kinds and degrees of visual development existing in the present, just as one can see mammals in various stages of adaptation to an aquatic environment (e.g., sea otters, seals, dolphins and whales). It is surmised that eyes evolved independently in the animal kingdom at least 40 times. Moreover, they are not all the same, but come in different kinds (e.g., bees can see ultraviolet light, while we cannot). Moreover still, having eyesight is not a matter of yes or no but of degree. Wherever vision evolved, it must have first begun as a light-sensitive spot on an organism (indeed some single-celled organisms presently have this), allowing the organism to respond to where the light is coming from. Increasing the fitness of its bearers, it would evolve gradually through time, typically increasing in complexity. The important point, as Richard Dawkins (1986) puts it, is that

> Vision that is 5 per cent as good as yours or mine is very much worth having in comparison with no vision at all. So is 1 per cent vision better than total blindness. And 6 per cent is better than 5, 7 per cent better than 6, and so on up the gradual, continuous scale.

(81)

If we agree that consciousness is a matter of degree and comes in different kinds, then we are in a position to try to explain it from an evolutionary point of view. Evolution by natural selection, it will be remembered, works incrementally on heritable variation, where every incremental step must be useful in some way to its possessor (Darwin 1859, ch. 6). Any trait that varies, is heritable, and confers nonrandom differential fitness is a trait subject to evolution by natural selection. Equally important is that the trait does not have to be physical or behavioral, it can be mental as well. Once this last point is admitted, the debate about consciousness gets really interesting.

The view of the philosopher John Searle (1984, 1992) is worth examining in some detail here, in spite of his many detractors (philosophers are a notoriously contentious lot, even more, I would say, than scientists, who are used to working in research teams). Searle's view is not the only interesting one on the topic of consciousness, as one might expect, but it is a good one to focus on at this point as it sets the stage for the rest of this chapter. According to Searle, there really is no mystery or problem of consciousness anymore, other than the details. Consciousness fits in perfectly with our modern understanding of physics/chemistry and evolutionary biology. For Searle, mind is neither reducible to the brain (nothing-but reductionism) nor a fiction to be eliminated altogether (eliminative materialism), but rather a macro-level property of the brain, much like liquidity is a property of H_2O molecules. Liquidity is not to be found as a property of any H_2O molecule itself, but emerges as a macro-level property of many H_2O molecules arranged in certain ways. So too with consciousness and neurons. Both levels are real. Moreover, the relation between the two, Searle points out, is not one of event-event causation, the familiar kind of causation that we find with billiard balls, for example. Instead, the relation is one of non-event causation. This is a difficult concept to grasp, but it works perfectly well for liquidity and many other properties, which are simple enough given some reflection, so the application of non-event causation to consciousness should at least be plausible. The relation between the two levels is also one of *supervenience*. Given that a chemist could infer the property of liquidity from a certain arrangement of H_2O molecules, the chemist, conversely, could not possibly infer that arrangement of H_2O molecules from the property of liquidity, or even that H_2O molecules are involved, since other molecules in other arrangements can also cause (non-event) the property of liquidity. In all of this, the details are a matter of physics/chemistry, but philosophically there is no problem. The same follows, in Searle's view, for consciousness. Given a particular brain state, a neuroscientist (when neuroscience is sufficiently advanced) should be able to infer the particular conscious state, but given a particular conscious state it should be impossible for the neuroscientist to infer the particular brain state, since different brain states can cause (non-event) the same conscious state.[6]

---

6   Interestingly, in an entirely different context Elliott Sober (1993) argues that "*All biological properties supervene on physical properties*" (73). For Searle, consciousness is a biological property. What is interesting about Sober's discussion is his focus on fitness, clearly a non-physical biological trait, and one that is arguably the result (at least in part) of non-event causation. To use Sober's example, a zebra might have the fitness value of 0.83 (specifically he says, "A cockroach and a zebra differ in numerous ways, but both may happen to have a 0.83 probability of surviving to adulthood"), but this biological property of the zebra, its overall fitness value, is a feature of a genotype in a particular environment; it is not a temporal effect like its eyes, which are an event-event product of its ontogenetic development.

According to Searle, biology fits in with this picture given that consciousness involves meaning and subjectivity. Famously, or notoriously, Searle argues that computers, no matter how complicated, could never have these features. He allows for the artificial production of consciousness, but such consciousness would have to be a property of biological entities of some sort, involving the right kinds of chemicals and cellular processes, not a computer. A recurring theme throughout Searle's writings is that consciousness cannot be produced by computers of any kind, since computers only compute (they apply rules to data). No matter how complicated they are, you can never get semantics out of syntax. Searle illustrated this point using his Chinese Room thought experiment (Searle 1984, 32–38; 1992, 200–212), in which an English-speaking person isolated in a room, applying rules from a Chinese rule book to questions passed to him in Chinese, would never get to understand Chinese, no matter how good he got at applying the rules from the rule book.[7]

Evolutionary biology takes its place in Searle's picture when the adaptive nature of consciousness is considered. Searle (1992) accepts Griffin's view that consciousness is widespread throughout the animal world. In fact he considers the evidence "overwhelming," although "We do not know at present how far down the evolutionary scale consciousness extends" (89). But Searle is no simple-minded Darwinian. He makes the point that it is a crude version of Darwinism to suppose that every organism trait must have been selected for (such as the human predilection for skiing). To this point one could add an important concept from evolutionary biology (important for this and other chapters in this book), namely, *pleiotropy*, the fact that a gene or gene complex (or a mutation) typically has two or more phenotypic traits. In cases of pleiotropy, if natural selection increases the frequency of one of the traits in a population, the other trait(s) tags along. For example, in many horned animals horn size and body size are correlated, such that if natural selection increases the body size (possibly as an adaptation to colder weather, known as Bergman's Rule), the increase in horn size is not naturally selected but merely tags along (Maynard Smith 1993, 64). In Chapter 4 we shall see another example, the case of male homosexuality, which the geneticist Dean Hamer argues has a partly genetic basis such that the genes responsible were favored by selection not because they promoted homosexuality in males but because they promoted earlier puberty in females (thus increasing their average number of offspring). Consciousness, however, along with its neurobiological basis, would seem far too complex to be explained merely as a pleiotropic trait evolved by tagging along. Quite rightly, then, Searle argues, as he must, that consciousness has many evolutionary advantages. He says,

> consciousness does all sorts of things . . . there are all sorts of forms of consciousness such as vision, hearing, taste, smell, thirst, pains, tickles, itches, and voluntary actions . . . in very general terms . . . "representation."
>
> (107)

---

7   See Hofstadter (1981, 373–382) and Gregory (1987) for criticisms, Barlow (1987) and Edelman (1992) for defenses.

But there is more, as indeed there must since representation can be accomplished unconsciously. Accordingly he says, "Consciousness gives us much greater powers of discrimination than unconscious mechanisms would have" (107). As evidence, Searle refers to case studies of epileptic seizure patients documented by the famous neuroscientist Wilder Penfield, in which patients while having a seizure were unconscious but continued with goal-directed behaviors that they had already learned, such as playing a particular piano piece. What only strengthens Searle's point is the fact that these patients while having a seizure do not sight read a piano piece or compose. In short, then, "one of the evolutionary advantages conferred on us by consciousness is the much greater flexibility, sensitivity, and creativity we derive from being conscious" (109). Indeed for Searle,

> Consciousness . . . is a biological feature of human and certain animal brains. It is caused by neurobiological processes and is as much a part of the natural biological order as any other biological features such as photosynthesis, digestion, or mitosis.
>
> (90)

One attractive feature of Searle's view is that he has consciousness as a macro-level trait. Even for those biologists who hold that natural selection works ultimately at the level of the gene, they routinely admit that genic selection is mediated through the phenotype, that natural selection selects genes via their phenotypic traits (Williams 1966, 25; 1992, 16; Dawkins 1976, 44–45, 235; 1986, 60). Just as conscious states supervene on neural states, so do phenotypes on genotypes (this follows from the synonymy in the genetic code, discussed in the Glossary). That Searle conceives of consciousness as an effect of non-event causation, whereas the genotype-phenotype relation is one of event-event causation, would seem to be a difference that makes no difference.

In spite of the attractive features of Searle's view, there are a number of problems. One is that he does not think that the human brain has rules of syntax built into it, syntactic circuitry one might call it. This is an issue I shall return to when we focus on Chomsky in the next chapter.

Another problem has to do with the conscious feeling of free will that we have. In Searle (1984) he argues that there is no room for free will in the modern scientific worldview,[8] and yet, he says,

> for reasons I don't really understand, evolution has given us a form of experience of voluntary action where the experience of freedom, that is to say, the experience of the sense of alternative possibilities, is built into the very structure of conscious, voluntary, intentional human behavior.
>
> (98)

---

8   Searle (1992, 90–93) says the same thing about God and life after death. Physics/chemistry, he says, holds that big systems are made up of smaller systems, the fundamental level being subatomic particles, while evolutionary biology explains living systems in terms of evolution by natural selection from the less complex to the more complex. In none of this is there room for God or mind-body dualism. Whether evolutionary biology really has no room for God is a matter we shall examine in Chapter 8.

In Searle (1992) he provides no answer, but it is an important question.

Connected with Searle's lack of answer is the fact that he seriously muddles the distinction between *selection for* and *selection of*. He says (1992), "Compare Elliot Sober's distinction between what is *selected* and what is *selected for* (1984, ch. 4)" (252 n. 7). We have seen the distinction briefly above, but let us look at it again. The distinction rests on pleiotropy and also on gene linkage (where an advantageous gene and a neutral gene are linked closely on the same chromosome). To repeat, of two traits x and y that are linked, say, by pleiotropy, if natural selection evolved only trait x, then trait y was not selected *for*; instead, trait x was selected *for* and trait y was merely a matter of selection *of* (Sober 1984, 100–102).[9]

This raises an interesting question: Could the feeling of free will be a consequence of selection in the sense of *selection of*? This becomes an interesting possibility given Harry Frankfurt's (1971) theory of free will. For Frankfurt, humans and many other animals have what he calls *first-order desires*, such as the desire for food, sex, or a cigarette. But only humans, he says, have *second-order desires*, desires *about* first-order desires, such as the desire to quit smoking (smoking being the first-order desire). For Frankfurt, we feel we act freely if we act not only in accordance with our second-order desires but if we also make our second-order desires our will. I'm not so sure that there are no animals other than humans that have second-order desires, desires about their first-order desires, but clearly second-order desires and a desire that they become our will would be adaptive for a species that is highly social and has consciousness, such as ourselves. Perhaps, then, the feeling of free will is simply a byproduct of the evolution of second-order desires and the corresponding will.

Aside from the problem of free will, a further problem for Searle is that he focuses on the individual. Consciousness, for Searle, is a feature of an individual brain, not of groups of brains.[10] The problem is that the best examples of consciousness we know of, consciousness in humans, evolved in close-knit groups rather than in roving individuals. Many have thought this fact is important for the topic of consciousness and should not be ignored.

The neurophysiologist Horace Barlow (1987), for example, focuses on the group rather than on the individual. In fact, he argues that consciousness is not a property of a single brain by itself but rather

---

9   Actually, even the concept of *selection for* is problematic, since it suggests that natural selection is teleological, which it is not (Stamos 2003, 204–205). Natural selection in itself is a "blind" process, its blindness aptly reflected in the title of Richard Dawkins's book *The Blind Watchmaker* (1986). In addition to being blind, almost all selection in nature is *selection against* (sexual selection being the possible exception if any)—it is "the grim reaper," as Dawkins nicely puts it (62). *Selection of*, then, is what remains from *selection against*, one kind of *selection of* playing a causal role in evolution (Sober's *selection for*), the other kind not playing a causal role (Sober's *selection of*).

10   This is in spite of his belief in what he calls "collective intentional behavior," which he says is not simply the sum of or reducible to individual intentional behaviors (Searle 1990, 1995). Intentionality refers to the *aboutness* of many mental states. The desire to drink, for example, is intentional because it refers to something beyond itself, same for the belief that something is wet. I suppose that Searle does not apply or extend consciousness to groups of brains because he distinguishes consciousness from intentionality. Not only are not all intentional states conscious states (Freud exploited this idea to the hilt), but not all conscious states are intentional states, such as many examples of pain, which just are but do not refer to something else (Searle 1992, 84).

is taught, awakened, and maintained by interactions with other modelled minds, and its characteristics in any individual depend to some extent upon these other minds . . . consciousness becomes the forum, not of a single mind, but of the social group with whom the individual interacts.

(373)

Barlow defends this remarkable thesis first by noting that the dawn of consciousness in a human individual begins roughly around the time it starts communicating with its parents. Next, he points out that when alone we do not lose consciousness but continue on as if we were addressing other individuals, even in our deepest introspections. The same is true even with straightforward sensations such as red. These are not "raw," says Barlow, but "carefully cooked," not only because our labeling of them comes from our social group but also because we want to communicate our sensations to others, for example when we say, "That strawberry is red enough to pick and eat" (370). Even the way we use the word "consciousness," says Barlow, supports his view, such as when we say a person who is sleepwalking is unconscious, which means we find the person incapable of communicating with us, or when we say we not accidentally but consciously did something like miss a dentist appointment, which means we were in a position to communicate our decision to someone else but perhaps did not (367–368). Granted, as Barlow recognizes, he has a problem with organisms such as bees (366). Honeybees, for example, are highly social and communicate to each other via behaviors such as the waggle dance (Griffin 1992, 178–194). And yet, for Barlow, the difference between them and us is so great that it warrants making a "qualitative" distinction (373), with us conscious and them not. But this does not mean for Barlow that only humans are conscious. He allows consciousness in higher animals that are social, but only in humans has consciousness flowered, with their ability to model other minds and their environment via language and to create culture. Thus, for Barlow, consciousness is a group adaptation, its "survival value" resulting from "the particular patterns of social behaviour it causes us to follow" (373), one of which is social morality (369).

And yet we are back to the problem that we found with those who made consciousness depend on language (syntax and semantics). Barlow does not go quite as far as they do, but his argument still suffers from the problem that we should not want to deny consciousness of pain, say, to a tiger or a mouse. Not only is pain more primitive than language but it is also more primitive than social groups: most animals (social and otherwise) seem clearly wired for pain, they exhibit pain behavior, and the adaptive advantage of consciousness of pain should be painfully obvious (excuse the pun) in that those who have less of it than their conspecifics are at a disadvantage in the struggle for existence and are therefore less likely to survive to adulthood and reproduce. (Consider the extreme case of an infant that for genetic reasons lacks the capacity for consciously feeling pain; it would not remove its hand when it touched a flame, or jump when it stepped on something sharp, or aggressively search for food when it is starving.) In short, there is nothing in the conscious experience of pain that requires or is dependent upon communication to others, although in social groups that can be of obvious help. Pain in itself is both raw and adaptive.

Arguably, then, Barlow goes too far when he says, "the ability to communicate . . . is the crucial test of consciousness" (368). But at the other extreme, Searle does not go far enough. One focuses on the group, the other on the individual. Perhaps the most balanced approach is to make a distinction something like the one made by the molecular biologist turned neuroscientist (and Nobel laureate at that) Gerald Edelman (1992), who distinguishes between what he calls *primary consciousness* and *higher-order consciousness*. Both are forms of consciousness. Primary consciousness is the sort of thing we find in Searle's account, consciousness aiding in such activities as discrimination and found in a high degree in many species of animals. Higher-order consciousness, on the other hand, involves introspective states and consciousness of being conscious, including self-consciousness, the concept of oneself as having a past, present, and future, the feeling of free will, and an inner modeling of the environment. While higher-order consciousness is different in kind from primary consciousness, both have degrees and both are for Edelman a matter of evolution by natural selection, with primary consciousness coming first and being necessary for the emergence of higher-order consciousness.

At the interface between the two levels one might bring in Popper's view that we saw in the previous chapter. For Popper (1978), the second stage in the evolution of consciousness is where an organism tries out behavior in its mind, where it precedes real behavior with imagined trial-and-error behavior. After that comes the evolution of conscious aims and ends. The evolution of language would be the next (and final?) major step. Here we create theories and conjectures and dissociate them from ourselves and subject them to scrutiny and criticism, so that unlike other animals which die with their theories we let our theories die in our stead.

Indeed, returning to Edelman, it is only with humans that we find "fully developed higher-order consciousness" (135). As he states more clearly, "Higher-order consciousness arises with the evolutionary onset of semantic capabilities, and it flowers with the acquisition of language and symbolic reference" (149). Only humans have this flowering, as we shall see in the next chapter. Moreover, the difference really does seem to be a substantial difference in kind—as Edelman puts it, "A conceptual explosion and ontological revolution" (150). At any rate, it explains why we have taken over the world. With language, in the full sense of syntax and semantics, our species passed through a major threshold, no longer, as many have pointed out, simply adapting to its environment, like other species, but adapting its environment to it. The reality is not that simple, of course, since all species alter their environments in some ways more or less, but it is still a basically correct way of putting the matter and not an exaggeration. Humans have managed to do things that no other species has come even remotely close to doing. And language was the essential element for the kinds of activity involved.

The question then becomes, "What is language and can it be explained by evolution?" This is the topic of our next chapter.

3

Evolution and Language

According to the SSSM, the mind-brain is a general-purpose learning device and language acquisition is simply a matter of learning capacity (i.e., innate intelligence) combined with stimulus-response conditioning. Humans readily acquire language because their species has the required learning capacity, while gorillas and chimpanzees have it less so and, lacking the correct vocal apparatus, can only be taught rudimentary sign language. Humans, on the other hand, can be taught virtually any kind of language. The only constraint is the degree of difficulty, learning capacity being a factor and vocal apparatus being another. On this view, if the Klingons of *Star Trek* fame came to earth and their language was not too difficult, humans could learn to speak Klingon. Philosophers have long held that the language learned by a person is completely acquired from outside the person, behavioral psychologists later codified this view of language, which received its culmination in B. F. Skinner's *Verbal Behavior* (1957), while science fiction books and movies have simply followed this rather common sense view.

The SSSM approach to language was completely undermined by the work of the linguist Noam Chomsky.[1] Beginning with a series of books and essays in the late 1950s and '60s (especially Chomsky 1957, 1965), including a now classic review of Skinner's *Verbal Behavior* (Chomsky 1959), Chomsky argued that language in humans is largely innate, so that in language acquisition the contribution from the environment is only partial. In arguing this, Chomsky did not mean that people are somehow born with natural languages such as English or Chinese already in them. Instead, he argued that all natural languages (of which there are roughly 5,000 currently in existence) are superficial phenomena. They are mere surface phenomena and are not as different from one another as one might think. What makes them superficial is not merely that they all have something fundamentally in common,

---

1   Linguistics is the academic discipline composed of historical linguistics (the study of the evolutionary history of natural languages; e.g., Lass 1997), sociolinguistics (the study of minute changes in the natural language of a language community; e.g., Labov 1966), comparative linguistics (the study of diversity in natural languages; e.g., Nichols 1992), and psycholinguistics (the study of the language faculty in the human brain, a field pioneered by Chomsky).

but rather that what they have in common, the common characteristics or linguistic universals, are innate in the human brain. This is what Chomsky called the *Universal Grammar* (UG).[2] In his earlier work, Chomsky conceived of the UG as a set of hundreds of rules of syntax, rules for making sentences. A common misconception is to think of the differences between languages as simply a difference in words. But what makes a language a language in the full-bodied sense is not words and their individual meanings (semantics) but the rules used to combine those words into sentences (syntax) and hence into new meanings. For example, the sentence "The dog bit the boy" and the sentence "The boy bit the dog" are exactly the same in terms of the words used. What makes the meaning of each sentence different from the other is not the words themselves but the word order, combined with the rules applied to those words that make them meaningful sentences. Chomsky argued that all natural languages have a lot in common in terms of their rules of syntax. The UG, then, can be conceived as the common denominator to the thousands of natural languages that constitute the numerator. As Sampson (1980) put it, "The essence of Chomsky's approach to language is the claim that there are linguistic universals in the domain of syntax" (131). As Sampson knew only too well, however, the essence goes much deeper than that. What Sampson says is true of Chomsky compared to linguists before him, but the real profundity of Chomsky's approach is his claim that the UG consists not only of these linguistic universals but of the fact that they are hard-wired into each and every one of our brains, as a language "organ" or "faculty" (Chomsky uses these terms interchangeably). To be accurate, then, it should be said that the essence of Chomsky's approach to language is his claim that there are linguistic universals in the domain of syntax and they are innate in the human brain.

The profundity of Chomsky's view becomes especially evident on the matter of language acquisition. For Chomsky, when we think a child learns a natural language such as English, granted the words and syntactic rules that set English apart from other natural languages are acquired from the child's environment, from those it interacts with linguistically, but the main contribution is from the child, from the "initial state" or UG existing already inside its brain, such that it was born with "unconscious knowledge" of language. The UG, moreover, works as a constraint on the kinds of languages humans can learn. Not only, for Chomsky, does the UG mean that the potential number of natural human languages must be finite, but if languages exist on other planets, such as Klingon, then it is quite possible that Klingons would have a different UG, in which case they would not be able to learn any of our languages and we would not be able to learn any of theirs (an unhappy implication for *Star Trek* fans).

---

2   Indeed, a recurring theme in Chomsky's writings is to go so far as to say that "A rational Martian scientist, studying humans as we do other organisms, would conclude that there is really only one human language, with some marginal differences that may, however, make communication impossible, not a surprising consequence of slight adjustments in a complex system. But essentially, human languages seem cast in the same mold" (Chomsky 1997, 121–122; see Chomsky 2000, 7). By parity of reasoning, Chomsky would have to say that a Martian geneticist studying organisms on Earth would conclude that there is only one biological species on Earth (or only a few), given the genetic code. For a critique of Chomsky's natural language nominalism, without denying the UG, see Stamos (2002).

In his later work, begun in the 1980s (Chomsky 1986), Chomsky began arguing for "minimalism" in the form of a hypothesis he calls "Principles and Parameters." (Chomsky refuses to call it a theory, as it is lacking in evidence, but rather a research program.) Here the idea is that a child during language acquisition learns literally *nothing* from its exposure to language users (unlike the earlier theory, where knowledge of language is contributed *mostly* from the initial state, the UG, and *partly* from the environment). Instead, the child's exposure to its circle of language users (usually its family) sets off switches in its brain, each one resulting in the set use of one or another syntactic rule already existing in its brain. To give a simple example, the English language is "head-first," meaning the verb comes before its object, as with "David ate apple pie" (subject-verb-object, SVO), whereas the Japanese language is "head-last," meaning the verb comes after its object, as with "David apple pie ate" (SOV)—indeed, as Pinker (1994, 111) points out, the two languages seem mirror images of each other, with the difference residing also in the position of adjectives and prepositions/postpositions. Neil Smith (1999, 82) suggests that it would not take a lot of these switches to account for the syntactic variety of natural languages. If a child's brain has to select from one out of a million natural languages, it can do this if it has only twenty switches in its brain and each of those switches is binary, since $2^{20} = 1,048,576$. With the switches set one way, we have English, with another way, Japanese. Probably a little more realistically, Robert Berwick (Green and Vervaeke 1997, 151–152) suggests that thirty-six parameters would be necessary to account for the basic structures of all human languages. But no matter what the number of parameters needed for this hypothesis, the point remains that the child does not really learn the language it acquires, not even partially, but merely has its switches set (whatever the number and their relation to each other) as it is exposed to its linguistic environment. This pattern of development, says Smith (1999, 83), following a suggestion by Chomsky (1988, 65), has a "close analogy" to embryological development, where the resulting organism results not simply from structural genes alone but also from regulator genes which switch on and off other genes at the appropriate times. The analogy shows, of course, that the Principles and Parameters approach is not enormously different from the earlier construal of the UG, since on the new approach there are still syntactic rules hard-wired into the brain which get switched on or off during language acquisition, while possibly there are other syntactic rules which are not a matter of switches at all but are necessarily in every human language.

Whether the UG is conceived in the older way or the newer way, the evidence for it is indeed quite stunning. Aside from the particularities claimed to be found in the UG, Chomsky's argument for an innate UG is based mainly on two closely connected arguments. His first argument is based on the fact that almost every sentence that a person utters or understands is a new combination of words, appearing for the first time in the universe. A language therefore cannot simply be a repertoire of conditioned responses. One might call this the "infinitely many sentences" argument. The second argument focuses on language acquisition in small children and Chomsky explicitly calls it the "poverty of stimulus" argument. A natural language involves not only a large number of words but also an extremely complex system of rules, and yet small children, with relatively underdeveloped intelligence, without any formal instruction, and experiencing only a small fraction of a language as it is used, rapidly acquire a natural language. Moreover, they correctly interpret the meaning of sentences they have never before encountered. Language acquisition, then, cannot simply be a matter of stimulus-response learning. Instead, the knowledge of language must already be in the child and it is brought to the surface by its exposure to a natural language.

Since Chomsky's original arguments, more intriguing evidence has been found in support of an innate UG. One class of evidence comes from what are called pidgins and creoles. When speakers of different languages are put together and they have to communicate to carry out practical tasks, as on plantations in the days of slavery (slave-owners sometimes mixed slaves from different backgrounds), a makeshift jargon of words called a *pidgin* typically develops. If small children, at the age of language acquisition, are exposed to a pidgin, they will naturally develop it into a full-blown language, rich with syntactical/grammatical rules. This brand-new language is called a *creole*, and some striking cases have actually been documented. One case involves sugar plantations in Hawaii in the 1890s. Immigrants were employed on these plantations from a variety of backgrounds. The children of these workers were tended to by someone who spoke to them in a pidgin. Remarkably, the children developed a creole from this simple exposure. A related case occurred in Nicaragua in 1979, the year the first schools for the deaf were created in that country. In those schools the children were drilled in lip reading and speech, with dismal results. But in the school buses and school yards, the children were developing their own sign language, a kind of pidgin, which quickly developed into a full-fledged sign language, a creole, by a younger class of children who were exposed to the pidgin of the older class of children, a creole with its own unique set of signs and syntactical/grammatical rules. The same sort of construction has even been documented in a single child, a deaf child lacking exposure to other similar children and only exposed to the bad American Sign Language (ASL) of its deaf parents (bad because they had been taught it in their mid teens) (Bickerton 1990, 169–171; Pinker 1994, 33–39). What is striking in each of the cases above and others of their kind is not only what these children add to the equation but that what they add is remarkably similar from case to case. As Pinker (1994) puts it, "The crux of the argument is that complex language is universal because *children actually reinvent it*, generation after generation" (32). This is not an entirely accurate way of putting the matter, since children in Chomsky's view do not *invent* let alone *reinvent* language (hence I would prefer something like *independently reproduce it*), but it does succeed in emphasizing Chomsky's dramatic point.

A very different class of evidence comes from what is known as *Specific Language Impairment* (SLI). Defects in specific grammatical abilities have been found to run in families in many cases. This does not in itself necessarily mean a genetic cause, but there is strong evidence that in some cases it is indeed genetic. Particularly striking is the case of a family in England studied by the linguist Myrna Gopnik. The grandmother had SLI. Of her five children, one was linguistically normal, as were all of the children from that child. The grandmother's other four children, however, had the SLI, and of their children, which totaled 23 in number, 11 had the SLI. In such cases one cannot simply blame the environment. Instead, quite apparently, the cause was a mutation in a single gene passed on from the grandmother, given the way it was transmitted in the family and that the family members either had the same SLI or not. Of further interest is that neither the sex of the children nor their birth order was correlated with the SLI, nor were their IQs (Pinker 1994, 48–50, 322–325).

A further class of evidence comes not from humans but from apes. It has long been known that apes such as chimpanzees and gorillas, much like dolphins and dogs and cats, exhibit remarkable intelligence. So why not teach them a human language? The problem is that they lack the required vocal apparatus. But chimpanzees and gorillas, unlike dolphins and dogs and cats, have hands. So why not teach them a sign language such as ASL? This is precisely what a number of researchers, trained in the school of behaviorism, attempted to do starting in the late 1960s and extending to the present, with chimpanzees such as Washoe, gorillas such as Koko, and bonobos such as Kanzi (where rather than signs a portable talking symbol board was used). These researchers made some amazing claims, which quickly captured the imagination of the public. They claimed not only that these animals were capable of learning signs and their meanings but that they were capable of making short and simple sentences, much like the average two-and-a-half-year-old human child. This research program came into disrepute beginning in the late 1970s, however, beginning with the work of Herbert Terrace, one of their own. Terrace and his research team found with his research chimp—which he fondly, and looking back now, ironically, named Nim Chimpsky—that although Nim put together words into strings, even long strings, they were not genuine sentences. Terrace reviewed his videotapes over and over again, but the conclusion seemed unassailable. Change the order of the wording in, for example, "give orange me give eat orange me eat orange give me eat orange give me you" and you do not get a change in meaning. Of course with a real sentence you do. These and further problems have seriously damaged the credibility of ape language studies ever since.[3] All of this is on Chomsky's side, since it seems that the apes cannot learn sign language not because of lack of intelligence or lack of physical apparatus but because their respective species lack a UG.

Aside from all of this extra evidence, even before it, Chomsky effected a veritable revolution in linguistics.[4] Not only did he set off an enormous research program that most linguists today subscribe to more or less—even though the discovery of what exactly the UG consists of is very far from settled, and much work remains in comparing different languages—but his detractors find it necessary to define themselves as anti-Chomskyans (a sure sign of what the dominant view is). In addition to linguistics, the Chomskyan revolution spilled out into other disciplines, known collectively as *cognitive science*, the research program devoted to studying how the mind works and which includes computer science, cognitive psychology, neuroscience, and (mainly because of Chomsky) linguistics.

---

3   See Terrace *et al.* (1979); Terrace (1987); Bickerton (1990, 106–110); Pinker (1994, 334–342); Smith (2002, ch. 14); Anderson (2004).

4   See Sampson (1980, ch. 6); Pinker (1994, 21–23); Smith (1999, 1); Antony and Hornstein (2003, 1).

My purpose here is not to defend the cognitive revolution, whether limited to linguistics or including the mind as a whole. Rather, the question for us in this chapter is whether Chomsky's language organ/faculty can be sufficiently explained by evolutionary principles, in particular natural selection. What is interesting is that Chomsky himself resisted an evolutionary explanation for the language organ and gave an alternative explanation instead. This prompted others to try to place Chomskyan linguistics on the foundation of evolutionary theory. We shall look first, though, at some arguments which attempt to show that given evolutionary biology Chomskyan linguistics cannot be right and must be replaced with a different model of language. These views, in turn, shall be critically evaluated particularly in the light provided by the most famous attempt at combining Chomskyan linguistics with evolutionary biology, namely, that of the psycholinguist Steven Pinker, who argues that the only reasonable explanation for the language instinct (Pinker prefers the word "instinct" because he believes that language bears all the marks of an instinct) is that it evolved slowly and cumulatively by good old natural selection. Finally, we shall then look at a very different attempt to fit Chomskyan linguistics into evolutionary theory, by the linguist Derek Bickerton, and we shall see if he succeeds over Pinker.

First, we need to look at the reasons Chomsky gave for why he thought the language organ/faculty was not the result of Darwinian selection. It should be pointed out beforehand, however, that Chomsky has expressed remarkably little interest in the origin of the UG. What interests him is the nature of the UG, not how it came into existence. Nevertheless, he has also expressed doubts that the UG originated by natural selection, saying the UG is an example of a "true emergence" and so cannot be explained by evolutionary theory (Pinker and Bloom 1990, 452). Much more plausible for Chomsky is the possibility that the UG emerged as a byproduct of "selection for bigger brains," where the natural selection was for processing abilities other than language. When hominid brains reached a certain critical mass (*Homo sapiens* being for Chomsky the only known species with a UG), not just for size but for complexity, unknown laws of physics kicked in resulting in the UG (474). In short, then, Chomsky has repeatedly resisted the possibility that the UG is an adaptation, evolved as all adaptations are by natural selection. This is even in spite of his easy acknowledgment that "language must surely confer enormous selective advantages" (487 n. 1)! Interestingly, Chomsky's antiselectionist stance on the UG was seconded no less by Stephen Jay Gould (Gould 1989a, 14), who as we have seen in Chapter 1 was very much an antiselectionist when it came to human nature. Indeed, Gould has been famous for trying to revive the "hopeful monster" view of evolution squashed by the Modern Synthesis, the view that a single mutation could occasionally produce a major change in an organism, immediately adapting it to a new way of life (Gould 1980b, ch. 18). Accordingly, Chomsky's UG, for Gould (and perhaps also for Chomsky), would be one such hopeful monster.[5]

Given Chomsky's rejection of natural selection as an explanation for the existence of the UG, some have taken the next step (and for them the logical step) of arguing that

---

5 For more on Chomsky's and Gould's rejection of natural selection as an explanation for the UG, see Pinker (1994, 355–364) and Dennett (1995, 384–397).

the UG itself must be rejected because the nature credited to it seriously conflicts with the way evolution works (evolution being the firmly established foundation science for any of the higher sciences claiming innate structures and behaviors). Chomsky's linguistic program, of course, has many detractors (see Antony and Hornstein 2003), but the only ones of interest for us here are those who argue against Chomsky from an evolutionary point of view.

John Searle, as we have seen in the previous chapter, takes evolution very seriously when it comes to the nature of mind. Part of his argument, as we have also seen, involves his claim that digital computers, no matter how complicated, could never have mind in general and consciousness in particular because they are purely syntactic machines and you cannot get semantics out of syntax. But Searle has gone a step further since his Chinese Room argument and has claimed that syntax is not something objective, out there to be discovered, but instead something in the eye of the beholder that we project onto the world. As Searle (1992) puts it,

> "syntax" is not the name of a physical feature, like mass or gravity . . . *syntax is essentially an observer-relative notion. The multiple realizability of computationally equivalent processes in different physical media is not just a sign that the processes are abstract, but that they are not intrinsic to the system at all. They depend on an interpretation from outside.*
>
> (209)

"Multiple realizability" refers to the claim common in computer science that the same function can be performed by physically different systems, that function is multiply realizable and therefore supervenes on a disjunctive base of hardware (i.e., different kinds of mechanical structures). Searle goes even so far as to claim that if a strange kind of machine were discovered we could not possibly reverse engineer it fully and discover what this part is for or that part, in particular we could not discover the syntactic rules built into the machine by its builders because nobody really builds syntactic rules into anything but instead reads them into things. As Searle puts it, "Computational states are not *discovered within* physics, they are *assigned to* the physics," and he adds that "This is a different argument from the Chinese room argument, and I should have seen it ten years ago, but I did not" (210).

Clearly if Searle is right on this matter, then the UG does not exist objectively inside our brains and is not coded for by our DNA. Instead, it is something that Chomskyan linguists read into human language and ultimately into the human brain. This would be a devastating criticism of the Chomskyan program if it could be maintained, and indeed Searle (1992) uses it explicitly against Chomsky's innateness thesis (220–221). However, it suffers from two serious difficulties. On the one hand, if his argument is an *a priori* one, an argument made prior to experience and imposed on experience, then his thesis is untestable and it leaves much to be desired as a result. On the other hand, if his argument is an *a posteriori* one and accordingly is open to testing, then it would seem already to have failed. I say this because if we look at the genetic code (Glossary), which is abstract but is carried in the relations between codons and amino acids, we do seem to have a *system of rules* down there that molecular biologists have *discovered* rather than imposed on the physical system. Specifically, there are rules for which codons code for which amino acids and rules of synonymy, with three codons coding for no amino acids whatsoever but functioning instead as stop codons, like periods at the end of sentences. Outside the genetic code but related to it, there are rules even more obviously syntactic, namely, rules for making proteins, as a change in the order of the codons in the RNA (or simply in the lettering of the DNA) can result in altered or in entirely different proteins, and hence in a change in the phenotype (what might be considered the meaning). Moreover, there are other rules in the genetic machinery such as DNA repair mechanisms, mechanisms where special enzymes either repair damaged DNA or correct mistakes in the copying of DNA (more on this in Chapter 8). Sometimes these rules are broken, often resulting in a mutation, but a broken rule does not negate the existence of a rule in the first place. And if there are rules existing objectively in the DNA world, then why not in the circuitry of the brain as well? It will do no good, of course, to *define* rules, syntactic or otherwise, as manmade and observer dependent, since such definitions, sometimes called "persuasive definitions," merely *beg the question*: they take for granted precisely what is in dispute. I am not saying Searle is necessarily guilty of this well-known fallacy, but he is if he takes his claim about syntactic rules to be an *a priori* truth.[6]

A very different kind of shot at Chomsky's UG is based squarely on evolutionary theory. Here I shall use the set of criticisms provided by Jagdish Hattiangadi (1987, ch. 9), not simply because he is a former teacher of mine, but because his criticisms are the best representatives of their kind and they lend themselves perfectly to a search in Pinker's writings for replies. In short, Hattiangadi argues that Chomsky's UG is essentialistic and therefore could not possibly have evolved by natural selection or other evolutionary mechanisms.

Before we get into this important criticism in more detail, however, to see if it works, we have to get clear on what essentialism means. Briefly, an essentialistic class of objects is a class defined by one or more properties which are said to be individually necessary and jointly sufficient for membership in the class. Think of the class of triangles, for example. A necessary condition for something to be a member of the class, to be a triangle, is that it has three sides. But that is not a sufficient condition for something to be a triangle. A two-dimensional acorn shape, for example, has three sides but it is not a triangle. For something to be a triangle, it has to have each of the properties that are *necessary* for something to be a triangle, the properties that when taken together are *sufficient* for it being a triangle. (As an interesting thought experiment, try to think of what all those properties are. It is not as easy as one might think.)

One problem with this definition of essentialism when applying it to things other than geometrical objects, such as three-dimensional physical objects, is that a class of objects might just happen to have one or more properties in common as a matter of chance. To avoid this consequence, essentialism is commonly thought

---

6 Searle (1992, 238), by the way, makes the same kind of argument against the objective existence of functions that he makes against syntax. See Dennett (1995, 397–400) for a biting reply very different in kind (but none the worse for it) than mine.

to be something deeper, to involve the concept of a *causal* essence, such that the individually necessary and jointly sufficient properties *cause* something to be what it is. A perfect example of this kind of essentialism is found in chemistry. Each chemical element has a causal essence which determines the chemical properties of the element and hence the kind of atom that it is. In the case of chemical elements, that property is the atomic number, the number of protons in the nucleus. The chemical element *gold*, for example, is defined by the atomic number 79. For an atom to be an atom of gold, it must have no more and no less than 79 protons in its nucleus. This number determines the number of electrons and hence the chemical properties of the atom. The number of neutrons in the nucleus, on the other hand, can change, resulting in what are called *isotopes* of the atom, but the chemical properties of the atom remain the same, the physical property of mass being different.

It used to be thought that biological species too have an underlying causal essence, Aristotle being the most influential example of this view (Stamos 2003, 102–111). Aristotle's influence extended for quite some time (Stamos 2007, chs 7 and 10). And then there was Darwin. Darwin, it has often been said, killed essentialism in biology. No longer were variations considered to be mere perturbations from a norm, but rather they were now considered to be the very stuff that natural selection operates on and hence the very stuff of evolution. Species and the populations that compose them were now no longer considered to have an underlying essence but to be statistical in nature. Modern biology, of course, has only confirmed this view with the study of genes and now the sequencing of DNA. Variation in a species is the norm. There is no underlying genetic essence. The only organisms that have the exact same DNA (barring mutations) are clones and identical twins.

This would seem to create a serious problem for Chomsky with his UG (keyword "universal"). As Hattiangadi (1987) points out, Chomsky supposes that the UG is shared by all humans, irregardless of racial differences, as the essence or "unchanging core" (170) of human language. The problem is that "The essentialistic theory is firmly and unalterably un-Darwinian . . . from an evolutionary point of view, *the universal grammar itself, if there is such a thing, could not have evolved*" (178). If it did evolve, it would evolve gradually and incrementally and statistically. Hence, it would not be essentialistic but variational, with racial differences as the human species left Africa and adapted to different geographies. But we do not find racial differences in language ability (an African baby can learn Chinese just as easily and as well as a Chinese baby, and so on for the different racial groups), and "Since we have no such inter-racial problems, 'universal grammar' could not have evolved, whatever it may be (and if it exists)." On the other hand, since such an essence could not evolve, and we know species evolve, either there is no such essence or it arose "in a single mutation that has come about very recently, say twenty or forty thousand years ago" (178–179). The problem now only worsens, says Hattiangadi. If the UG arose as the result of a single mutation, we should still find some people who do not have it, but of course we do not. Secondly, the first humans who got UG would have been no more rational than their fellow human beings who did not have it and so would be at a selective disadvantage compared to their fellows, since they would have been concentrating on developing a natural language rather than on adaptively useful things, so that "it is quite incredible that they survived." And if language counterbalanced this by being adaptively useful, it is "even harder to find an advantage that men might have found for the first beginnings of language, if they were already rational anyway" (179). Third, there is the problem of vocal anatomy. The human language ability is not only in the brain but in the mouth and neck, more specifically in the tongue and dental arch and in the location of the larynx, all of which are necessary for articulate speech and all of which are wrong in our ape cousins. But they are not wrong in ancestral humans going back over a million years, even though they had smaller brains. Hence, on Chomsky's view, "We must have spoken for many million years before we found out what we were saying." To this we may add that the probability of the UG *and* the right vocal anatomy resulting either from a single mutation or a simultaneous combination of mutations is vanishingly small, next to zero if I may put it that way.

To all of this, Hattiangadi adds some further difficulties (177–178) which he calls psychological but they are still from an evolutionary point of view. Briefly, if language acquisition is a matter of the UG, then an adult ought to learn a language as easily as a child, but an adult does not. Secondly, language relearning (following brain damage) should be no more difficult in an adult than in a child, but it is actually more difficult. Finally, if language and IQ are highly independent, as would have to be the case on Chomsky's view, then correlations between the two would go against Chomsky and "we could test this" (Hattiangadi, however, suggests no specific kind of test).

As for the commonalities in structure found between human languages, Hattiangadi admits that there are such (he calls them "trivially true"). But he does not think essentialism is needed to explain them. Instead, he appeals to a common evolutionary origin of human languages and to subsequent branching evolution under similar circumstances; in other words, his simpler explanation is that human languages have "some common *hereditary characters* which have evolved under very similar but not identical *selective pressures*" (175).

What is odd is that Hattiangadi seems to involve himself in essentialism when giving his own explanation of human language. He says, "What is innate to man and common to all men is a system of expectations which are brought to the world at birth" (175). Hattiangadi was a student of Karl Popper and it shows in his theory of human language evolution. For Hattiangadi, following the main ideas of Popper that we examined in Chapter 1, the common expectations that we are born with are basically theories that get discarded or modified in a Popperian *modus tollens* way. To support this view, Hattiangadi appeals to the Sapir-Whorf hypothesis, according to which language embodies theories and points of view, sometimes with remarkable differences such as the Hopi language, which has no past or future tense or words for units of time.

How well do Hattiangadi's criticisms fare? Although the landmark attempt at combining Chomskyan linguistics with evolution was made by Pinker and Bloom (1990), and I shall use that source a little when looking for replies, it is more useful to turn to Pinker's book on the subject (1994), aptly titled *The Language Instinct* (a useful summary of which is Pinker 1998). I have no idea whether Pinker ever read

Hattiangadi (1987), but as his criticisms are rather generic it turns out that with a little searching we can find answers to most if not all of them in Pinker's writings.

Beginning with the claim that linguistic universals can be explained by common origin and common selection pressures, Pinker has a long and detailed reply involving evidence from pidgins and creoles (233–240; see also Bickerton 1990, 118–122, 169–171). These phenomena support the idea of an innate UG since they should not exist if language universals are merely the result of common origin and common selection pressures. What needs to be added to this is what Pinker calls "One of Chomsky's classic illustrations . . . the process of moving words around to form questions" (40). This is a very specific rule-governed process, which displays remarkable uniformity given the many possible ways a statement could be turned into a question. Interestingly, Chomsky (e.g., 2000) uses the uniformity of such rules to argue not only against the view that language acquisition in children is inductive but also against the view that language universals are merely historical characters surviving a common origin. He says,

> languages have not "drifted" to incorporate this "simplification" of the rule of question-formation over many millennia. The problem, in short, is one of poverty of stimulus, and speculations about genetic kinship of languages have nothing whatsoever to do with it, in this and innumerable other similar cases.
>
> (56)

Interestingly, this sort of empirical evidence, repeatedly marshaled by Chomsky, goes far in undermining criticisms such as we have seen earlier in this chapter by Searle, the claim that syntax is observer relative, a view Hattiangadi (1987) seems to share when he argues against the existence of "universal elements of grammar" on the grounds that "contexts influence meaningfulness" (177).

Next, there is Hattiangadi's criticism that Chomsky's UG entails little or no correlation with IQ. Pinker actually cites evidence that supports the lack of correlation between the two, in further support of Chomsky, namely, the evidence from specific language impairment and linguistic idiot savants (48–53, 322–325).[7]

As for the view that languages embody theories and these can be highly variable, Pinker (1994, 59–64) points out that subsequent anthropological research has refuted Whorf's claim that Hopi language has no past or future tense and no units of time. Instead it is, he says, another SSSM myth. Indeed, for Pinker, "there is no scientific evidence that languages dramatically shape their speakers' ways of thinking" (58).

Next, for the claim that if Chomsky's UG is real then adults ought to acquire a natural language as easily as a child but they do not, Pinker (1994, 290–296) points out that Chomsky's UG bears a characteristic common to many adaptive instincts, namely that it is age dependent, having a "critical period" in ontogeny. One example he gives is the instinct to breed. Gazzaniga (1992, 56–57) provides the further example of bird songs, which are limited to a critical period in the youth of birds and

---

7   See Badcock (2000, 244–253) for evidence from cases of Williams and Down syndromes.

are species-specific (i.e., they only learn the songs of their own species, with variation in the songs being fairly limited). I would think a further good example is the instinct of imprinting found in goslings.

Perhaps of greatest interest and importance is the claim that the UG is essentialistic and therefore misconceived. Remarkably for someone attempting to place Chomsky's UG into a traditional evolutionary framework, Pinker (1994) makes some statements which seem to smack of essentialism. Pointing out that anatomically modern humans go back roughly 200,000 years and migrated out of Africa roughly 100,000 years ago, he finds it "hard to believe that they lacked language" and that language began roughly 30,000 years ago (a common view given the dating of cave paintings and other evidence of abstract thought, a view that we shall return to near the end of this chapter). Instead, he says, "The major branches of humanity diverged well before then, and all their descendants have identical language abilities; therefore the language instinct was probably in place well before the cultural fads of the Upper Paleolithic emerged in Europe" (353). "Identical language abilities" certainly sounds essentialistic, especially if the focus is on the UG (as it is with Pinker). Pinker even says, again sounding very essentialistic with regard to his concept of the UG, that "Babies become human at three months when their larynx descends to a position low in their throats" (354).

Chomsky himself can also be found using language about the UG that certainly looks essentialistic. For example, in one of his early books (Chomsky 1968) he says,

> The study of universal grammar . . . tries to formulate the necessary and sufficient conditions that a system must meet to qualify as a potential human language, conditions that are not accidentally true of the existing human languages, but that are rather rooted in the human "language capacity" and thus constitute the innate organization.
>
> (24; quoted in Hattiangadi 1987, 170)

Then again in one of his recent books (Chomsky 2000) he says, "the initial state is a common human possession. It must be, then, that in their essential properties and even down to fine detail, languages are cast to the same mold" (7). And yet in the very same book he says of the "faculty of language" that it is "a 'species-property' that is shared among humans to close approximation" (2). In fact, Chomsky can often be found saying the same thing about the UG, even in his earlier writings. For example, in an interview conducted by Bryan Magee (1978) Chomsky states that the language organ is "roughly uniform for the species" (184). In a book published close to the same time (Chomsky 1980), again in reference to the language organ, he says, "we may suppose that there is a fixed, genetically determined initial state of the mind, common to the species with at most minor variation apart from pathology" (187). And then again in a recent book (Chomsky 2002) he says, "This language organ, or 'faculty of language' as we may call it, is a common human possession, varying little across the species as far as we know, apart from very serious pathology" (47).

Perhaps one might then call Chomsky's UG *almost essentialistic*, if that makes any sense. In any case, essentialism is arguably not the problem that Hattiangadi and others make it out to be, and here Pinker and Bloom (1990, 476–477) have an excellent passage explaining why. In reply to the claim by Lieberman (1989, 203–205, 223) that Chomsky's UG is essentialistic and therefore not in touch with modern evolutionary biology, Pinker and Bloom reply that it is common in the literature of physiology and anatomy to speak of a structure as a single kind of system common to all members of the species (unlike racial variations, I might add), a perfect example being the structure of the human eye, with variations typically considered as deviations from the norm. "This is because natural selection," they say, "while feeding on variation, uses it up." Moreover, they point out that, beyond the norm, variations are well known, such as specific language impairment and the heritable deficit studied in a family by Gopnik (see Pinker 1994, 326–331, for more on the essentialism issue).

To all of this I would add the consideration that modern evolutionary biology has not killed essentialism, even though it is true that no species has a genetic essence. To me the most striking piece of evidence is the genetic code. Although known to be chemically contingent, it used to be called "universal" until some minor deviations were found to exist in some single-celled organisms such as yeast and some organelles such as mitochondria. Hence the code, though uniform in all plant and animal species, is now called "standard" rather than "universal." It is also known that, although chemically arbitrary, it is far from random (Crick's "frozen accident"), in that it has been fine-tuned by natural selection in terms of minimizing errors (Freeland and Hurst 1998). The point is that even though the genetic code is standard rather than universal, it is still an essence (and in the deep sense of causal), the same in every human being and other animal that has it, even though it evolved incrementally by natural selection. In the early history of the code it would have been easy to change, as organisms were relatively simple, while in more complex organisms codon-amino acid change would typically be deleterious to the organism and selected against, so that the standard genetic code must have become fixed very early in the evolution of multi-cellular life (Maynard Smith and Szathmáry 1999, 39–46). The idea here is not that of a species-specific essence, of course, but it still harms the criticism that Chomsky's UG could not have evolved by natural selection because it is essentialistic. It is quite possible that the UG (which might better now be called the SG, following the lead of molecular biology) evolved to the point where it became fixed in the human species, deviations such as SLI being selected against in early hunting-gathering groups but still recurring now and then as deleterious mutations typically do.

The claim that the UG is essentialistic and therefore cannot be real has taken some further twists that are worth considering. Gerald Edelman (1992), for example, famous first for his earlier work on the Darwinian nature of the immune system and later for his theoretical model of the brain which he calls "neural Darwinism," begins with the now common mantra that "population thinking deals a death blow to typological thinking or essentialism" (239), that "Biology (particularly Darwin's work) shows essentialism to be false" (234). He then adds that psychologists have shown, to such an extent that "it has been generally confirmed" (236), that humans do not naturally categorize objects into classes essentialistically but instead clustorally (my term), where membership in a class depends not on an object having each of the characteristics that define the class but only some of them (none of them being individually necessary let alone jointly sufficient) and even then as a matter of degree, so that membership in a class is not a yes-or-no affair but instead is fuzzy, a matter of degree. Adding an evolutionary spin to this, stating that "categories are determined by bodily structure and by adaptive use as a result of evolution and behavior" (239), Edelman then argues that this feature of the human mind-brain counts against Chomsky's UG because "A lot of categorizing must be done in order to speak" (242), so that Chomsky's UG "does not agree with the empirical facts concerning categorization" (243).

This is a remarkable chain of inference against Chomsky's UG. It is comparable to someone arguing that all plants and animals on earth cannot possibly have the same genetic code because Darwin killed essentialism in biology. Now imagine the reverse, someone arguing that humans cannot possibly classify objects clustorally because the genetic code in them is essentialistic and the genetic code along with their DNA built their brains. In each case, the reasons given are totally irrelevant. Returning to Edelman's argument, it might indeed be the fact that humans naturally and habitually classify objects clustorally, but this has nothing to do with whether the UG is essentialistic. The fact is also that humans, in particular chemists and geometers, not only can but do categorize objects in the classical essentialistic way. Does this now mean that the UG is essentialistic, or at least partly essentialistic? Not at all. Whether the UG is or is not essentialistic is entirely irrelevant to human categorizing, whether the objects categorized are atoms or animals, since an essentialistic or non-essentialistic UG is perfectly compatible with both essentialistic and clustoral modes of classification.

Even more remarkable are the arguments made against Chomsky's UG by the linguist George Lakoff and the philosopher Mark Johnson, arguments made again in the name of anti-essentialism. Lakoff and Johnson (1999) argue throughout their book that mind is what they call "embodied" (16), that like Searle substance dualism is not only false but the matter of the mind really matters for the nature of mind. To this they add by now the common mantra that biology (along with cognitive science and neuroscience) conceptualizes human nature "in terms of variation, change, and evolution, not in terms of a fixed list of central features ['the classical theory of essences']" (557). Chomsky's UG, however, is "an essence" (472), an essence that is twofold since "The capacity for 'language' defines the essence of human nature, and universal 'syntax' defines the essence of language," so that the UG "could not have evolved through natural selection. Chomsky's Cartesian perspective rules out such a possibility" (476).

But Lakoff and Johnson go much further than this recurring theme. Relying on Edelman's theory of neural Darwinism (480), the most extreme theory on brain plasticity to date (see Edelman 1992, ch. 9), they make two claims that, if true, make Chomsky's UG impossible (on any account of origin). First, Chomsky's UG, as a language "organ," would have to be conceived as a structural or functional module in the brain. However, they claim that "From the perspective of neuroscience, Chomsky's idea of 'syntax' is physically impossible." This is because the UG requires in the brain a "neural module, localized or distributed, *with no neural input* to the module. But this is physically impossible" (480). Indeed, they seem to be against neural module theories of the mind-brain wholly and completely (38). The main problem with their view is that for mainly empirical reasons (namely, the study of behavior, brain damage, brain surgery, and brain scans) it has become commonplace in cognitive science and neuroscience that the brain is not a singular multi-purpose organ but is instead, like the rest of the body, highly if not massively modular, divided into a number of specialized units (= *modules*, like the components of a stereo system, though the comparison can easily mislead, as a brain module can be distributed throughout various regions of the brain) which carry out specific functions and which were evolved by natural selection.[8] The vision cortex, the motor cortex, short- and long-term memory, unconscious memory, and Broca's and Wernicke's areas for language are among the most common examples given for brain modules, the latter two having been known for quite some time. That Lakoff and Johnson think that the theory of the UG, as a syntax module of the brain, requires "no neural input" (see 475–476) is certainly not a claim made by Chomsky, and moreover it is by no means required by module theory (any more than stereo system theory requires "no signal input") *unless one presupposes extreme brain plasticity to begin with!* And that is the fundamental problem with their argument: quite simply, it begs the question.

The second claim that Lakoff and Johnson make, based again on an extreme model of the plasticity of the brain (probably further than Edelman himself would want to go), is that the brain is so plastic, with neuronal connections dying off in the first few years and new ones forming as a result of experience, that "much of what is given at birth is not present five years later. But what is given at birth is supposed [by Chomsky] to be innate and thus something that cannot be lost. The neural facts don't fit the philosophical theory of innateness" (507). Linguistic universals, instead, are "universals of common experience, which occur after birth" (508). Plasticity of the brain, let us be more specific, refers to the fact that the connections between the 11 billion or so brain cells (neurons) that make up our brains have up to 50,000 connections (synapses) between any two of them and these connections can change in response to environmental experience. The issue (see Gazzaniga 1992, ch. 2) is whether the brain is basically plastic, such that neuronal change is due to growth and reorganization of neurons and their synapses in response to experience, with the organism's genome coding not for a primary repertoire of circuits but rather only for *constraints* on the primary repertoire (Edelman), or whether the brain is basically static, with pre-existing genetically determined circuits activated or inactivated by experience, allowing for some change in neurons and their connections but not a whole lot (Chomsky). What is remarkable about Lakoff and Johnson's claim (again) is that they take a theoretical model of the brain (plasticity such as Edelman's) and argue as if it were a fact, whereas the reality of the matter is anything but, given that experts continue to debate both sides (see furthermore the reply to Edelman in Chomsky 2000, 103–104).

---

8   See Pinker (1997, 21, 23, 27–31); Anderson and Lightfoot (2002, ch. 10); *Scientific American* (August 31, 2002, devoted exclusively to the mind); Dror and Thomas (2005).

What needs to be remembered in all of this, before we let ourselves get carried away arguing over neuronal models, is, firstly, the enormity of species-specific instinctual behaviors exhibited by animals, from the waggle dance of honeybees and the intricate webs spun by spiders to the dams built by beavers and so on, and secondly, returning to human language, the nature of the evidence supporting Chomsky's theory of the UG, especially poverty of stimulus, creoles (these two forming what Pinker calls the "reinvention" of the same language rules), and specific language impairment. This evidence speaks powerfully in favor of Chomsky's thesis, but typically it is ignored by critics with alternate models of language. Moreover, it need not be the case that the neuronal circuits that make up the UG are in place when the baby is born. Instead, it would make more sense that they gradually form in the brain (directed by the genetic program in the genome) as many other phenotypic traits are formed, not from conception to birth but later in growth within a specific time frame (like teeth, or feathers on birds).[9]

After making our way through all of these objections and seeing why they fail, we are now in a much better position to return to our original question: Can Chomsky's UG be reconciled with Darwinian evolution by natural selection? Or to put it more strongly, is natural selection the best explanation of the UG? Steven Pinker is the strongest advocate of the *yes* side, no matter how the question is put, and it remains to look at the basics of his answer as well as the alternative view championed by Derek Bickerton.

Pinker (1994, ch. 11) uses an interesting analogy to get his point across, namely, the trunk of the elephant. There are a number of features of the trunk of the elephant that make it an excellent analogy to human language. It is extremely complex, for a start, made of roughly 60,000 muscles, it is amazingly multi-functional, and no other living species has anything like it. Now imagine what critics might say, critics who have trouble with a Darwinian point of view: (i) Some, says Pinker, might deny the uniqueness of the trunk, (ii) some might deny its complexity, (iii) some might argue instead that it is a byproduct of the evolution of the largeness of the elephant's head, (iv) some might deny that such an amazing thing could evolve by natural selection, either because they cannot imagine incremental steps (again, natural selection requires that every stage confer some sort of advantage resulting in a statistical increase in reproductive success) or because the organ does more than what its ancestors in the wild would need, and (v) some might argue that it is the result of a single dramatic mutation.

In each case, says Pinker, the critics would be wrong, not only for the trunk of the elephant but likewise for the human language instinct. For a start, the closest living relative of the elephant is the hyrax, which shares roughly 90% of its DNA with elephants. It has long nostrils, but nothing like the trunk of the elephant. Humans and chimpanzees share much more DNA, something like 98–99%, and yet as we have seen from ape language studies, chimpanzees (along with the other great apes) have nothing or next to nothing like the human language faculty. We really are unique in this respect.

---

9   For more on this implication of the UG, see Pinker (1994, 288–290, 315–317; 1997, 34–36) and Anderson and Lightfoot (2002, ch. 9).

Second, the complexity of the trunk of the elephant is a given (to anyone willing to research into its anatomy, that is). But many will balk at the idea that the human language faculty is really all that complex as Chomsky and Pinker and others make out. But here again, this is only because one has not done the necessary research. The reason why many will be unconvinced *before* they do the research is because we commonly think of language simply as a vocabulary of words. The rules we use to put them together into meaningful sentences are something we take for granted. And the reason is obvious: we use them unconsciously. A big part of the science of linguistics is to bring these rules to the surface. And the complexity is truly staggering. To see this, one has only to try to follow a linguistics book on syntax (e.g., Chomsky 1965), even an introductory one (e.g., Pinker 1994, chs 4, 5, and 6; Yule 1996, chs 9 and 10). (Personally, I find studying molecular biology easier.) Linguists typically take a sentence and use a labeled tree diagram to break the sentence into its parts and to show the rules used. A very simple sentence such as "The monkey ate a banana" would appear as in Figure 3.1 (Adapted from Yule 1996, 105).

The example is not simply one of the subject-verb-object (SVO) rule of syntax. In each case the article comes before the noun, which is another rule. And from a simple example like this, it only gets worse (or better, depending on how one looks at it). Every human language contains phrase structure rules (rules for positioning nouns, proper nouns, pronouns, noun phrases, verbs, adverbs, verb phrases, articles, adjectives, prepositions, and prepositional phrases), recursion rules (rules that reapply so that a sentence can have other sentences in it, or a phrase other phrases in it), transformational rules (rules for turning a sentence into another sentence, such as turning a sentence into a question), lexical rules (rules for which words are to be used for sentence constituents such as N), and tense rules (rules for past, present and future tense). The UG would contain the human language universals of these rules. But it would also include more. As Pinker points out, the human language instinct is "composed of many parts" (362): hundreds of rules of syntax, a complicated morphology (a system for building bigger words from smaller words, like the word "talkative" from "talk"), a huge lexicon (a storehouse of intuitive concepts and categories), a modified mammalian vocal tract, phonological (speech sound) rules and structures, speech perception/comprehension, parsing algorithms (rules for analyzing an ambiguous sentence into its grammatical constituents, as with "Dr Ruth will discuss sex with Dick Cavet"), and learning algorithms.

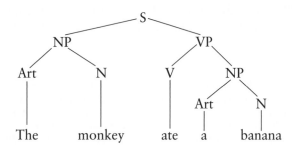

Figure 3.1 S, sentence; NP, noun phrase; VP, verb phrase; Art, article; N, noun; V, verb.

Third, just as some might argue that the trunk of the elephant resulted as a byproduct of the evolution of the largeness of the elephant's head, some might argue that the language organ evolved as the result of the evolution of the largeness of the human head. This, as we have seen earlier, has been Chomsky's preferred view. But for Pinker this argument fails in both cases, and for some of the same reasons. First, it makes no sense why evolution would select for larger brains *per se*, since they create greater pain and mortality in childbirth, are more difficult to balance, and are metabolically greedy, reasons that in themselves mean that larger brains are maladaptive and would be selected against. The non-selectionist has it all backwards. If larger brains evolve, it must be as a byproduct of selection for something else, adaptive reasons that outweigh the maladaptiveness of bigger brains. For Pinker, this can mean only one thing: "Selection for more powerful computational abilities (language, perception, reasoning, and so on) must have given us a big brain as a by-product, not the other way around" (363).[10] Second and finally, language ability in humans does not covary with brain size, as it should if language is a byproduct of brain size. The linguistic abilities of dwarfs and hydrocephalics (an accumulation of fluid in the brain, causing great enlargement of the brain and mental handicap) are the extremes of evidence here (363–364).

Fourth, some might deny that not only the trunk of the elephant but also the human language faculty could evolve by natural selection. In the case of the trunk of the elephant, homologies, says Pinker (350), suggest that it evolved from the fusion of the nostrils and upper lip in the common ancestor of the elephant and hyrax, followed by incremental steps. This would be a standard case .of evolution, with the initial lineage branching into a number of lineages as it spread geographically, including extinction along the way (e.g., mastodons and mammoths). Moreover, that the trunk of the elephant can be used for activities that we see in circuses does nothing for the claim that these functions could not have evolved. Human hands did not evolve for playing the piano or strumming the guitar, but these activities evolved culturally due in part because human hands evolved for the functions they did, the cultural functions being exaptations of a sort.

The same is equally plausible for the human language faculty. The primate brain had a single origin and evolved gradually in a branching fashion, with much extinction along the way, resulting in only one living species, *Homo sapiens*, with a highly evolved language instinct. The stage at the beginning of the human ascent to true language is now usually called *protolanguage*, following Bickerton (1990), language with little if any structure, "language that lacks most of the distinguishing formal properties of language" (118), language as it would have been found in *Homo erectus* and as it is

---

10   Bigger brains, indeed bigger anything, could result simply as a matter of cold weather, known as Bergman's Rule. (Body size tends to be larger in the colder regions of a species' range, since larger body size helps regulate body temperature. Larger body size would therefore be the result of a selection process, homeothermia being the selection pressure.) But Pinker's "must" still stands, for Bergman's Rule does nothing to explain the increase in *proportion* of human brain size to body size.

currently found in ape language studies, children in the two-word stage, and pidgins.[11] Here is where paleoanthropology and archeology have tried to contribute, making inferences from the bones and artifacts of *Australopithecus*, *Homo habilis*, *Homo erectus*, early *Homo sapiens* (which began roughly 200,000 years ago in Africa), and Neanderthals (sometimes classified as *Homo sapiens neanderthalensis* but more commonly now as *Homo neanderthalensis*—I shall assume the latter—which evolved roughly 300,000 years ago in Europe). For language to be possible, says Pinker, the right vocal machinery would have had to evolve along with the language organ in the brain, and it is possible to infer from bones the kind of larynx (vocal tract) an animal had (animals have a larynx high in the throat, but in humans it is low in the throat, which is required for the full range of human vocalization, vowels, and consonants). Moreover, a jaw of a certain kind of shape is needed in order to give the muscles that move the tongue proper leverage.[12] In addition, the language organ is located in the cerebral cortex of the brain, so perhaps inferences could be made from the skulls (size, shape, and faint imprints of the wrinkle patterns). Tools could also contribute as evidence. Language is social, so that its selective advantage would have to be social. (For example, think of a hunting party and compare the advantage of having syntax and semantics to having semantics alone, including gestures. Indeed, to really get the point, try going through a day, or better a week, using only words and no sentences. The difference is dramatic.) Evolving gradually over time, then, each specific feature of the language instinct would be the result of a series of advantageous mutations that conferred to its host a statistical increase in survival and reproduction. A mutant with an improved feature, says Pinker, would not be completely understood by those in his group, but he would be partly understood as a result of "overall intelligence" (365), so that there need not be selection against the mutant. Thus the mutant could

---

11   The problem of the origin of life is similar in some respects to the problem of the origin of language, in that DNA is too complex to be the original genetic molecule, so that it had to have evolved ultimately from a simple kind of genetic molecule no longer in existence.

12   Interestingly, a fairly recent discovery of a teenage Neanderthal shows that it had a kind of bone in the neck thoroughly modern in form and consistent with a larynx low in the throat; also, Neanderthal jaws are sufficiently similar to humans to allow for the right kind of leverage of the tongue. Given that the split in hominids that eventually led to Neanderthals on the one hand and modern humans on the other occurred roughly 600,000 years ago, this has led Aiello (1998), for example, to conclude that Neanderthals could have been capable of full human speech but also that it is possible they had none of it at all, given "the virtual lack of convincing evidence for explicit symbolic thinking in Neanderthals" (23). Instead, he suggests that the parts of the vocal anatomy necessary for human language "may have been simple consequences of fundamental and morphological and lifestyle changes that accompanied the evolution of *Homo ergaster* (early African *Homo erectus*) almost two million years ago" (24). The idea, then, is that the vocal anatomy evolved for reasons other than modern language and that its use for the latter is an exaptation, a view seconded by Tattersall (2000, 62). But still, we have to wonder why the larynx descended in the throat of these hominids (if indeed it did that far back) while it did not descend in any other animal lineage known to biologists. Moreover, if the reason was not language, we have to wonder what the adaptive advantage of it was, for an adaptive advantage it surely must have been, one strong enough to more than compensate for the disadvantage of having the larynx descend low in the throat, the disadvantage being the capability of choking to death on food.

reproduce and pass on his genes, including the mutation, but the mutation could also occur by chance in other members of the population. Suppose the mutation confers in the host the ability to linguistically make some sort of fine distinction. This creates a selection pressure, says Pinker, to evolve in that population the matching system. Consequently anyone else who got the mutation would completely understand the mutant and be favored by natural selection. And on and on for thousands upon thousands of years, leading eventually to the modern language organ (the UG) in contemporary humans. Intermediate stages, says Pinker (366), are easy to imagine, more complicated rules being built from less complicated rules.

As for the claim that language does too much, that it does far more than our ancestors in the wild needed, so that it could not have evolved by natural selection, Pinker (367) replies that our ancestors in the wild were not simply grunting cavemen but hunter-gatherers who dealt with a complexity of information, including complicated tool making and detailed knowledge of the biology and ecology of the plants and animals they depended on. In addition was the importance of the ability to communicate precise information about time, space, objects, and who did what to whom. (Indeed, think of the importance of story-telling for these social groups, in addition to gossip, making alliances and other social relations, and long-term planning.) All of this required complicated language. Pinker further points out that our hominid forebears would be competing against other similar hominid groups, resulting in a kind of "cognitive arms race" (367), escalating until only one group or kind of group was left. What adds to all of this is the fact that recently existing stone age cultures studied by anthropologists, such as the Tasmanians, the various groups in New Guinea, or the bush people of Botswana, have fully modern languages (Mellars 1998, 97).

For Pinker, then, all things considered, the conclusion is "inescapable" (362) that language evolved incrementally by natural selection. As Dawkins (1986, ch. 11) argued, natural selection is the *only* real explanation for biological, adaptive complexity. And this is what allows Pinker (359, 364) to respond to the kind of objection that we saw in Chapter 1, the just-so evolution by adaptation story-telling exposed and parodied by Gould (1980a). It is the adaptive complexity of the language instinct, the adaptedness of incremental steps, and the convergence of evidence that makes the explanation by natural selection anything but just-so.

But there is one final objection that we have not yet dealt with, namely, that a more plausible view is that the human language faculty resulted from a single dramatic genetic event, a "big bang" as it is sometimes called. Derek Bickerton is the key linguist here. He characterizes the debate as gradual evolution versus "catastrophic" evolution. For Bickerton (1990), who by the way like Pinker is a Chomskyan,

> syntax must have emerged in one piece, at one time—the most likely cause being some kind of mutation that affected the organization of the brain. Since mutations are due to chance, and beneficial ones are rare, it is implausible to hypothesize more than one such mutation.
>
> (190)

Undaunted by the complexity of language, Bickerton in the pages preceding this quotation argues that the field of generative grammar (Chomskyan linguistics) began with supposing the UG to contain a long list of rules, but that as the field has progressed and matured it has increasingly simplified the nature of the UG. The UG is still fairly complicated, but not nearly to the extent that it was in the 1950s and '60s. Banking that this trend will continue, Bickerton thinks that it is quite possible that language (the UG) evolved from protolanguage in a single step. He thinks this might have happened roughly 200,000 years ago, at the same time when anatomically modern humans first emerged in Africa, possibly from a mutation in a single woman (175–177). To Pinker (1994), on the other hand, Bickerton's suggestion that "a single mutation in a single woman . . . simultaneously wired in syntax, resized and reshaped the skull, and reworked the vocal tract" is outright "jaw-dropping" (366).

As we have already seen above (note 12), there are some paleoanthropologists who believe that the human vocal anatomy evolved for reasons other than language and preceded the evolution of language, a view that if true (I for one am not convinced) would certainly aid Bickerton's thesis (though as we shall see below, he does not use it). That leaves only syntax. Part of the problem is that Pinker and Bickerton do not have the exact same idea of the UG. Even though they are both Chomskyans, Pinker retains the more complex form of the UG, Bickerton the hopefully less complex form. This is something not worth trying to resolve here. What is worth addressing is that Bickerton has since had the benefit of the publications by Pinker and Bloom (1990) and Pinker (1994, 1998) and yet he has not backed down from his basic thesis. Instead, he has only defended it more (Bickerton 1998), though now no longer requiring a single macromutation but minimally only a recombination of genes. The evolution of protolanguage to syntax is still "a single step" (341), "as abruptly as a creole emerges from a pidgin" (355), but now in the sense that syntax is the linkage into a "single circuit" of a number of already existing computational parts, "some protolanguage-related, some not" (344). Sexual selection would kick in to spread the new gene combination (or macromutation) throughout the population, resulting in a relatively rapid "speciation event" (353). The "birth of syntax" would then create enormous selection pressure "for an improved vocal apparatus even at the cost of maladaptive effects on ingestion" (342).

Interestingly, Bickerton (1998, 354–357) gives five independent lines of evidence and argument which together he believes point toward the truth of his catastrophist model of language evolution and away from the gradualist model of Pinker and others. We shall finish this chapter by examining those lines of argument and seeing how well they stand.

Bickerton's first argument appeals to what he calls a "cognitive explosion" that occurred "only *after* the emergence of anatomically modern humans in South Africa about 120,000 years BP [before the present]." His numbers have become a little off, as the number given is still around 200,000 years ago. At any rate, for Bickerton the only plausible explanation for the "cognitive explosion" is the sudden birth of language in a small *Homo erectus* population, possibly he says in South Africa, a population like other *Homo erectus* populations having only protolanguage, with the result that not only a new species was born, *Homo sapiens*, but one with full-blown language, a change that gave *Homo sapiens* an enormous competitive advantage. Had the evolution of language progressed gradually, as Pinker holds, then, says Bickerton, the evidence should reveal a gradual increase in technology and culture. But the evidence speaks otherwise.[13]

Bickerton might have done better to place the birth of language much later, not at the beginning of *Homo sapiens* but, as the paleoanthropologist Richard Klein (2002) supposes, just before the first really big cognitive explosion, known as the "Upper Paleolithic revolution." This revolution has been variously described as a "symbolic revolution," a "cognitive revolution," a "cognitive explosion," and a "behavioral big bang." The revolution occurred in Europe roughly 40,000 years ago, the time *Homo sapiens* migrated into Europe. For Klein, the brain mutation responsible for language occurred roughly 50,000 years ago and probably in Kenya, allowing our species to spread out and conquer, resulting in the Upper Paleolithic revolution. During the Middle Paleolithic, Europe and the Middle East were occupied by Neanderthals, beginning roughly 300,000 years ago. Roughly 200,000 years later, in other words 100,000 years ago, *Homo sapiens*, which evolved in Africa roughly 200,000 years ago, migrated into the Middle East from Africa and for roughly 60,000 years there was very little difference between the two hominid species in technology and culture, which were stone age and relatively simple. Then, at around 40,000 years ago, *Homo sapiens* migrated into Europe, having undergone at the time a profound change, in which they produced advanced weaponry, kept records on bone and stone plaques, formed long-distance trade networks, expressed themselves symbolically through art and music, and buried their dead in some cases with goods and fancy accompaniments (indicating social stratification and a belief in life after death).[14] Only 10,000 years after the arrival of these *Homo sapiens*, the aboriginal hominids, the Neanderthals, sadly went extinct, their extinction beginning in the Middle East and reaching completion in Europe.[15]

This line of evidence, however, has three serious problems with it, not only for Klein's argument but by extension for Bickerton's as well. First, as Pinker (1994) points out, *Homo sapiens* left Africa roughly 100,000 years ago and spread to various

---

13   Indeed, Bickerton (1990, 175–177) argues that language would be a necessary prerequisite for technological progress but not a sufficient cause, competition with other hominids, such as Neanderthals, being the remaining driving force.

14   Interestingly, my Ph.D. supervisor and friend David Johnson (2003) has taken this evidence a step further by arguing that the Upper Paleolithic revolution was the first of two great revolutions that produced the modern objective mind, the other revolution being the Greek revolution that occurred not in Asia Minor but in mainland Greece between 1100 and 750 BC and which ushered in philosophy and science (Thales being not the first philosopher-scientist but one in a long succession).

15   The consensus is that their disappearance was one of short-term replacement (Mellars 1998, 91). I use the word "sadly" above not only because of the loss in biodiversity, but from a moral point of view. As Tattersall (2000) puts it, "In light of the Neanderthals' rapid disappearance [keep in mind they survived and thrived in Europe for roughly 260,000 years] . . . and of the appalling subsequent record of *H. sapiens*, we can reasonably surmise that . . . interactions [between the two] were rarely happy for the former" (61). Such is human nature.

parts of the globe, not just the Middle East and Europe. They arrived in Australia, for example, roughly 60,000 years ago. And yet, says Pinker,

> The major branches of humanity diverged well before then [the Upper Paleolithic], and all their descendants have identical language abilities; therefore the language instinct was probably in place well before the cultural fads of the Upper Paleolithic emerged in Europe.
>
> (353)

Second, there is growing evidence that the Upper Paleolithic revolution was not really a revolution, that the kind of culture and technology that shows up so dramatically there is instead part of a continuum with major features already existing many tens of thousands of years previously in a number of places in Africa. Kate Wong (2005) brings together the growing evidence from a small but increasing number of archaeologists and paleoanthropologists, all of which points to the conclusion that "modern human behavior emerged over a long period in a process more aptly described as evolution than revolution" (91). The kinds of evidence are harpoons carved from bone from 80,000 years ago, red ochre (a pigment possibly used for symbolic purposes) and the grindstones for making it from 285,000 years ago, stone blades from 510,000 years ago, but most striking of all the recent discovery in a cave at Blombos in South Africa of bone tools, finely cut stones, bones of deep-sea fish, red ochre, engraved bones, and long strands of pearl-like beads, all from 75,000 years ago. The growing evidence is one of "fits and starts" (93), not a "cognitive explosion," precisely as one should expect if the gradualist model is true.

Third, as Mellars (1998, 104) points out, not only has the intensity of archaeological research for the time range between 50,000 and 100,000 years ago been more intensive in Europe than in Africa, but we have to make a distinction between what he calls "cognitive potential" and "behavioral performance" (108). That groups of humans outside of Europe and before the Upper Paleolithic failed to provide the kinds of evidence that we equate with modern humans does not mean they were incapable of the behavior responsible for those kinds of evidence, including full-fledged language. Lack of evidence is not evidence of a lack. Instead, the growing physical evidence out of Africa suggests to Mellars "strong evidence for the emergence of fully modern cognitive and language capacities in this area at a substantially earlier date than in Europe" (108).

Interestingly, the questions parallel in a striking way a more basic phenomenon in evolution. A species might evolve in a particular locale by gradual Darwinian evolution and expand its range into a new locale. In the fossil record, the species in the new locale looks like it arose instantaneously, especially if there is no fossil record for where it originally evolved. But that is simply an illusion of the fossil record. The cause of what we see is gradual evolution, migration, and an imperfect fossil record, not a genetic big bang (Darwin 1859, 464–465; Hoffman 1989, 126–128). In the case of humans and language, the same could very well be the case. The language faculty could have evolved in a population of humans facing a unique situation in a particular locale. The behavioral advantage (including possibly a time lag in greater technology) allowed the population to grow and migrate into new locales, creating the impression of a revolution in those new locales. But just as with evolution in general, there is no need to suppose here a genetic big bang.

Bickerton's remaining four arguments can be dealt with more summarily. His second argument in the list is that there is an absence synchronically (at the same time) of intermediate stages, living fossils one might say, between protolanguage and language (the UG), not even in the pidgin-creole interface, which occurs in a single generation. Third, neither in cases of language impairment nor even in language acquisition do we find stable intermediate stages between protolanguage and language (the keyword here is "stable"). Fourth, if there were intermediate stable stages diachronically (over time) between protolanguage and language, then "human ingenuity" ought to be able to reconstruct some of them, but this has yet to be done. Fifth and finally, the various parts of UG appear to be interdependent and indissociable, such that "syntax could not possibly work as a computational device unless its major defining properties were intimately linked and had emerged simultaneously" (357).

For Bickerton, then, the "burden of proof" lies not upon the shoulders of the catastrophist, but rather on those of the gradualist. I should say, however, that it lies on the shoulders of the catastrophist. This is partly because the essence of Darwinian evolution includes gradualism, so that any case of catastrophism needs to be argued for as a special case. Moreover, all of Bickerton's four arguments above, when examined from a modern evolutionary point of view, make some rather standard errors. For a start, the lack of presently existing stable intermediate stages between protolanguage and language is not a good reason to believe that there never were any such intermediates. (The evidence from creoles is irrelevant, since the humans with creoles have only a modern UG.) This point becomes strikingly obvious in the case of the standard genetic code, which no molecular biologist doubts had intermediate stages (including the possibility of relatively stable ones) even though today we find none. Moreover, the fact (if it is a fact) that no linguist has been able to reconstruct a stable intermediate stage fares no better, since the same could be said, again, with respect to the genetic code. (Indeed, the appeal to lack of evidence as evidence is known as the *appeal to ignorance* fallacy.) Bickerton's fifth and final argument reminds me of the argument of intelligent design creationists, who take a particular structure such as the human blood clotting system (coagulation) and argue that it is "irreducibly complex" because all the parts of the system are necessary for its functioning and any missing part would result in a nonfunctioning system (hence no evolution by intermediate stages). In the case of the human blood-clotting system, this has been shown to be untrue partly by looking at other species with functioning intermediate stages (see Miller 1999, 133, 152–158). In the case of language, we do not have the benefit of contemporary species with various stages of a UG, but Bickerton's argument is no less a poor one because of it.

In sum, the human capacity for language seems best explained by standard evolutionary principles, criticisms against Chomsky's UG notwithstanding. But enough of that. All the talk about slow and gradual, about *erectus*, and about explosions brings us to our next big question, namely, sex.

4

Evolution and Sex

In his 1972 movie, *Everything You Always Wanted To Know About Sex But Were Afraid To Ask*, Woody Allen deals with a number of romantic and sexual topics, often with great humor and even deep insight into human nature. Among the topics explored in the series of vignettes are whether aphrodisiacs work; sexual perversions such as whips, bondage, and cross-dressing; love and sex with sheep; and women who achieve orgasm only while having sex in public places. None of these topics shall be explored in this chapter. What we shall explore, instead, is the deeper significance of sex from an evolutionary point of view. The four topics covered are whether the different mating strategies of men and women have an evolutionary basis, whether men have an evolved rape instinct, whether there is a genetic and ultimately evolutionary basis to homosexuality, and whether the general repugnance for sex with immediate kin (sibling-sibling, parent-child) has an evolved basis. These are core topics that could well be used in a sequel to Allen's masterpiece, and I can imagine them being done with great humor and insight, some vignettes looking at the topics from an evolutionary point of view, some from a SSSM point of view, each a parody. But alas, although humor can sometimes be more insightful than straightforward analysis, I shall leave the humor and the sequel to moviemakers—but would I love to make that movie!

There is actually one vignette in Allen's movie that is relevant to this chapter, the vignette titled "Are the findings of doctors and clinics who do sexual research and experiments accurate?" The vignette deals with a mad scientist played by John Carradine. Unfortunately, that is where the relevance ends. Not only does Carradine's character not touch on the topic of evolution, but there is nothing inherently mad in developing evolutionary explanations of gender-different mating strategies in humans, as well as of rape, homosexuality, and incest avoidance. Such theories, of course, might turn out in the end to be seriously mistaken. But since they involve one of the best confirmed theories in the history of science (the basics of evolutionary biology), they need to be taken seriously and assessed in the manner that all scientific theories are assessed, in terms of testability, simplicity, coherence with what is already known, fruitfulness, and so on.

The first topic we shall examine is the difference between men and women when it comes to mating. The SSSM, of course, is that, whatever the differences, they are purely arbitrary and culture bound. This is a popular view in academia. What used to be popular, both inside and outside of academia, was the view of Freud, who traced most of our sexual behavior to early childhood psychodynamics, males with their Oedipus complex, females with their penis envy. Freud is no longer popular inside of academia today, while outside of academia one is likely to read popular books such as John Gray's *Men Are from Mars, Women Are from Venus* (1992), probably the most famous book of its kind. The main problem with Gray's book is that he attributes the causes of the mating differences between men and women (which he stresses are statistical) primarily to the differences in hormones. From an evolutionary point of view, this kind of analysis simply will not do, since it focuses only on *proximate* causes, not on *ultimate* causes (a common distinction in evolutionary biology). We want to know, first of all, if the hormonal differences have a genetic basis, and if they do, what it is that ultimately explains the differences in genes.

The main pioneer on the ultimate causes of differences between males and females was—by now it should be no surprise—Charles Darwin. Darwin was struck by the many obvious differences between males and females of many species, such as the manes of male lions, or the big bright plumage of peacocks (peahens are plain brown). For Darwin, it made little sense to consider such differences adaptations to the environment, since the males and females generally inhabit the same environment. Moreover, what Darwin called secondary sexual characteristics are in many cases maladaptive to some extent (e.g., the feathers of the peacock make it more conspicuous to predators). Consequently, Darwin developed a supplementary theory to natural selection, which he called *sexual selection*. This is the kind of selection that explains *sexual dimorphism*—physical differences in males and females—not in terms of adaptation to the environment but in terms of competition for mates. Male lions have the manes because manes are an intimidation factor in male-male combat for the harem; peacocks have the large bright plumage because mating in peafowl is decided by female choice.

Darwin's theory of sexual selection was not well received in its time, even by his own followers, and it was only in recent decades that it has gained respect in professional biology (Cronin 1991; Andersson 1994). Although now recognized as a powerful theory, it still does very little to explain mating differences in species such as humans. This is where modern evolutionary psychology comes in, the premier evolutionary psychologist on human mating differences being David Buss. Although Buss has an entire book on the topic (Buss 2003), we shall focus here mainly on his *American Scientist* publication (Buss 1994). In that article, Buss provides nine hypotheses on human mating derived from modern evolutionary theory, which he then tests using in part the results of the largest cross-cultural study to date, conducted by himself and colleagues between 1984 and 1989. This study, comprised really of 37 studies conducted in 33 countries, covered all major religious, racial, and ethnic groups, and involved surveying a total of 10,047 people.

An important part of the background theory is the addition to sexual selection theory made by Robert Trivers (1972), who developed the concept of *minimal parental investment.* What this means, at bottom, is that the sex (male or female) that has more of a minimal investment in producing offspring is going to be more selective in mate choice, while the sex that has less of a minimal investment in producing offspring is going to face more competition in mate choice from members of the same sex. Applied to human females, the minimal parental investment for passing on their genes is nine months plus years of nurturing and caring, while in males it is but a minute or two of sexual intercourse. This is a major asymmetry between males and females, one which entails that females instinctively ought to be far more selective in mate choice than males.

There are further asymmetries that entail this conclusion. With females, the *capacity* for spreading their genes is far less than it is for males. A female can have at most maybe 15 children in one lifetime (keeping in mind that in our evolutionary past the average human life expectancy was maybe 25 or 30 years), while in males the number is potentially in the thousands. Connected with this asymmetry is a further asymmetry dealing with *age* (although for the reason just given, it would not mean as much in our evolutionary past). Both males and females begin their reproductive years at roughly the same age, but males are capable of producing children well into their 40s and 50s, even into their 80s, while the female capacity ends at menopause, which begins at around 50 (with the increasing risk of birth defects and infant mortality beginning long before that, as well as the increasing risk of mortality to the mother). The final asymmetry, and it is again a big one, involves *paternity confidence.* Females know that the baby inside of them is their own, while males can never be perfectly sure that they themselves are the father (DNA paternity tests, of course, are entirely irrelevant to the evolutionary picture).

According to Buss, the fact of these reproductive asymmetries, combined with the fact of evolution, entails nine hypotheses about human mating that are open to testing. It is important to note that these hypotheses are *derived* from the background theory/knowledge; they are not merely consistent with it. (I refer here, of course, to the complaint of Gould and Lewontin examined in Chapter 1.) Rather than examine all nine hypotheses, I shall focus on what I consider to be the five most salient.

Buss's first hypothesis is that "*Short-term mating is more important for men than women*" (220). By *short-term mating* Buss refers to what we might call "casual sex," "one-night stands," or "brief affairs" (also "experimental unions" according to some of my Modes of Reasoning students), whereas *long-term mating* refers to a committed relationship, what many would call a "real" relationship. Buss tested the above hypothesis by surveying 148 college students (prime candidates for such a survey because of their raging hormones). He found that (i) males expressed a greater desire for seeking a short-term mate than a long-term mate, while females expressed the opposite, (ii) males expressed a desire for far more sexual partners in their lifetime than did females, and finally (iii) males expressed a willingness to engage in sexual intercourse far sooner after meeting someone than did females. Buss supplements this evidence with that of a study conducted by two other researchers where college students were approached by an attractive member of the opposite sex and after a short introduction were asked either "Would you go out on a date with me tonight?," "Would you go back to my apartment with me tonight?," or "Would you have sex with me tonight?" Of the females asked, 50% agreed to the first question, 6% to the second question, and 0% to the third question. Of the males asked, 50% agreed to the first question, 69% to the second question, and 75% to the third question (the remaining 25% usually cited an excuse such as a fiancée or an unavoidable obligation that night).

Buss's fourth and fifth hypotheses are so closely related that he puts them together, the one hypothesis dealing with the attitude of males in relation to fertility in women, the other with the attitude of males in relation to reproductive value in women: "*Men seeking a short-term mate will solve the problem of identifying fertile women, whereas men seeking a long-term mate will solve the problem of identifying reproductively valuable women*" (221). *Fertility* refers to the probability that a female is currently able to conceive a child (this would be higher in, say, a 24-year-old than in a 14-year-old). *Reproductive value*, on the other hand, refers to the probable number of future children produced by a female (this would be higher in, say, a 14-year-old than in a 24-year-old). What these hypotheses both mean is that males would be sensitive to physical features in a female indicating age. They would be attracted to signs of youth and successful child-bearing such as smooth skin and rosy cheeks, full lips, full and firm breasts, fairly wide hips, and would be turned off by signs such as wrinkles, gray hair, sagging skin and breasts. This is because the former features are correlated with high fertility and high reproductive value while the latter features are correlated with the opposite. One needs to keep in mind that just as biology is statistical, these instincts would be statistical too (so one can forget about the aberrant teenage male who prefers old women). Throughout our evolutionary past, those males who had less or none of this instinctual sexual attraction to youthful beauty would tend to have less or no sex with women of high fertility and high reproductive value, and so would tend to have less or no success in passing on their genes compared with those who had more of the instinct. It all boils down to what biologists call *reproductive success*. Genetic traits that increase reproductive success would be favored by natural selection, in this case genetic traits that create in men an instinctive sexual attraction primarily to young women.

As for anthropologists and others who claim that standards of female beauty vary greatly from culture to culture, Buss replies that this is really only true when it comes to plumpness and thinness, that men in all cultures are turned on by signs of youth and turned off by signs of old age.[1] (What females are primarily turned on by is another matter, dealt with in the last hypothesis below.) Indeed I have to wonder, and unfortunately Buss does not address this, that if Buss is right, then we ought to find, cross-culturally as we do here in the West, that females are mainly the ones who like to dance, wear make-up, and enhance themselves with adornments and sexy clothing (such as, lately, push-up bras and low-riding jeans), all in the interest of youthful and sexual attractiveness. This, it seems to me, would be an important cross-cultural test, especially when it comes to make-up. At any rate, Buss in his survey of

---

1   Feminists also tend to argue that the concept of female beauty is culturally relative and socially constructed. Naomi Wolf, for example, in *The Beauty Myth* (1990), claims that politics rather than biology wholly determines the concept of female beauty, namely, the power politics of patriarchy. For an extended reply from the perspective of sociobiology, see Alcock (2001, 136–143).

college students found that while both sexes rated physical attractiveness important, males rated it significantly more important than did females. His cross-cultural study produced the same result. That both sexes should hold physical attractiveness important should not be a surprise from an evolutionary point of view. Physical attractiveness is a sign of "good genes." And again, those throughout our evolutionary past who did not discriminate, or who actually preferred mates with signs of aging, tended to pass on their genes less than those who did discriminate. Nor should we be surprised that males place a higher premium on physical attractiveness than do females. This follows alone from the reasons given earlier, and there are further reasons that we shall examine below. What surprised Buss is that among college students males rated physical attractiveness more important when it came to seeking short-term mates than when it came to seeking long-term mates. There is nothing in his background theory that would entail this result. Perhaps the result, then, is the product of culture. Unfortunately, Buss's cross-cultural study did not repeat the same question, so it could not possibly offer corroboration. Returning to college students, although Buss, as already stated, found that females did not place as high a premium on physical attractiveness in a mate as did their male classmates, they did, however, just like them, rate physical attractiveness more important when it came to short-term mates compared with long-term mates. This is what we should expect, given, as we shall see, that socioeconomic status in males is the key feature of attraction in females seeking long-term mates. Interestingly, Buss corroborates his results about the differences between males and females when it comes to physical attractiveness by referring to a cross-cultural study on divorce conducted by Laura Betzig, who found that a woman's old age is significantly more likely to result in divorce than a man's old age.

Buss's sixth hypothesis is that *"Men seeking a long-term mate will solve the problem of paternity confidence"* (224). As stated earlier, there is an enormous asymmetry between males and females when it comes to knowing who the parents are, females knowing that the baby is theirs, males never being entirely sure (an asymmetry exploited with both hilarious and tragic results in the *Maury Show*, where suspected fathers are subjected to DNA paternity tests, the mothers sometimes being "one thousand percent sure" that so-and-so is the father, often to find out that he is not, usually to the delight of the suspected fathers). Cross-culturally the lack of paternity confidence in males has resulted in a number of strategies designed to increase confidence, such as chastity belts, harems guarded by eunuchs, laws against female adultery, wife abuse, and even murder (see Ghiglieri 2000, 149–154). Interestingly, in his cross-cultural study Buss found that the desirability of chastity in a mate is highly variable, with males in roughly only two-thirds of the countries valuing chastity in females more than females value it in males, the other third revealing no difference between the sexes. But this only involves sex *before* long-term mating has begun. Sex *after* long-term mating has begun is an entirely different matter. Among the college students surveyed by Buss, fidelity was ranked first by males seeking long-term mates, whereas females typically ranked it third or fourth. Buss also cites the research of Laura Betzig, who found that internationally a wife's infidelity was more likely to result in divorce than a husband's infidelity. And then of course there is the prevalence of cross-cultural strategies for paternity confidence listed above.

The topic of infidelity raises the enormously interesting question of differences in jealousy between males and females. Buss argues that there is a "qualitative" (i.e., instinctual) difference (again, it is going to be statistical) and that it follows from the facts of evolutionary theory and reproductive asymmetries. Because of lack of paternity confidence, males should be more upset by *sexual* infidelity (a sexual affair) on the part of their spouse than by *emotional* infidelity (an emotional but not sexual affair). On the other hand, because females do not lack paternity confidence and because they instinctively place a high value on economic support from their spouse (examined in the next hypothesis), females should be more upset by *emotional* infidelity on the part of their spouse than by *sexual* infidelity (since emotional infidelity is much more threatening to the loss of economic support). In a study on college students conducted by Buss and a number of colleagues, they found that the majority of males indeed reported that they would be more upset by sexual infidelity than by emotional infidelity, whereas the majority of females reported the reverse. This result was corroborated in another study by the same researchers in which they placed electrodes on 60 men and women to test their physiological responses (frowning, sweating, heart rate) to the two infidelity scenarios, males showing more distress at the thought of sexual infidelity, females the reverse.

Among psychologists, interestingly, the topic of asymmetry in jealousy has generated a large debate.[2] Needless to say, I have no intention to enter into the many details of this debate, except to point out that the psychologists on the con side provide an alternative and simpler hypothesis based on mere differences in beliefs, what they called the *double-shot hypothesis*. If men believe that women are unlikely to have sex unless emotionally involved, then this explains why men are more stressed by the prospect of sexual infidelity on the part of their spouse; sexual infidelity alone suggests to them both kinds of infidelity. On the other hand, if women believe that men are likely to have sex when they are emotionally involved but need not be emotionally involved to be sexually involved, then this explains why women are more stressed by the prospect of emotional infidelity on the part of their spouse; emotional infidelity alone suggests to them both kinds of infidelity. Hence the sexual asymmetry in jealousy.

Neven Sesardic (2003, 440–441) provides the interesting reply that this model can be tested against the evolutionary one. First, one could ask both men and women which of the two scenarios they find more upsetting, emotional infidelity without sexual infidelity or sexual infidelity without emotional infidelity. If women typically choose the first scenario and men the second, then the double-shot model loses credibility and the evolutionary model gains support. Alternatively, one could ask them which part of infidelity they find more upsetting when it is *both* emotional and sexual. Again, if women typically choose the first part and men the second, the double-shot model loses credibility and the evolutionary model gains support. As Sesardic goes on to point out, not only does the double-shot hypothesis fail to explain

---

2   See, for example, the series of pro and con papers published in the journal *Psychological Science*: Buunk *et al.* (1996); Harris and Christenfeld (1996a); DeSteno and Salovey (1996a); Buss *et al.* (1996); DeSteno and Salovey (1996b); Harris and Christenfeld (1996b).

why men and women would have the different beliefs ascribed to them, but it has also failed subsequent testing along the above lines. I would add the need for a good cross-cultural study, a topic that gets messy in the *Psychological Science* debates.

Buss's ninth and final hypothesis is that *"Women seeking a long-term mate will prefer men who can provide resources for her offspring"* (227). This instinctive trait follows from the enormous investment females need to put into reproduction in order to pass on their genes, from the handicap of pregnancy to the time and resource handicap of nurturing and raising a child. Again, the reference is primarily to our evolutionary past. Females who instinctively were less discriminating in long-term mate choice with regard to economic resources, or who lacked the instinct altogether, would tend to pass on their genes less than those who had it. Accordingly, we should expect females to be attracted more to males who exhibit attributes such as ambition, above-average earning capacity, professional degrees, material wealth, and high social status (such as fame, as this usually translates into material resources). And indeed in a group of 108 males and females, Buss found not only that females desired the above attributes in a long-term mate far more than did males, but that females desired those attributes more in long-term mates than in short-term mates.[3] This was corroborated in Buss's cross-cultural study, where he found that in 36 of the 37 cultures examined females seeking a long-term mate expressed a much greater interest in the socioeconomic status of the prospective mate than did males, the one exception being Spain. In his Figure 15, Buss shows the percentages for Japan, Zambia, Yugoslavia, Australia, and the USA, which are relatively the same. However, even though Buss says the results were "strikingly consistent around the world," one has to wonder why Spain would be the exception. A further anomaly concerns his Figure 11. One might expect women in the USA to give the lowest rating to the importance of social status in men compared to the other countries shown (Brazil, West Germany, Estonia, Taiwan), because of the USA's greater relative wealth and

---

3   Interestingly, Buss's seventh and eighth hypotheses concern females and short-term mating. One might think that females would have evolved an aversion to short-term mating, given the reproductive costs to them. But this does not necessarily follow. For a start, both sexes evolved an enjoyment for sex. This enjoyment is part of the proximate cause of sexual relations, the ultimate cause being reproductive success. Indeed, the female orgasm has been shown to increase the retention of seminal fluid (Ehrlich 2000, 191–192). But aside from shared enjoyment, females are going to have different proximate causes of short-term mating than males. For males at a conscious level, short-term mating is typically going to be an end in itself, sex purely for the enjoyment of it. For females at a conscious level, on the other hand, short-term mating is often going to be a means to an end. For a start, as Buss points out, short-term mating affords females protection against other males who are aggressive but undesirable, hence the greater desirability of muscle and strength in a short-term mate. Short-term mating also affords females the quick acquisition of economic resources, which are important both for the female and her offspring. Gifts from the males are one way this is achieved, bartering is another—hence the pervasiveness throughout the past and present of female prostitution, which has always greatly overshadowed male prostitution. Short-term mating also affords a female the opportunity to evaluate a prospective long-term mate, stinginess at the short-term mating stage, for example, being a bad sign. And of course short-term mating is a means of trapping a male into a committed relationship (getting a piece of the man on the up, so to speak). Given the reproductive asymmetries between human males and females, we should not expect to find any culture in human history where these roles are reversed, and arguably we do not.

social programs, and yet women in that country gave it by far the highest rating (compared to the above four countries). Perhaps there is something about American culture that accounts for this. At any rate, in further support of his ninth hypothesis, Buss cites Laura Betzig, who found that a man's inability to provide for his wife and children was internationally a significant cause for divorce. (I am reminded here of the true story made into the movie *Catch Me If You Can*, where the main character's mother leaves his father for his father's banker and starts a new family with him because her husband fell into financial difficulties.)

In one of his closing paragraphs, after stating that "a woman's physical appearance is the most powerful predictor of the occupational status of the man she marries," Buss adds on a more hopeful note that

> Some adaptive problems are faced by men and women equally: identifying mates who show a proclivity to cooperate and mates who show evidence of having good parental skills. Men do not look at women simply as sex objects, nor do women look at men simply as success objects. One of our most robust observations was that both sexes place tremendous importance on mutual love and kindness when seeking a long-term mate.
>
> (227)

This might well be, and yet what the evolutionary psychology of human mate choice indicates is that, at bottom, "Love ain't nothing but sex misspelled" (to use the title of a book of short stories by Harlan Ellison). In terms of songs, it means that females want to hear "Everything I do, I do it for you" (Bryan Adams), while males cynical with experience cannot help but hear "What have you done for me lately?" (Janet Jackson). Indeed there's nothing like science to take the fun out of a date or the joy out of a wedding day.

Or maybe not. In reply to Buss's nine hypotheses, one might argue that in general he places too much reliance on evidence from college students and not enough on cross-cultural studies. But the fact remains that there is further evidence supporting Buss's instinct theory. Cross-culturally, men tend to marry younger women, women older men, and both prefer it that way, the average age difference being somewhere around three years (Buss 1994, 223; Badcock 2000, 174; Alcock 2001, 141). Older age in men generally means higher socioeconomic status, younger age in women greater fertility and reproductive value. Studies on sex fantasies also corroborate Buss's thesis. Not only do males fantasize about sex more often than do females, but their fantasies tend to be plainly sexual, whereas the fantasies of females tend to involve affection and commitment (Badcock 2000, 175). Indeed, one has only to look around to see that males are by far the main consumers of pornography, no matter what their cultural background, with almost all videos and strip clubs catering to the causal sex desires of males and featuring mainly young, physically desirable females. Finally, there are studies which reveal that when presented with a line-up of the opposite sex clothed in outfits representing a spectrum of social status from high to low, the outfits were irrelevant to the males, whose choices were based on physical features, whereas the choices of the females were based largely on the outfits, the clothes making the man (Hamer and Copeland 1998, 173).

Although impressive, it still might be the case that all of the above evidence from Buss and others simply follows from the fact of male dominance and the oppression of women in our own and all other cultures. This is a big part of the reason that the cultural anthropologist Marvin Harris rejects sociobiological and evolutionary psychological approaches to innate mating differences between men and women. According to Harris (1989),

> if they were really free to choose, women would choose as many partners as men choose when they are free to do so . . . What has kept us from recognizing this truth is that women have never been as free to choose the option of multiple sexual partnerships as men have.
>
> (254)

Indeed according to Harris, "there are very few, if any, societies in which women's sexual freedom is not curtailed more than men's" (255). In support of this thesis about what women would choose if they had the same freedom as men, Harris cites as evidence "quasi-polyandrous female-centered households . . . common among the urban poor in many parts of the world" (253), where men cannot earn enough to support a family so that the women are more promiscuous than women in monogamous households. But then Harris goes on to point out that these women typically expect gifts from their lovers! He even quotes an African tribeswoman, from a hunting-gathering tribe where the males visit a lot of different camps within the tribe, who says, "A woman should have lovers wherever she goes. Then she will be well provided for with beads, meat, or other food" (261).

Harris's approach is part of what is known in evolutionary psychology as the *structural powerlessness hypothesis*. Females seek males with power and economic resources because females are deprived of these by males, whereas males do not seek females with power and economic resources because males are the ones who control these in the first place. Hence there is nothing innate about the differences in mating desires between males and females; it is all cultural. Aside from the fact that Harris's first piece of evidence above actually corroborates Buss's theory about short-term mating in females, the structural powerlessness hypothesis fails to explain why males are cross-culturally dominant. Harris (1999, 27, 105) thinks that male dominance is no more natural than female dominance and is simply cultural. But it is highly unlikely that male dominance is simply an accident of history. Moreover, the structural powerlessness hypothesis cannot avoid the fact that males not only have to compete with other males for power and economic resources, but they also have to compete with other males for the genetic resources of young, fertile females. Both facts entail haves and have-nots among males, with the have-nots always outnumbering the haves. The structural powerlessness thesis cannot avoid the conclusion, then, that over thousands upon thousands of years the differences in reproductive strategies between males and females are not going to be merely cultural but innately biological.

Either way, the structural powerlessness hypothesis is testable and has actually been tested. If the theory is true, it entails that the socioeconomic status of males will be less important to females who enjoy high socioeconomic status and independence than to females with low socioeconomic status and independence. Interestingly, in study after study, females with high socioeconomic status and independence continued to place a high emphasis on the socioeconomic status of a prospective mate (the exact reverse of males with high socioeconomic status and independence). This suggests to the evolutionary psychologist Bruce Ellis (1992) that "female preference for high-status males is the product of a psychological mechanism that operates whether a woman's own SES [socioeconomic status] is high or low" (274). For Buss (2003), not only does the above and other similar evidence "directly contradict" (46) the structural powerlessness hypothesis, but that hypothesis fails to account for why males are dominant in the first place with regard to power and economic resources, why males exclude power from other males as much as from females, why males are more power and status driven than females, why males evolved bigger and stronger bodies than females, and why males prefer young female mates. Only evolutionary psychology, he says, "accounts for this constellation of findings" (47).

Not to be outdone, Harris (1989) has a further kind of evidence in aid of his thesis that females would be just as promiscuous as males if culturally they were allowed to be. He says, "there is nothing in the behavior of our closest primate relatives that supports the idea of innate female sexual coyness" (253). Citing the example of female bonobos, he points out that they are as sexual as the males, not only heterosexually but homosexually, with females often rubbing their clitorises against each other for multiple orgasms. Although bonobos look much like chimpanzees, they are a little thinner, are designed a little better for upright walking, are less aggressive and violent, and are hypersexual, using sex not only for reproduction but for social bonding. In some ways, then, they are more like us than any other modern ape. But Harris is misleading when he says they are our closest primate relatives (actually he lumps them with chimpanzees in the above statement). Chimpanzees and bonobos are the closest relatives of each other, having diverged from a common ancestral stem roughly a million years ago, while that stem and humans diverged from a common ancestral stem roughly 6 million years ago (that stem and gorillas diverged from a common ancestral stem roughly 9 million years ago, while that stem and orangutans diverged from a common ancestral stem roughly 15 millions years ago). Given that chimpanzees, gorillas, and orangutans share behavioral features not shared by bonobos (male dominance, rape, etc.), it is likely that the anatomical and behavioral features of bonobos not shared by chimpanzees, gorillas, and orangutans evolved separately in bonobos. In other words, any similarities shared between humans and bonobos that are not shared with chimpanzees, gorillas, and orangutans are *analogies*, not homologies. Harris's argument fails, then, even if he has the anatomical and behavioral facts of bonobos right.[4] Just because female bonobos are

---

4   Given that the old picture of peace-loving chimpanzees fell to later observations of war and killing between male chimpanzees of different communities, members of each community being kin related, and that bonobos are relatively rare and less studied and severe fights between male bonobos from different communities have now been observed and documented, Ghiglieri (2000, 174–175) suggests that the current picture of peace-loving bonobos is premature if not wishful thinking.

as sexual as male bonobos, assuming that they are, that provides no good evidence to think that female humans would be just as sexual as male humans if cultural restraints were lifted.

Harris's mistake goes even deeper. G. C. Williams (1966) makes the point, based both on evolutionary theory and on studies of animal behavior in the wild, that "promiscuity, active courtship, and belligerence toward rivals are not inherent aspects of maleness" (186). This might seem a point in Harris's favor, but it is not. What determines the above characteristics in the sex of a species, says Williams, is which sex contributes "the greater amount of material and food energy to the next generation" (185). It is "almost always true," he says, that it is the females. From the viewpoint of evolutionary homologies, this is going to be significant for human evolution. Now what about those relatively few species where it is not the females but the males who make the main contribution to the next generation? True to expectation, says Williams, it is the females who exhibit the traditional male aggressiveness, territoriality, and mating rivalry. To use two of his examples, female seahorses transfer their unfertilized eggs into the males, where the eggs are fertilized, develop, and eventually born, and true enough it is the females who exhibit "the traditional masculine aggressiveness in courtship and general promiscuity." The same is true in species of birds where the males assume the main role of incubating the eggs and feeding the chicks. Does any of this also apply to the human case? Obviously not. We know that humans evolved in small hunting-gathering groups, that males are generally bigger and stronger than females, that patriarchy is and was probably always the norm, and that in the few remaining hunting-gathering groups such as the Eskimos there is a sharp division of labor, with the males doing the hunting and the females remaining behind to do other things.

So we are back to Buss, with human males being more promiscuous and competitive for mates (which makes me think that maybe highly advanced androids would be the best solution to the "war between the sexes," keeping in mind that clichés are not necessarily false). At any rate, the comparative method brings us to our next topic, namely, rape. Rape is a cross-cultural fact, a human universal, with men being by far the majority of the perpetrators (Brown 1991, 161, 188; Ghiglieri 2000, 103). According to those of a SSSM persuasion, rape is not something innate but instead is caused by social factors such as broken families, poverty, the commercialization of sex and violence, peer pressure, and of course patriarchy. Feminists tend to focus on the last of these, the consensus view receiving its now classic treatment in Susan Brownmiller's *Against Our Will* (1975). According to Brownmiller, rape is not a crime of sex but of violence, it is not sexually motivated but is motivated instead by hate and power, specifically the hatred of women by men and their desire to exercise power over them. Rape, in her own words, is "man's basic weapon of force against women, the principal agent of his will and fear" (14), such that "any female may become a victim of rape" (348). On this view, women are monogamous not only because they are forced to be but also because it affords them protection against rape by other men. Men rape during war, then, because women have lost their protection.

While rape is clearly a sensitive and emotional topic and one needs to have a good measure of sympathy and empathy for any group that suffers under the tyranny of hate, oppression, and violation, rape is also a topic that needs to be examined by the logic of evidence rather than by the force of emotions. Feelings are not arguments. As Nietzsche so aptly put it in *Human, All Too Human* (1878, §15), "strong belief proves only its own strength, not the truth of what is believed" (23). What this means is that we have to get interdisciplinary, taking into account not only the psychology and statistics of rape, but also cultural and anthropological data and, most importantly of all, evolutionary biology.

But even before we get into all of that, the idea that rape is about hate and power rather than sex seems extremely odd right from the start. Rape by a male, after all, requires an erection, and an erection requires sexual arousal. Moreover, if men really do hate women as much as radical feminists say, then why do not men simply beat them up? Why rape at all? When men really do hate, as with racial and ethnic minorities, they simply beat, bash, and kill. On the other hand, there is reason to believe that sex is intimately connected with violence. What supports the feminists here is the very language we use for sex. The word "fuck," for example, although the 16th-century origin of the word is unknown, is related to the Middle Dutch word *focken*, meaning "thrust," and the Swedish word *focka*, meaning "strike, push" (Webster's). Hence we say "fuck off" or "I'm fucked," along with words such as "screw" and "bang." All of these words connote violence and harm.

Be that as it may, from an evolutionary point of view a very real possibility is that rape is an adaptation that evolved in males because it increased reproductive success. A further possibility is that it is simply a byproduct of other adaptive behaviors, such as the male desire for sexual variety and inexpensive sex, or even more simply the use of aggression to get what they want. Both theories are explained and defended in the controversial book by Randy Thornhill and Craig Palmer, *A Natural History of Rape* (2000), the biologist Thornhill defending the adaptation theory, the anthropologist Palmer defending the byproduct theory. While Thornhill and Palmer provide a book-length treatment of both theories, I want to focus mainly on the adaptation theory, using not the book by Thornhill and Palmer, however, but instead the chapter on rape by the anthropologist Michael Ghiglieri in his book *The Dark Side of Man* (2000). What makes Ghiglieri's chapter preferable is his use of statistics, which is more impressive than the use of statistics in the Thornhill and Palmer book.

Ghiglieri's argument has basically three parts: one looking at the statistics of the female victims, another at the statistics of the male perpetrators, and the third looking at rape in the animal world.

Beginning with the female victims, Ghiglieri brings together an enormous amount of rape statistics, largely from the FBI and other US government sources, which I shall simply summarize here. For a start, it is estimated that 13 to 25% of all US women will experience rape within their lifetimes, an estimate based not simply on reported rapes but on the estimate (fair in virtually everyone's mind) that for every rape reported 5 to 20 go unreported. More importantly for Ghiglieri's thesis, of 95,769 rapes reported in the US in 1996, only 60 resulted in the deaths of the victims. Taking unreported rapes into account, that means that rapists kill fewer than 1 in 10,000 of their victims. Equally important, 90% of the victims came from an age group that comprises less than one-third of the female population, specifically the ages between

12 and 35, which corresponds quite closely to the years of female fertility. Even more significantly, 77% of the victims were between the ages of 16 and 24, the most fertile and sexually attractive tenth of the female population and, not accidentally, the age group preferred by the pornography industry for its male consumers. It is important to notice also that, as Ghiglieri stresses, the idea that women want to be raped is a myth proved so by the fact that the overwhelming majority of rape victims reported various degrees of trauma from the experience.

Turning now to the male perpetrators and the 1996 statistics, roughly two-thirds were strangers to their victims, 44% of arrested rapists were under 25 while most of the rest were in their late twenties and thirties, and roughly only one-third were under the influence of alcohol or drugs. The socioeconomic scale of the perpetrators is also important. Ghiglieri found that the majority were low in the socioeconomic scale (relatively uneducated, unemployed, underemployed, low incomes), a fact corroborated, he says, by the statistic that blacks accounted for 42% of arrests for rape. Moreover, two-thirds of the rapists had prior arrest records, 94% having begun crime at 15 and rape at 18. Interestingly, criminals who used guns were less likely to rape than criminals who did not use guns, a statistic Ghiglieri attributes to the fact that criminals who use guns tend to take in more money and hence have the money to court women by traditional means. Also significant is the amount of force used in rape. A number of studies have concluded that most rapes involve no more than *instrumental force*, the minimum amount of force needed to force and control the victim so as to perform the rape. In one US study of 479,000 rape victims, it was found that only 58% were physically injured and the average medical expense of those injured was $115. The conclusion about instrumental force is corroborated by another study according to which 61.7% of the rapists said they did not intend to use force at all. What also corroborates Ghiglieri's thesis about males are the studies he cites which found that males typically get more sexually aroused by films and pictures of damsels in distress than by women who are sexually willing.

Given all the above, the idea that rape is motivated mainly by power, control, and hate, what Ghiglieri calls "the central dogma of rape in sociology, psychology, and other social sciences" (92), just does not fit the facts, he says. If the dogma were true,

> we should see three trends. First, men would rape older, more powerful women more often. (They do not.) Second, rapists would come in all ages and from all walks of life. (Again, this is not the case.) Third, when socialization changes, rape should change. (It does not.)
>
> (101)

For Ghiglieri, instead, the statistics of rape indicate that rape evolved as a "biological adaptation" (103), a trait favored and evolved by natural selection since it increases in males the odds of producing offspring and therefore passing on their genes. But it is also what he and others call *condition-dependent*, an instinct that depends on certain conditions for it to be triggered. Given that males are more promiscuous than females, that females are sexually motivated by socioeconomic status in males, and that there are never enough desirable females for males, this leaves males, says Ghiglieri, with basically three strategies, which they can use either individually or in combination, namely, "honesty, deception, or coercion" (104), honesty if they have the socioeconomic status desired by females, deception if they do not, and coercion if deception does not work. Hence, says Ghiglieri, "The condition that leads to rape is the failure by a male to win the resources and status needed to attract a female" (103).

As perverse as it might sound, if Ghiglieri is correct it follows that rape is what Buss calls a *short-term mating strategy*. What adds to this view are the statistics of war. Agreeing that rape is a universal in war and that in situations of conquest the females have lost their defenders, Ghiglieri argues that the reasons why men rape are not so simple as hate and victory. First, most of the soldiers are young males and have yet to produce children, they are uncertain about surviving the war and hence are uncertain about their future fatherhood, they are constantly meeting young unprotected females, and rape in war is rarely punished. Although statistics about rape and reproduction during war are scarce, Ghiglieri cites the case of West Pakistani soldiers who in 1971 raped 200,000 to 400,000 Bengali women over a nine-month period, which health officials estimate produced at least 25,000 children. Hence for Ghiglieri, "rape is a massive reproductive victory," a chance to "plant your seed before you go" (91), so that "rape during war may be an instinctive male reproductive strategy" (92).

Ghiglieri's final category of evidence comes from animal behavior in the wild. According to Brownmiller (1975), "No zoologist, as far as I know, has ever observed that animals rape in their natural habitat, the wild" (12). If this statement was naive in 1975, it was only more so in 2000, when the books by Ghiglieri and Thornhill and Palmer were published. As Ghiglieri points out, biologists devoted to studying animal behavior in the wild have documented numerous cases of rape of females by males, not only in gorillas and chimps and orangutans but in many species such as mallard ducks, where females are sometimes gang raped to the point of drowning. Thornhill and Palmer (2000, 144) provide a very long and impressive list of publications by biologists. Ghiglieri focuses on the case of orangutans, where one-third to one-half of the observed copulations in the wild were by males too young to have established themselves as attractive to females but large enough to force themselves on the females (95–97). Indeed, rape is rampant in the animal world. Of course one can always define "rape" in such a way that only humans are capable of committing rape, but a *question-begging definition* cannot possibly settle an empirical matter. Define the word "rape" as one may, the fact is inescapable that throughout the animal world males routinely force themselves sexually on unwilling females.

This is not to say that the rape instinct thesis does not have problems when applied to humans. First, there are cultural differences. As Ghiglieri points out, a woman is 8 times more likely to get raped in the USA than in Europe, and 26 times more likely than in Japan (84). But then nobody denies that culture is a factor. With regard to the above differences, they could be the result of differences in the economies, in rape laws and prison times, in different senses of entitlement and degrees of patriarchy, and in the percentage of rapes reported, while the latter in turn could be the result of differences in intimidation faced by females. Cultural differences like the above do nothing to negate the possibility of a rape instinct in males.

Statistics, however, raise a deeper problem. Ghiglieri uses a lot of statistics and they are absolutely essential for his argument, but we cannot simply assume that they are all reliable. As is well known, statistics can mislead for a variety of reasons, such as poor sampling procedures, sampling bias, and unclear terms and questions used in the gathering of data. Certainly one cannot discount Ghiglieri's statistics as Richard Lewontin (1999) does in his book review of Ghiglieri's book, when he states, "I begin to doubt my own species identity, having never engaged in, or fantasized about, any of these activities [rape, murder, etc.] whether drunk or sober, asleep or awake" (738). The problem is that Lewontin might simply be an exception, like the proverbial 95-year-old who smoked a pack a day all his life and never got lung cancer. One does not refute statistics simply by citing a single contrary case.

Interestingly, the psychologist Dorothy Einon (2002), in a hostile book review of Thornhill and Palmer (2000), provides a number of statistics and statistical claims that she believes completely undermine the rape adaptation thesis. In sum, she says, "There is no evidence that the distribution of rapists is significantly screwed [sic] towards the lower social classes"; 29% of rape victims are below 11 years old and 62% are between 16 and 24, while the next highest range is between 12 and 15; fertility is extremely low for the first two years following puberty, it reaches a peak around 30, stays high throughout the 30s and then declines rapidly after 45, so that "The average forty-year-old woman is more fertile than the average sixteen-year-old"; 89% of victims know the rapists; the average woman is capable of breeding over a 25 year period and has 6 viable children, "There is thus a 1:1506 chance of a random rape resulting in conception"; "Assuming a 28-day menstrual cycle, in one in every 349 rapes a man would have a 50% chance of fathering a child"; if sperm competition is involved "it is likely that the average rapist would need to rape about 5[00]–600 times before he had a 50:50 chance of fathering a child"; "in many societies illegitimate children are killed or neglected"; in small hunting-gathering groups rape would not likely go undetected and so would likely be punished; females would tend to outnumber males in our evolutionary past because of death to males from accidents, fights, and diseases; and finally, like all female great apes, which have an average interbirth interval of about 4 to 6 years and are "rarely fertile," women before the advent of feeding bottles "like great apes, were probably fertile for a few days every 2–4 years" (448–451).

I shall leave it to the reader to decide about all of these conflicting statistics, except to make three points. First, pay especial attention to Einon's "thus" above; it does not at all follow (this is because producing an average of 6 children over a 25-year period does not automatically mean that there were only 6 possible days during that period for getting pregnant). Second, in a study of 782 healthy couples, Dunson et al. (2002) found that nearly all pregnancies occurred within a 6-day fertile window of each month, the specific-day probability of fertility declined with age for women from their late 20s on, and women 19–26 were twice as likely to get pregnant as women 35–39. All of this, of course, supports Ghiglieri and casts doubt on Einon's use of statistics. Third, if rape is primarily a weapon of hate and violence, it follows that the best strategy for a female facing rape is to not fight back. As everyone knows with regard to bullies, standing up and fighting back or trying to run away only further incites the bully to violence. One should not fight back or run away unless one knows one has a good chance of succeeding. On the other hand, if rape is mainly about sex, it follows that the best strategy of a female facing rape is to scream and fight back. Much hinges, then, on which theory of rape is basically correct. Interestingly, Ghiglieri cites a study which found that only 27% of females facing rape who fought back were actually raped, while 56% of females facing rape who did not fight back went on to be actually raped (99). In other words, fighting back doubles a female's chance of escaping rape. And of course this is what we should expect if rape is indeed mainly about sex.

Statistics do, however, haunt Ghiglieri's thesis in a very different way. The major problem is the prevalence of rape when the victims are little girls or old women, since these victims are either too young or too old to produce offspring. Again, 90% of the victims are 12 to 35 (from a biological point of view, a child becomes an adult at puberty). But still, 10% of the victims fall outside of that range (assuming the statistics are correct). Ghiglieri attributes the rape of little girls and old women (he tells us that less than 5% of rape victims are over 50) to two factors, first that they are typically more vulnerable to rapists, and second that the men who do this sort of thing are crazy. He says, "As with men who rape children or old women, those who choose anal or oral penetration or other forms of rape with no reproductive potential are mentally ill" (100). Perhaps they are mentally ill (personally I think those who rape children and old women are), but they cannot be so by definition (for then again we would have the problem of a question-begging definition). The real problem is that the concept of mental illness is a culturally fluctuating one (homosexuals, e.g., were officially considered mentally ill by psychiatrists until around 1980). I suggest, then, that Ghiglieri would do better simply to view the rape of little girls and old women as the rape instinct misdirected, since evolved instincts are never perfect and are inherently blind (as, e.g., when the instinct in cats for chasing mice and birds is misdirected when chasing the end of a string). As Alcock (2001) points out,

> Evolutionary biologists have developed several potential explanations for maladaptive responses, including the by-product hypothesis (in which the fitness-reducing action is a byproduct of generally adaptive proximate mechanisms) and the novel environment hypothesis (in which evolved proximate mechanisms generate maladaptive reactions in an evolutionarily novel environment).
>
> (218–219)

The byproduct hypothesis makes more sense here, especially given that, as I'm sure Ghiglieri would agree, most rapists are simply not consciously aware that reproduction is the real reason for why they commit rape. If rape is an adaptation, the rape of little girls and old women, as with oral and anal sex in general, would simply be a byproduct of that adaptation. (The moral issue, of course, is quite another matter!)

But then this raises the possibility that rape itself is not an adaptation but simply a byproduct of other adaptations. David Buss (2003, 273–274) favors this view, even though he cites a recent study which found that the pregnancy rate per incident from the penile-vaginal rape of reproductive-age women is 6.42%, which is surprisingly high given that it is only 3.1% per incident among consensual couples trying to get pregnant (the difference is almost certainly due to the main age range of the targets of rapists). Given the level of promiscuity of males and their desire for casual and inexpensive sex with young, fertile females, Buss thinks the simpler theory is still the byproduct theory, although he is open to the direct-product theory. Especially interesting is that he thinks progress cannot be made until we distinguish between different kinds of rape (date rape, war rape, stranger rape, homosexual rape, etc.). Unfortunately, Buss does not provide any suggestions for tests to help decide between the two theories, even if different kinds of rape are specified, and indeed one has to wonder what would constitute a good test. I suspect that the needed testing will eventually come from geneticists.

The possibility that rape might be a byproduct rather than a direct product of evolution brings us to our next topic, homosexuality. Right from the start, however, homosexuality would seem to defy an evolutionary account, since homosexual behavior entails zero reproductive success. If there are genes for homosexuality, then, one would think they would have been weeded out by natural selection long ago. But this does not necessarily follow, as we shall see from two theories that attempt to explain the universality (albeit a universal minority) of male homosexual behavior and attraction.

The first theory is the kin-selection theory of homosexuality proposed by E. O. Wilson (1978). In direct opposition to what many religions teach, Wilson suggests that homosexuality is "normal in a biological sense," that it "evolved as an important element of early human social organization" (143). In support of this thesis, Wilson points out that homosexuality is a human universal and that homosexual behavior is common in the animal world from insects to mammals, in the latter often as a "bonding device." Wilson also points out that in primitive societies studied by anthropologists (societies still in the hunting-gathering stage) homosexuals often play roles such as seers, shamans, matchmakers, peacemakers, artists, advisors, and keepers of tribal knowledge. In addition, Wilson points out that male homosexuals have been found to have higher average IQ scores than male heterosexuals. All of these are "clues" (146), not proof, but the biological basis of homosexuality for Wilson becomes "more consistent with the existing evidence" (147) when combined with the theory of *kin selection*.

This theory was developed in the early 1960s by William Hamilton (1996), who was puzzled by the existence of sterile castes in social insects. Since the members of these castes are sterile, they obviously cannot pass on their genes. How then could such a clearly maladaptive state evolve by natural selection? The answer, said Hamilton, is kin selection. The groups are not merely groups but kin groups, groups that share more genes in common than non-kin groups. It follows that if a gene or gene complex exists that lowers *individual fitness* (in the reproductive sense of "fitness") but increases the *inclusive fitness* of the individual's close kin to a degree sufficient to overcompensate the loss in individual fitness, then that gene or gene complex will be favored by natural selection. Returning to social insects, although the sterile members have zero individual fitness, they are highly altruistic to the members of their kin group, sacrificing their labor and even their very lives for the good of the group. This results in the sterile members passing on their genes, in effect, through their fertile kin by a kind of proxy.

This theory, which we shall return to in Chapter 7, fits in quite nicely with the gene-centered view of evolution promoted by G. C. Williams (1966) but especially by Richard Dawkins in his now classic *The Selfish Gene* (1976). Genes, of course, cannot be selfish in the literal sense, but it helps to understand evolution by looking at genes *as if* they were selfish, selfish in the sense that they will do whatever they can to make copies of themselves. Altruistic sterile castes make evolutionary sense not because these insects really care more about their kin than about themselves, but because their genes evolved in a manner that made them behave as if they do. What aids this theory is the idea that genes are not physical chunks of DNA or RNA but instead are units of information, units that are nonphysical but which require a physical medium (Williams 1992, 10–13). Partly because of the synonymy of the genetic code (Glossary), it follows that genes are units of information. Just as the same story can be carried by a book, a magnetic tape, or a brain, the same genetic information (genes) supervenes on physically different strings of DNA or RNA, strings with different sequences of genetic letters. What matters ultimately is the message, not the medium. Natural selection is ultimately for types, not tokens of types (Rosenberg 1994, 93–97).[5]

For Wilson (1978), then, genes for homosexuality in humans could have evolved via kin selection. Freed from reproductive roles, homosexuals could pass on their genes by proxy through their close relatives, especially if the homosexuals held social positions that gave them a high standing in their society. Indeed, for Wilson, given the connection between kin-selection theory and altruism, "Homosexuals may be the carriers of some of mankind's rare altruistic impulses" (143).

There are, of course, some problems with this theory. Richard Lewontin (1991), for example, claims that sexuality in humans is not divided into two classes but is a continuum, that it has varied historically by social class, that there exists no direct evidence of a genetic basis for sexual preference, and that there is no evidence for "helpers at the nest" in human societies, so that the kin-selection theory of homosexuality "is a made-up story, from beginning to end" (77). We shall return to these points, some below and some in the next chapter. Of more immediate concern are the problems raised by David Buss (2003, 251–252). For one, Buss cites a modern study on male homosexuals which found that they were not more altruistic toward their close kin but instead were more estranged from them. But of course this could simply be a matter of cultural noise. More importantly, Buss points out that though the kin-selection theory explains altruism, including whatever altruism is involved with the genes for homosexuality, it does not at all explain the homosexual orientation itself. Only humans, after all, have obligate homosexuality (homosexuality in animals is bisexuality). In other words, kin-selection theory does not explain why homosexuals as altruists are homosexuals. I suppose that Wilson would have to appeal in reply to

---

5   To illustrate the difference between types and tokens, there is one word in "sex sex" if by "word" one means *type* and two words if by "word" one means *token*. Each of the two words is a token of the type.

pleiotropy, where the homosexual orientation would be the hitchhiker. But then why are not the most altruistic people in the world, such as Albert Schweitzer, typically homosexuals?

According to Wilson (1978), "The kin-selection hypothesis would be substantially supported if some amount of predisposition to homosexuality were shown to be inherited" (145). This, however, does not necessarily follow, as the work on homosexuality by the behavior geneticist Dean Hamer clearly shows. The work leading up to the genetic study by Hamer *et al.* (1993) is quite an interesting piece of detective work, but it is the popular exposition by Hamer and Copeland (1998, 182– 199) that I shall refer to. First, based on the Kinsey scale, where 0 is for exclusive heterosexuality and 6 is for exclusive homosexuality (one's number in the scale is determined by four measures: attraction, fantasy, behavior, and self-identification), most males are at either end of the scale, much like being right-handed or left-handed, whereas females are distributed through the scale, much like height. Second, males tend to be consistent on all four measures at any one time, whereas females on average are less consistent. Third, males on average tend to stay pretty much the same throughout their lives, whereas females are much more likely to flip-flop. Fourth, in monozygotic (identical) twins, if one male twin is gay, the other twin has a higher probability of being gay, which is lower if the twins are dizygotic (twins from different eggs) but still higher than the statistics for unrelated people. With females, on the other hand, if one monozygotic twin is a lesbian, the probability that the other twin is a lesbian is no different than if they are dizygotic twins. Moreover, the sister of a lesbian has a 6% chance higher than normal of being a lesbian, whereas the daughter of a lesbian has a 33% chance higher than normal of being a lesbian. Since sisters share more genes in common than mothers and daughters, these percentages make no sense genetically. In other words, the evidence suggests that lesbianism is entirely cultural. More evidence suggests, however, that male homosexuality is roughly 50% genetic and 50% cultural. Not only do homosexual males have more homosexual brothers, but they also have more homosexual cousins and uncles. Moreover, the homosexual relatives are concentrated on the mother's side. To geneticists, this can only mean one thing: genes on the X chromosome, the sex chromosome inherited from the mother. (Females have XX chromosomes, receiving one each from their mother and father, whereas males have XY chromosomes, the X coming from the mother, the Y from the father.)

Hence, Hamer and his colleagues went searching for "genetic markers" in the X chromosomes of homosexual males (genetic markers are gene regions correlated with certain phenotypic traits but not necessarily causes of those traits). Looking at 40 pairs of homosexual brothers and 22 different markers, they found something highly significant at the end of the long arm of the X chromosome, called Xq28. In that region, 33 of the 40 pairs of homosexual brothers were the same for a series of five closely spaced markers, a result that shows 83% sharing, which should have been roughly 50% if there was no genetic connection to sexual orientation. Even though the media hailed the result as the discovery of "the gay gene," Hamer and his colleagues always stressed the modest nature of the results and the need for further research. In a second study, they used seven new families plus four families from the previous study, looking now not only at homosexual brothers but also at the heterosexual brothers of the homosexual males. Most of the heterosexual brothers had different genetic markers from their homosexual brothers, sharing only 22% of those markers, while the homosexual brothers shared 67% of the same markers, still significantly higher than the 50% level.

Hamer suggests that the genes responsible for male homosexuality might make enzymes that control sex hormone metabolism in the developing brain or they might make proteins that build specific brain circuits or they do something not at all yet known. While the initial results are significant, Hamer is clear to stress their tentative nature and the need for further research (research money, after all, is far more likely to go to cancer research and the like). Nevertheless, Hamer feels confident enough to provide a reply to the social constructionists, pointing out that females are encouraged more than males to be in touch with their feelings and males are ridiculed for being effeminate not because of the accidents of cultural history but "because that's the way men and women really are . . . In this way culture would support the biology, and the nature holds up the nurture" (192).

Equally interesting is the way Hamer's research undermines Wilson's kin-selection theory. What Hamer and his colleagues found is that females who had "the gay version of Xq28" began puberty roughly six months earlier than those without it, suggesting that selection was for the increased reproductive span ("reproductive value" in Buss's terminology). In other words, the homosexual expression of "the gay gene," confined as it is to males, is merely a pleiotropic byproduct of evolution by natural selection, not an adaptation in itself. Of course, it is possible that both kinds of selection (kin selection and natural selection proper) have operated in human history when it comes to male homosexuality—we need not necessarily choose only one—but Hamer does indeed offer the simpler theory in terms of Ockham's Razor (Glossary).

Some are not convinced by this kind of research, however. Edward Stein (1998), for example, argues that genetic researchers on homosexuality are guilty of what he calls *essentialism*, "the view that our contemporary categories of sexual orientation can be applied to people in any culture and at any point in history" (427). In fact, he says, "current scientific research [on sexual orientation] *assumes* essentialism" (437) (his italics), and he specifically mentions the research of Hamer.[6] Stein is not

---

6   Stein's focus, however, is on the work of Simon LeVay (1992), who studied the hypothalamus of the brains of 41 corpses (mostly males, most had died from AIDS, and most were known or presumed to be either homosexual or bisexual). The hypothalamus is the region of the brain that controls appetite, thirst, body temperature, insulin, sex hormones, and the pituitary gland (which secretes various hormones responsible for growth and other bodily functions, such as adrenaline). LeVay found that a specific part of the hypothalamus of the brains of homosexual and bisexual males was significantly smaller than that of the brains of heterosexual males. My own problem with LeVay's study is that he jumps from correlation to causation. Following the work of the neuroscientist Antonio Damasio (1994), it is possible that the homosexual activity affected the size of the hypothalamus, rather than *vice versa*, analogous to the possibility that the mental activity of Einstein resulted in the pronounced altering of his brain (see Diamond *et al.* 1985), since "Activity in the hypothalamus can influence neocortical activity, directly or via the limbic system, and the reverse is also true" (119).

opposed to essentialistic categories *per se*. He agrees that blood types are not a social construction and would objectively apply to all other cultures whether they were aware of it or not. But genetic researchers on homosexuality, he says, do impose on their research socially constructed categories, namely, heterosexual, homosexual, and bisexual. The reason he says this is that there are other cultures, such as the ancient Greeks of Periclean Athens, that would not recognize these categories of sexual orientation. In that culture, only males were citizens, citizens were allowed to penetrate non-citizens (slaves, children, women, foreigners) but were not allowed to be penetrated by non-citizens, and gay relations were typically asymmetric with regard to age.

Beginning with the latter claim, it is not at all certain that the ancient Athenians would not have recognized the categories of heterosexual, homosexual, and bisexual. The same goes for other extinct cultures. First, and Stein admits this, the information available about sexual orientation and practices is fragmentary and incomplete. But even so, from what we do know about the ancient Athenians, the general acceptance of male homosexuality appears to have been confined to the upper class. More importantly, it is well known that marriage between males was unheard of, that strict homosexuality was scorned throughout all of Greece, that the ancient Greeks had brothels filled with female prostitutes (*pornai*), that in addition to their wives men often had mistresses (*hetairai*), and that male homosexuality had its critics who considered it unnatural, the most famous being Plato (*Laws* 836e–841d). Of apparently even greater significance is the fact that in ancient Athens especially, as David Cohen (1997) argues, social restrictions were such that the basic human need for courtship was seriously thwarted. Satisfying the need for sex, of course, was not a problem, given slaves and prostitutes. But courting desirable females was an entirely different matter. The problem was that females, especially young females, were closely guarded and had virtually no freedom in public. Consequently, they could not be courted. Young males, however, did enjoy public freedom, and before puberty were considered to be more female than male. Adult males, then, simply redirected their courtship instinct toward young males, who were no longer the proper object of pursuit once they reached puberty. Hence, as Cohen puts it, "the heterosexual norm was the standard by which roles and behavior were judged" (164). The upshot of all of this is that the ancient Athenians might well have recognized our categories for sexual orientation, in spite of their culture and politics.

With regard to Stein's criticisms of LeVay and Hamer and other genetic researchers on homosexuality, it is no more certain that they impose on their research our contemporary categories for sexual orientation. Scientists have to use terms where they know there are continuums and scales (we all do, as with the weather). Words by their very nature imply a partitioning of ideas about the world into distinct categories, but that is something about words, not necessarily about the ideas or the information they are used to convey. In the case of Wilson but especially of Hamer, they are concerned with finding a genetic and ultimately an evolutionary explanation for homosexual behavior and orientation (all the while admitting other influences). They are not concerned so much with categories as with influences and degrees. Interestingly, both Wilson and Hamer appeal to the Kinsey *scale*. Would this scale have been inapplicable to the ancient Athenians? Stein is silent here. The problem is that Stein, like so many others, does not take biology seriously enough. Wilson and Hamer, along with all biologists, understand completely that biology is statistical, even though the way biologists use language does not always indicate this (a fact that would only mislead those who know little biology). Surprisingly, Stein claims that *both* essentialism and constructionism are *"empirical* theses" (his italics) and that "If nativism [innateness] is true, then constructionism is false, and essentialism is true" (436). Although Stein does not tell us what he *would* accept as empirical evidence of innateness, I find it interesting that he himself would fall prey to dichotomous thinking (the equivalence of a mortal sin for social constructionists). Sexual orientation can be *partly* innate, a matter of *degree*. It need not be *either* innate or environmental. Such is the nature of instincts, especially in humans.

So much for homosexuality, and rape and human mating strategies before it. If neither appear a sure thing for an evolved instinct in humans, surely incest avoidance between immediate kin (sibling-sibling, parent-child) does. On the one hand, there is in most family situations the availability of having sex with immediate kin as opposed to more distant kin or non-kin, while on the other hand, most people strongly prefer distant kin or non-kin and experience enormous repugnance and emotional disgust at the mere thought of having sex with immediate kin (just try the thought experiment right now!). If we look at this coolly and objectively, it must seem extremely odd, given our obvious instinct to have sex. Hence the temptation to attribute incest avoidance to an instinct.

A standard example of this kind of argument is provided by E. O. Wilson. According to Wilson (1978), incest taboos "are among the universals of human social behavior" (36). Of course, just because something is found in all human cultures does not automatically mean that it has a genetic basis. Making fire is a human universal, and so is whistling, but it would be rather silly to say we have genes for making fire and genes for whistling. But Wilson does not simply make his argument based on universality. That is only the first step in his sociobiological explanation of the incest taboo. His next step is to focus on what is called *inbreeding depression*, the fact that the offspring of immediate kin are statistically less viable than the offspring of non-kin. This is observed not only in plants and animals but in humans. Wilson provides the example of a research study conducted on a number of women in Czechoslovakia who had sex with their fathers, brothers, or sons. Of the 161 children that resulted from these unions, 15 were stillborn or died within their first year and more than 40% suffered from serious mental and/or physical defects. On the other hand, of the 95 children produced by these same women from non-incestuous unions, only 5 died during the first year of life and none of the rest had any serious mental or physical defects. This, of course, is not simply an observation but follows instead from genetic theory, as the offspring of close kin are going to have more homozygous genes of harmful recessive alleles. Given the selection pressure against close incestuous unions resulting from inbreeding depression, and that none of the human societies throughout evolutionary history had a modern understanding of genetics and were unlikely to do the kinds of studies needed to establish inbreeding depression, Wilson argues that an instinctual sexual repugnance gradually evolved in our pre-human past, manifesting itself in human culture as an incest taboo. Although an instinct, it nevertheless can be and often is overcome (typically accompanied in humans with a sense of shame) in situations where access to outbreeding is severely limited. In further support of this instinct theory, Wilson cites the example of Israeli kibbutzim, where children from various parents were raised as siblings, and where neither marriage nor heterosexual activity between children raised from birth in the same kibbutz occurred, even though such unions were not opposed.

Richard Dawkins (1989) notes that "Selection for active incest-avoidance could be as strong as any selection pressure that has been measured in nature," and he is surprised that many anthropologists oppose such a "strong Darwinian case" (294). However, anthropologists are not all that opposed. The most famous anthropologist on human incest avoidance is Edward Westermarck, who proposed in 1891 that incest avoidance is an imprinting kind of instinct that is gradually developed when young children are raised as siblings. Hence, incest avoidance is known as the *Westermarck effect*. Interestingly, Westermarck also proposed that the cross-cultural taboo on incest developed not for those who had the instinct but instead for the relatively few who were deficient in the instinct.

It is interesting to compare recent writings by anthropologists on this topic. Donald Brown, for example, in his book on human universals (1991), devotes the entire fifth chapter to the subject and favors the instinct theory. In addition to the evidence and arguments we find in Wilson, Brown adds that we need to distinguish between incestuous marriage and incestuous sexual relations, since the former does not entail the latter (hence we have to be careful about historical cases of royal couples, for example, who may have married only for political-economic reasons, not for sexual-romantic ones). In addition to the example of Israeli kibbutzim, where he cites a study which concluded that there was not a single case of sexual intercourse between adults who had been raised together from childhood, Brown cites other cases that are equally compelling, such as the Chinese practice of "minor" marriages, where, until recently, Chinese parents often adopted a female child so as to make her the wife of their son. The anthropologists who studied this practice found that minor marriages were roughly 30% less fertile, both sexes in minor marriages resorted more than others to extramarital sex, and minor marriages ended more than others in separation and divorce. Brown also cites evidence from parent-child incest. The fact that father-daughter incest is much more common than mother-son incest is fully consistent with the instinct theory, he says, given that traditional social roles entail less opportunity for the imprinting effect to develop between a father and a daughter. Finally, Brown cites widespread evidence of inbreeding depression among mammals as well as instinctual incest avoidance.

The anthropologist Marvin Harris (1989), on the other hand, in spite of all of the above, is "convinced" that the incest taboo is purely cultural, "another manifestation of the principle of exchange" (198). He points out, first, that in small human pre-agricultural groups inbreeding depression would result in the gradual elimination of harmful recessive alleles "because such societies have little tolerance for infants and children who are congenitally handicapped and impaired" (200). Second, he points out that many purely cultural phenomena are universal or nearly universal, having been passed from culture to culture or reinvented simply because they were useful, such as boiling water and cooking food. Third, he points out that many male rulers of ancient kingdoms and empires married their sisters, such as in ancient Egypt and among the Incas and Hawaiians. Especially interesting is that for about the first 300 years AD in Roman-controlled Egypt, brother-sister marriages were not uncommon among the general populace and were considered perfectly normal. Fourth, the fact that minor marriages in China were regarded as inferior, even humiliating, explains the lower fertility, the extra-marital sex, and the failure rate of those marriages. Fifth and finally, among the Israeli kibbutzim, Harris says that out of 2,516 marriages 200 of them were between children raised in the same kibbutz such that, "although they were not in the same class for six years," five of the marriages were "between boys and girls who had been reared together for part of the first six years of their life." These five marriages, says Harris, "actually disconfirm the [instinct] theory" (201). For Harris, then, incest taboos evolved purely culturally throughout human history because of the advantages afforded to small hunting-gathering groups that exchanged their sons and daughters with neighboring hunting-gathering groups, advantages such as alliances against common enemies and the sharing of food in times of scarcity. The incest taboo developed culturally as a way of ensuring such exchanges, incest reflecting a breakdown in exchange. With the advent of agriculture and urbanization, marital exchange continued to persist for economic and political reasons, while parent-child incest became further discouraged since it threatened the integrity of the family organization.

There are some serious problems with Harris's view. Starting backwards, his theory of cultural evolution for the incest taboo fails to explain, first, the widespread and independent recurrence of repugnance at the thought of incest by people raised in modern societies that do not have the same pressures that small groups did in the past and, second, the general intensity of that repugnance. Next, Brown (1991) cites a study which found that father-daughter incest tends to occur when the mother was either unable or unwilling to fulfill her mating role, so that father-daughter incest actually "allowed the family to *maintain* its functional integrity" (125). Next, even if there were five marriages between adults who were raised together as children in the same kibbutzim, this hardly "disconfirms" the instinct theory, especially since we are not told how long the "part" of their lives was and what their ages were. But even if it was fully the first six years of their lives, biology is statistical, and you cannot disconfirm a statistical theory with such a small sample. Next, Brown adds some important details about the incestuous marriages in Roman-controlled Egypt that are missed by Harris. For example, according to the Roman censuses taken periodically at the time, 15–21% of the marriages were brother-sister, which of course is high. However, of the five sibling marriages cited by the modern historian that both Harris and Brown refer to, the age differences were 7, 8, 4, 8, and 20 years, most of which are not conducive to producing the Westermarck effect. Moreover, as Brown points out, we just do not know enough about the child-rearing practices of these Egyptians to be able to say their culture disconfirms the instinct theory (126–127). Finally, while it is true that inbreeding depression in a small population can eventually result in a population with less harmful recessive alleles because of selection against deleterious homozygotes, and has actually been observed in some species of plants and animals (Futuyma 1998, 620–621; Freeman and Herron 1998, 626–627), it is highly unlikely that human hunting-gathering groups would remain genetically isolated long enough to enjoy this effect. But even if they did, their homozygosity would prove maladaptive, given that a population that has less genetic diversity has a poorer capacity to respond to changes in the environment such as modified or new diseases. (Genetic diversity is, after all, the reason that sex evolved in the first place.) A population low in genetic diversity, as David Raup (1991, 124–125) points out, is ripe for extinction.

According to Harris (1989),

> In this age of sexual liberation and experiment . . . brother-sister mating is probably on the verge of becoming just another "kinky" sexual preference of little interest to society, providing that incestuous siblings use contraceptive safeguards and seek genetic counseling.
>
> (206)

I doubt that this will ever happen, both for evolutionary and for recent historical reasons (it did not happen, e.g., in the hippie "free love" generation of the 1960s), but it would certainly make for a farcical vignette in my proposed sequel to Woody Allen's masterpiece, which I suggest be given the title *Everything You Never Wanted To Know About Sex And Wish You Hadn't Asked*.

5

Evolution and Feminism

Many of the questions that concern feminism overlap with other questions in this book, such as epistemology in the first chapter, mating and rape in the previous chapter, and race in the next. But many of the questions are unique in their own way. Unlike the races involved in the racial rights movements, for example, which fight for justice and equal rights for minorities, women are not a minority, nor have they ever been. Instead, they have always constituted roughly half the human population. The discrimination and injustice faced by women, moreover, are both cross-cultural and world historical, with a history that, unlike human races, is as old as humanity itself.

Indeed, anyone who applies some real empathy to the matter cannot help but feel overwhelmed by the burden of what Simone de Beauvoir (1949), the matriarch of modern feminism, called "the second sex" or "the other." I would hope that every reader will agree that women have been and still are seriously discriminated against and ought to get fair treatment in every relevant area of society. So what has evolution to do with any of this? As we shall see in this chapter, many say it is irrelevant, while many others say it needs to be taken seriously into account.

What is remarkable is that, especially in recent years, courses in colleges and universities devoted to feminism, typically called "Women's Studies," have shifted from issues of justice to an assault on biology. There are two fronts being fought here. On the one side, there is the denial that gender differences are biological. Instead, the argument repeatedly made is that gender is a social construction, in line with SSSM thinking. This is a topic we shall return to in this chapter. On the other side, there is the repeated claim that the science of biology, in particular evolutionary biology, is rotten to the core with sexism in the form of male chauvinism, so that it is not to be trusted. That will be the final topic we shall examine in this chapter.

Of course, there is no *one* feminist point of view, and *feminism*, as the term is used, denotes a wide continuum of perspectives. But this is not reflected in women's studies courses. Instead, a very narrow perspective has come to dominate, and the metaphors of wars and fronts are not misplaced. The battle, indeed, is for the minds of the young, specifically the minds of young women. It is all highly reminiscent of the approach to ideas known as *memetics*, according to which ideas take on a life of their own and evolve strategies for their spread, much like viruses, with our brains as the unwitting hosts. I shall save that approach for Chapter 8 on religion and shall not belabor it here, but after reading that chapter one might well return to the topics in the present chapter with a fresh perspective.

At any rate, the single best book on the topic of anti-biology in women's studies is *Professing Feminism* (2003) by Daphne Patai and Noretta Koertge. Patai is a professor of literature, Koertge a philosopher of science, the previous editor of the journal *Philosophy of Science*, no less. Both are feminists of the traditional sort. What is remarkable about their book is its exposé of women's studies programs, including firsthand accounts of what they call *biophobia* in these programs. While feminism was originally about equality and women's rights, according to Patai and Koertge it has become derailed in academia to include indoctrination into an aberration of true feminism. And indeed *indoctrination* is the word Patai and Koertge use in their subtitle. Sometimes the indoctrination includes viewing men as the enemy. But more generally, it includes an anti-science stance, in line with the SSSM, such that not only is science viewed largely as a male social construction, but gender differences themselves are viewed completely as a social construction. In these courses there is hardly any encouragement of what could be called the love of learning. Instead, the attitude promoted is that of hostility. Hostility toward men. Hostility toward science. Evolutionary biology is especially denied any explanatory power, at least when it comes to human gender.

Indeed, it is quite remarkable what is taught and driven home in these courses time and again. For example, it is commonly taught that men in general are not inherently physically stronger than women in general (Patai and Koertge 2003, 138). The idea that they are inherently stronger is simply a social construction, a product of a male-dominated society. In a truly egalitarian society, women would be bench pressing, on average, the same weight as men in the local gym. (Does this mean that in a feminist society they would be on average even stronger?) Another example concerns the pain of childbirth (139). It is often taught in women's studies courses that women have been brainwashed by patriarchy into believing that pain during childbirth is natural. Accordingly, they are taught that the pain of childbirth is a social construction, that there would be no such pain in a truly feminist society. In connection with this idea, they are also often taught that there was little infant mortality before childbirth was taken over by the male science of modern medicine. Yet another example, one that we shall return to in detail, is the idea that the male-female dichotomy is a false dichotomy, that instead of two genders there are three or more or even a continuum.

What Patai and Koertge (2003) object to is not the exploration of new ideas that challenge mainstream thought (indeed such challenges characterize the history of science), but rather

the pedagogical practice of presenting unsubstantiated ideas to students ill-prepared to examine them, and dressing these notions up as well-founded and properly documented feminist correctives to "malestream" prejudice. Equally deplorable is the habit of branding any disagreements with these ideas as demonstrations of a lack of genuine commitment to feminist aims.

(140)

This latter point is where indoctrination really shows its face. Patai and Koertge provide a chapter devoted entirely to "Proselytizing and Policing in the Feminist Classroom." Here they document case after case of women's studies professors using their classrooms to recruit students into the life of feminist activists. A party line is established, contrary opinions and teachings are summarily rejected, the enemy is clearly identified and denounced, and sympathizers with the enemy are branded as traitors. As a result, feminist professors of the more traditional sort have increasingly ceased to teach in women's studies courses. As a prime example, Patai and Koertge cite the case of Alice Rossi, past president (1983–1984) of the American Sociological Association and a pioneer within feminism of a biopsychosocial approach to human behavior. According to Rossi, with this new wave of women's studies professors and students genuine debate became impossible. "It was all settled in their minds," she said.

You didn't have to know any biology, you didn't have to know anything about genes, you didn't have to know anything about how your hormones worked . . . they just came back at me all the time with the same argument—"Well, there's no reason to deal with anything biological because that only takes away from the *real* factors. The real things that determine things are social and political."

(147)

What is more, these student recruits attend classes outside of women's studies, where either of their own accord or of the prompting of their women's studies professors they shout down attempts to teach views that are contrary to their easily acquired ideology. (I've experienced this myself when I taught a course on some of the topics covered in the present book.) Patai and Koertge cite the case of a more traditionally-minded professor in women's studies who one day was simply trying to discuss a British feminist writer's description of women's complicity in wars. "That's bull!," retorted a women's studies major, who continued to assert that men were entirely to blame for war and women were merely the victims of patriarchy.

With this brief introduction, one has then to wonder, what is it about evolutionary biology that many feminists, teachers and students alike, find so threatening?

The common denominator has got to be what evolutionary biology claims about the evolution of sex, in particular the evolution of male-female differences. We have seen some of this in the previous chapter. The threat to many feminists lies not simply in the evolution of male and female *physical* differences in terms of genitalia. The threat goes much deeper, to size and strength differences and to behavioral differences. Often, quite misleadingly, the differences are said to imply a kind of *essentialism* or *determinism*, as if that were enough for their dismissal. I have dealt with these confusions in Chapters 4 and 1 respectively. Quite simply, biology is statistical, and genes are not the only influence on our behavior. But even once those misconceptions are themselves clarified and dismissed, the objections remain. In all of this, in short, the underlying fear is that *is* entails *ought*. This is a dichotomy that we shall return to in Chapter 7, at a more theoretical level. For the present, we need to examine this fear and show that it is largely if not totally misplaced, given what is known about evolution. (I shall leave the question of whether evolutionary science is itself objective for the very last.)

The best place to begin is with the claim that gender is a social construction. There is no uniformity on what this claim means. Sometimes it means that gender *roles* are a social construction. Sometimes it means that *all* gender differences are a social construction, including the male-female sex dichotomy. Sometimes it means a bit of both. In any case, a confusion that we should immediately dismiss is the confusion between matters of biology and matters of justice. These are separate issues, but all too often one does not find them separated in feminist literature. For example, in a book titled *Justice, Gender, and the Family* (1989), Susan Okin, a professor of political science, not only states that "The rejection of biological determinism and the corresponding emphasis on gender as a social construction characterize most current feminist scholarship" (6), calling them "findings" (8), but that

> the sharing of roles by men and women, rather than the division of roles between them, would have a . . . positive impact because the experience of *being* a physical and psychological nurturer—whether of a child or of another adult—would increase the capacity to identify with and fully comprehend the viewpoints of others that is important to a sense of justice.
>
> (18)

The call is for a society that "minimizes" gender. Minimizing gender is a matter of justice, for Okin, especially at the level of the family unit (where justice begins). Indeed, so confident is Okin in the "findings" of feminism that *gender* is defined by her as "*the deeply entrenched institutionalization of sexual differences*" (6). The problem is that biology is being determined by what is perceived to be the demands of justice, namely, the minimization of gender. Biology and justice are not kept separate. *Is* is not kept separate from *ought*. The unstated fear is that if gender differences are really a matter of biology, then there can be no justice. Hence, gender differences need to be no more than cultural and historical. But what if they are not? What if the causes of the "deeply entrenched institutionalization of sexual differences" are fundamentally evolutionary? What if, in other words, culture basically reflects biology here? Does that mean the end of justice?

No one doubts, of course, that there is much flexibility in gender roles. After all, women have amply shown that they can make equally good doctors, lawyers, scientists, and even firefighters. But that is not what is at issue here. Rather, it is whether there are fundamental genetic differences between men and women, statistical differences, both physically and behaviorally, differences rooted in evolutionary history, differences that, again albeit statistically, predispose women to being one way and men to being another. We can argue about what *ought* to be the case for as long as we want. But we should not let this affect our understanding of what *is* the case, especially if what *is* is deeply biological. To think otherwise is to embrace an ideology that, no matter how good its intentions, is doomed to failure. That is why the communist experiment, to name a major example, failed in the Soviet Union, and why it is destined to fail in its remaining strongholds. All along it has been based on an erroneous theory of human nature. Arguably the same fate awaits the feminist experiment based on the social construction of gender. Specifically, if the common denominator in gender differences that we see virtually all over the world is simply a matter of culture and historical contingency, then feminist approaches to social change have a realistic chance of success. But if that common denominator is largely biological via evolution, then the feminist experiment hasn't got a snowball's chance in hell, as the saying goes, any more than the communist experiment. The only hope in that case is to enact gender-specific laws (such as sexual harassment laws; see Altman 1996), laws that come down harder on one gender than on another given that gender's predisposition to unwanted social activity, laws that force gender equality (e.g., political representation), laws that attempt to counterbalance innate differences. These laws must not be temporary but permanent, however. If we are indeed fighting biology developed over millions of years of evolution, we have got to recognize that we cannot realistically hope for a fully egalitarian society, one that "minimizes" gender. The only hope for that would be full-scale genetic engineering, a hope dampened by the dream-turned-nightmare warning of *Jurassic Park*.

This brings us to the topic of *sexual selection*, the very mention of which is anathema to many feminists much like the mention of devil worship is to fundamentalists. Discussed briefly in the previous chapter, sexual selection was Darwin's solution to the problem of explaining sexual dimorphism in sexually-differentiated species. For example, male lions have a mane, female lions do not. Male elephant seals are many times larger than the females. Male peafowl, which we know as peacocks, are the ones with the big brightly colored plumage, while the females, the peahens, are relatively plain brown birds. And so on in example after example throughout the animal world. Darwin could find no explanation in natural selection for these differences, since males and females of the same species inhabited the same environments. So he developed sexual selection theory to account for the differences. Here, selection and adaptation are not in terms of the environment, but rather in terms of mating. Darwin focused on basically two kinds of sexual selection. One is male-male combat, where males literally battle it out for acquisition to females. This explains why male lions have a mane. Fights are often won wholly or partly due to intimidation, and a mane aids the intimidation factor by making a lion appear larger than it really is. Male-male combat also explains why male elephant seals are much larger than the females, since the males have harems and battle other males for acquisition to them. With birds and many other species, however, the males do not literally battle each other for females. Instead, they compete via display, with the females choosing their mates based on appearance or dancing.

Sexual selection theory was never favorably received even by Darwin's fellow Darwinists. Indeed, it suffered for roughly 100 years. But since the late 1970s, however, it has enjoyed a veritable renaissance, and is today a well-established theory in modern biology, the evidence having grown so overwhelmingly in its favor, the real debate only being over the details.[1]

One has to keep this in mind when reading feminist authors who reject sexual selection theory on ostensibly *biological* grounds. Foremost here is Anne Fausto-Sterling. What makes her particularly interesting is that she is both a professor of biology and a professor of women's studies. Because of this dual role, she is often cited as an authority by feminist writers and sociologists. With regard to sexual selection, while Fausto-Sterling in her discussion of it, confined as far as I know only to *Myths of Gender* (1992), does not outright reject it, she does take pains to show its sexist origin, in particular Darwin's belief that males are evolved females such that "females represented the less variable, more juvenile or primitive states of the species" (180). More to the point, when it comes to sexual selection in humans, she argues against minimal parental investment theory developed, as we have seen in the previous chapter, by Robert Trivers and which has since become a mainstay in sociobiology and evolutionary psychology. According to Fausto-Sterling, one first has to demonstrate that the trait to be explained by parental investment theory is "really under genetic control" (188). Another problem is that "one cannot use knowledge about the genetic variance of a particular trait in a currently existing species to make accurate inferences about the trait's evolutionary origins" (188). Since there is no fossil record for behavior, the problem, she says, remains. Indeed, she goes on to say not only that "Human sociobiology is a theory of essences" (195) but that "*Human sociobiology is a theory that inherently defies proof*" (199). There is no way to tell, she says, what behavioral traits are controlled by genes—whether the adaptive product of natural selection or the mere product of "random genetic events"—and what behavioral traits are the product of learned behaviors made common through shared history.

This approach to knowledge, however, suffers from what the 18th-century Irish Bishop and philosopher George Berkeley called "we have first raised a dust and then complain we cannot see." If one wants, one could take what Fausto-Sterling says and apply it to all of evolution, the way creationists do. But in so doing, all that one has done is satisfy one's predetermined agenda. In order to accomplish this, one has to ignore the basic nature of scientific knowledge, in particular knowledge of evolutionary biology, which, as I have argued in Chapter 1 on knowledge and shall argue again in Chapter 8 on religion, is best characterized as *inference to the best explanation*. There is not one unequivocal arrow that points to evolution as a fact, as "proof" (this is a bogey term)—how could there be?—but instead millions upon millions of arrows which when viewed collectively point unequivocally to one and only one conclusion and not at all to any other. That is the Darwinian program in evolutionary biology, including sociobiology and evolutionary psychology. One cannot accept it for evolution as such but not for sexual differences in human

---

1   See Cronin (1991, 1992); Spencer and Masters (1992); Andersson (1994); Futuyma (1998, 325–326, 586–594); Pinker (2002, 344–351).

evolution. To do so exposes one's underlying agenda, which is a political agenda, not one motivated by love of knowledge.

When it comes to sexual selection, the modern evidence is overwhelming not only for sexual dimorphism but for sexual selection as the only plausible explanation. For example (and I give here only one), Futuyma (1998, 345) discusses experiments on long-tailed widowbirds, where the males have long tail feathers and mating is based on female choice. In one experiment, some of the males had their tail feathers partly cut off and attached to the tail feathers of other males. The males with the artificially short tail feathers mated less than normal, while the males with the artificially long tail feathers mated more than normal. Modern biology is filled with thousands of similar examples and experiments. But what is more, the evidence perfectly fits sexual selection theory and no other. As discussed in the previous chapter, sexual selection is the logical outcome of basic asymmetries almost completely universal between males and females (the rare exception is sex role reversal in some birds and fish). Females produce large and few gametes (egg cells), while males produce small and numerous gametes (sperm cells). This asymmetry entails a conflict in reproductive strategies. A male can mate with many females and suffer little if at all in terms of reproductive success (fitness) if he happens to mate with a genetically inferior female, while a female can only have one male to fertilize an egg and suffers a high cost in reproductive success (fitness) if she mates with a genetically inferior male. Because of this basic and profound asymmetry, males find themselves competing against each other for access to females, while females, especially genetically superior females, face the opposite problem, an embarrassment of riches in terms of which male they are going to mate with. The results in terms of evolution are going to be innate sex differences in both physical and behavioral traits, which is precisely what we see.

With humans, of course, culture muddies the waters to some degree, but only to a degree. The same forces that explain sexual dimorphism in lions and in peafowl have operated and continue to operate on humans. Anyone who has gone to a bar or to an Internet dating site can view the numbers game played by males and see a very different strategy used by females. Here culture reflects biology. The same goes for strip clubs and prostitution, where the consumers are mainly males. Moreover, throughout evolutionary history, given that humans evolved in small groups and competed against similar groups, there had to have evolved a division of labor. Not only do the reproductive asymmetries between males and females entail this, but also the very nature of group competition. Hence it is not simply because of culture that we find males bigger and stronger in general than females, or that males are typically the ones with facial hair, or that soldiers in wars are almost always males, or that most sex and violent crimes are committed by males. To pass all of this off as the mere product of cultural contingencies can only be accomplished by ignoring the enormous amount of evidence to the contrary from biology.

We are not finished with Fausto-Sterling's attack on sexual dimorphism in humans, however. While she does not attack sexual selection theory directly, when applied to humans she attempts to undermine it by using an argument for which she has by now become quite famous. In short, she argues that the male-female dichotomy is a false one. (If the dichotomy is false, then so is the "di" in sexual dimorphism. And if there is no sexual dimorphism, then there is nothing for sexual selection theory to explain.) In the case of feminism in general, Patai (Patai and Koertge 2003) argues that for radical or new-wave feminists

> It is important to attack the notion of biological dimorphism because the fact of dimorphism is essential to our history of sexual reproduction. And sexual reproduction, from a biological point of view, confirms the normalcy, indeed the ordinariness, of heterosexuality, which various famous feminists have interpreted as the "institution" at the root of women's oppression. Biology presents a particularly intractable problem for this sort of feminist analysis.
>
> (316)

Hence these feminists feel a need for rejecting the biology, either by attacking science as a whole as a misogynist institution, or by attacking the particular facts and theories upon which sexual dimorphism is based. We shall return to both of these themes in this chapter. Patai, indeed, goes so far as to say that social constructionism has become the new orthodoxy in feminism because of a perceived need to see gender roles as social constructs rather than as biological determinants in order to have hope for their agenda. Hence, she says, their "denigration of 'facts,' 'logic,' and 'rationality'—they often [sic] dismissed as forms of 'masculinist' linear thinking'—is, I believe, in large part driven by feminists' opportunistic desire to leave themselves free to make any claim they wish" (306).

This might all be true, and yet to reject an argument because of an alleged ulterior motive (whether that motive is real or not) is to commit the fallacy known as *circumstantial ad hominem*. The keyword here is "argument." It is perfectly legitimate to reject a position, testimony, or statement on the basis of an ulterior motive (we do this in court cases all the time, and well we should). An argument, however, is an autonomous entity, which has to be judged according to criteria that have nothing to do with the motive of the person who gave the argument (in short, acceptability, relevance, and sufficiency of premises). In other words, a good argument can come from a person with a bad motive, a bad argument can come from a person with a good motive, and so on.

In the case of Fausto-Sterling, she has a very interesting argument against sexual dimorphism in humans that needs to be examined and assessed, and the assessment of it should have nothing to do with whatever her motive for putting it forward. In brief, her argument is based on the existence of what are commonly known as *hermaphrodites*, but the politically correct term now is *intersexuals*. According to Fausto-Sterling in her book *Sexing the Body* (2000), roughly 1.7% of the human population is composed of intersexuals (53). This is an overall average, which varies among human populations. Since intersexuals "quite literally embody both sexes," she claims that their existence "weaken claims about sexual difference" (8). Instead, what we really have is a continuum. "While male and female stand on the extreme ends of a biological continuum," she says, "there are many other bodies that evidently mix together anatomical components conventionally attributed to both males and females" (31). In the West we divide the continuum into two categories, but that is a social decision. (Presumably other societies could divide the continuum into, say, six categories.) Hence the male-female dichotomy is a social construction, and a false one at that. What nature gives us is "no either/or. Rather, there are shades of difference" (3). But there is more. Surgeons, she says, remove parts of intersexual babies so as to force them into either the male or female mold. Hence, "sex is, literally, constructed" (27).

There are many problems with this argument, both biological and conceptual. One problem concerns her meaning of *intersex*. In an earlier publication, where, not completely seriously, she argues for five sexes rather than a continuum, Fausto-Sterling (1993) uses the term *intersex* to refer to individuals with XY chromosomes but predominately female anatomy, to individuals with XX chromosomes but predominately male anatomy, and to individuals with mixed genitalia.

According to the psychologist Leonard Sax (2002), this is a fair meaning, except that it does not generate the 1.7% that she claims exists in *Sexing the Body*. To arrive at that percentage in her book, she uses a different definition of *intersex*: "bodies having mixtures of male and female parts" (257 n. 4). This definition, like its antecedent, seems to presuppose what it is used to deny, namely, that there are male and female parts. But more than that, it is surprisingly vague, since it is not clear whether it refers only to phenotypic parts or also to genotypic parts.

In a publication contemporary with *Sexing the Body*, Fausto-Sterling (Blackless *et al*. 2000) defines *intersex* as any "individual who deviates from the Platonic ideal of physical dimorphism at the chromosomal, genital, gonadal, or hormonal levels" (161). What I find interesting about this definition is the strong implication that anyone who believes in sexual dimorphism is a Platonic essentialist, believing that male and female are fixed abstract essences existing outside of space and time!

The main problem with her two more recent definitions, however, is simply that they are much too wide. According to Sax (2002), not only is a definition of *intersex* based on deviation "from the Platonic ideal" not clinically useful, but when a clinically useful definition is in fact used then the five most common conditions claimed by Fausto-Sterling to be intersex do not turn out to be intersex conditions at all. More specifically, when a clinically useful definition of *intersex* is used, such as Fausto-Sterling's (1993) definition, or Sax's (2002) "those conditions in which (a) the phenotype is not classified as either male or female, or (b) chromosomal sex is inconsistent with phenotypic sex" (175), then the 1.7% disappears and is replaced by something much smaller, namely, 0.018%.[2]

---

2   According to Sax (2002), true cases of intersexuality include *mosaics* (individuals with XX sex chromosomes in some cells and XY in others and the phenotype includes either a penis or vagina or both), *complete androgen insensitivity syndrome* (individuals with exclusively XY sex chromosomes but no male genitalia), and *congenital adrenal hyperplasia* (individuals with exclusively XX sex chromosomes but either male genitalia or mixed genitalia combined with masculinized behavior). These conditions result in the roughly 0.018%. Fausto-Sterling, however, also includes five other conditions to get her 1.7%, which according to Sax no clinician would regard as a true intersexual. These are (i) *late-onset congenital adrenal hyperplasia*, which accounts for 88% of Fausto-Sterling's 1.7% (genitals are normal at birth and roughly only 10% of XX females get a mildly enlarged clitoris, starting usually in their early twenties, while XY males usually show no symptoms at all, except for roughly 50% of them which experience abnormal balding before the

Continued on next page

The difference is between roughly 50,000 true intersexuals in the United States and roughly five million. That is a significant difference. What is more, as Sax (2002) points out, "The most original feature of Fausto-Sterling's book is her reluctance to classify true intersex conditions as pathological" (177). In *Sexing the Body* (2000) she muses that we may well come one day to view intersexuals as "especially blessed or lucky," perhaps as even "the most desirable of all possible mates, able to pleasure their partners in a variety of ways" (113). For Fausto-Sterling, at any rate, the entire continuum should be considered both "natural" and "normal," with "complete maleness" and "complete femaleness" as only "the extreme ends of a spectrum of possible body types" (76). For Sax (2002), on the other hand, "*natural* and *normal* are not synonyms. A cow may give birth to a two-headed or Siamese calf by natural processes . . . Nevertheless, that two-headed calf unarguably manifests an *abnormal* condition" (177). Here the word *abnormal*, of course, must be taken not only in the sense of "statistically rare," but also "defective." (Again, the two senses are not the same, as what is defective can be statistically rare or statistically common.) That true intersexuality is a pathological condition, a genetic defect or disease, is evident from the kinds of cases that compose true intersexuality, all of which involve various hormonal imbalances and often involve sterility, groin pain, tumors, and cognitive disorders. If we readily classify problems such as hemophilia, diabetes (Type 1), and asthma (which I have) as genetic defects or diseases, then why not true intersexuality as well? The reason can only be political correctness, not biological correctness. Unfortunately, political correctness does not make the problems go away. In the future, heavy breathing from asthma might be considered socially desirable as a sexual turn-on, but it is a physical defect nonetheless and it needs medical treatment.

From an evolutionary point of view, perhaps the most serious problem with Fausto-Sterling's definitions of *intersex* is that they ignore gamete production. Surely if anything defines sex from a biological, evolutionary point of view, it is the production of either sperm or egg cells. Sexual reproduction evolved from asexual or vegetative modes of reproduction because, at bottom, it increases variation, the greater the genetic variation the greater the raw material upon which natural selection operates, and also the lower the probability of extinction in a changing environment.[3] What this ultimately means for our present question is that, to borrow the title of a

---

age of 50), (ii) *Klinefelter syndrome* (individuals with exclusively XXY sex chromosomes, such that they have normal male genitalia but a tendency to sterility), (iii) *Turner syndrome* (individuals with only an X sex chromosome, such that they have normal female genitalia but most are infertile and grow to roughly half a foot shorter than they should be, (iv) chromosomal variants other than the two previous (individuals with exclusively XYY or XXX sex chromosomes, such that men with an extra Y chromosome are physically indistinguishable from normal XY men, are typically fertile, but usually have a lower than average IQ, while women with an extra X chromosome are physically indistinguishable from normal XX women, are typically fertile, but also usually have a lower than average IQ), and (v) *vaginal agenesis* (individuals with exclusively XX sex chromosomes and normal ovaries but the outside third of the vagina failed to develop and was replaced by roughly an inch of fibrous tissue).

3   The whole matter, it turns out, is far more complicated, but arguably still basically true. See, e.g., Futuyma (1998, 606–613).

recent book by Joe Quirk (2006), *sperm are from men, eggs are from women.* A true intersexual from this point of view, then, is going to be a human with the machinery to produce both sperm and eggs, whether fertile or not. Given this definition, I would predict that the number of true intersexuals, or true hermaphrodites in this sense, is going to be extremely small, certainly much less than 0.018%. And indeed, it turns out, the condition is extremely rare among intersexuals however defined (Parvin 1982; Krob *et al.* 1994).

But even if we ignore all of these problems—even if we ignore whether true intersexuals compose roughly 1.7% of the population or only 0.018% or even less, whether they are normal in the good sense of "normal," and whether they produce sperm and eggs—the problem remains that Fausto-Sterling is guilty of a conceptual error of the most basic kind. Night grades into day and day grades into night, and yet night and day are not extreme ends of a continuum. They are basic categories in spite of dusk and dawn. Moreover, biology is full of such categories, categories that are not fully discrete, such as ectotherm and endotherm, adult and infant, sexual and asexual. Even when we focus on something like species, where in roughly 15% of species situations biologists cannot agree on whether what they are dealing with is a species or a subspecies/variety, irregardless of whether their species concept is based on reproductive isolation or morphological discreteness or something else, biologists do not take the 15% of what they call "messy situations" and argue that species are not real (see Stamos 2003, 83–84, 331–332). In the case of intersexuals, 1.7% is certainly far too small to deny the categories of male and female as basic, natural, and normal.

One also has to wonder about the practical implications of Fausto-Sterling's (2000) suggestions, not just in terms of surgery. She clearly (110–113) would like to see the categories of male and female disappear from birth certificates, drivers' licences, passports, and apparently even sports events. But what would she recommend for public washrooms, changerooms, and prisons? All women, feminists included, should be outraged at the implication that these should all be shared, for it is nothing but a misconceived recipe for a dramatic increase in sexual harassment and rape. Indeed, with the denial of the male-female dichotomy, feminism becomes its own worst enemy.

What Fausto-Sterling (2000) does for the sexual categories of male and female she also does for sex hormones. In a chapter devoted to a fairly detailed account of the early history of hormone research, she concludes by advocating the abandonment of the concept of sex hormones. Testosterone and estrogen are found in both males and females, furthermore they affect many tissues and organs such as bones, blood, nerves, kidneys, liver, and heart. The history of the division of these hormones, she argues, *superimposed* on humans the male–female sex hormone dichotomy. The dichotomy is not from nature. Hence, she says, we need to free ourselves of the "sex hormone straitjacket" and "agree to call them *steroid hormones* and nothing else . . . They are, in short, powerful growth hormones affecting most, if not all, of the body's organ systems" (193).

To agree to this suggestion, however, one has to ignore a mountain of sex hormone research to the contrary, especially the research on animals and humans spread out over the past few decades. To be sure, males and females produce both testosterone and estrogen, and these hormones do not just affect primary and secondary sexual characteristics. The most we get from Fausto-Sterling (2000) is the bland acknowledgment that testosterone and estrogen "are present in different quantities and often affect the same tissues differently in conventional males and females" (193). Aside from the acknowledgment of conventional males and females, this statement masks an enormity of research that serves to justify the standard male-female dichotomy, both physical and behavioral (keeping in mind, of course, that biology is statistical). For example, in a study and review of testosterone levels in male birds, Wingfield *et al.* (1987) confirm the role of testosterone in dominance relationships and in territorial aggression. In a variety of monogamous bird species, challenges to males from other males resulted in an increase in testosterone, while parental behavior in males was either accompanied by or slightly preceded by a reduction in testosterone. When a testosterone implant was given to these males, mating and aggressive behavior increased while paternal behavior decreased.

Not surprisingly, testosterone in humans has been studied to a far greater degree, and the results have been basically harmonious with what has been found in animals. Not only do males have roughly ten to twenty times more testosterone in their bodies than do females, but it especially kicks in during puberty, resulting in a deepening of the voice, growth in facial and bodily hair, a dramatic increase in muscle strength, and also a dramatic increase in aggressive impulses. The fact that the vast majority of violence in culture after culture throughout the world is committed by young adult males is not merely the result of cultural obstacles faced in the transition to manhood. The British psychologist John Archer (2006) provides a long review, a meta-analysis, on the enormous amount of research conducted on testosterone and aggression, in particular on male aggression, and the overall evidence is unequivocal. Although it is not a simple matter of strict covariance (aggressive behavior does not strictly covary with the level of testosterone in an individual), nevertheless what is known as the *challenge hypothesis* (originally developed and corroborated for birds and later corroborated for other animals including primates such as chimpanzees) has withstood the test of time when applied to human males. According to this hypothesis, testosterone levels rise in situations that require aggression, especially in reproductive contexts (taken broadly to include not only direct competition for mates but also dominance disputes, mate guarding, and territorial protection). Accordingly, studies on human males have repeatedly found that males in the presence of attractive females (whether physically present or in films) experience an increase in testosterone, that males in competitive situations, situations either directly or indirectly related to reproduction (matters of reputation, challenges to masculine honor such as insults, sexual jealousy, competition in sports, competition for socioeconomic status, etc.), experience an increase in testosterone (there is a qualification here that I shall address below), that males experiencing an increase in testosterone from external sources (e.g., a testosterone injection) experience an increase in libido, that inmates convicted of violent crimes have on average higher testosterone levels than inmates convicted of nonviolent crimes, that males who tend to beat their mates have a higher level of testosterone, that males willing to discuss problems with their mates and who are more sensitive to their needs have lower levels of testosterone, and that males involved in parental care experience a decrease in testosterone. All in all, the evidence favors a reciprocal model between testosterone and behavior. Higher testosterone men may be less able to enter into long-term committed relationships, may be more inclined toward promiscuity (short-term mating), may be more inclined toward risk taking, and may be more inclined toward violent, criminal behavior, whereas the opposite behaviors in turn tend to reduce testosterone levels.

Testosterone differences also cut across racial lines. For example, it has long been known in the medical community that, according to studies conducted in the United States, black men have on average a 15% higher level of testosterone than white men, and accordingly almost twice as high a level of prostate cancer (Ross *et al.* 1986). Similarly, black women have on average a higher level of testosterone than white women, and correspondingly a lower level of osteoporosis (Perry *et al.* 1996).

Sometimes, when it comes to the relation between testosterone and aggression, the question is raised over which one is the cause of the other. Social constructionists like to exploit any problems here. The cultural anthropologist Marvin Harris (1989), for example, refers to primate studies on dominance and hierarchy where testosterone levels were found to be highest *after* a successful aggressive encounter and lowest after losing, and also evidence from human studies where males showed an increase in testosterone *after* winning a sports event or receiving an MD degree and a sharp decrease *before* going into surgery or *before* going on patrol during a war. Hence he says the correlation between high levels of testosterone and high rank is not a matter of the former causing the latter. This prompts him to conclude that

> the primate evidence does not indicate the existence of a hormonal barrier that would prevent women from learning to be more aggressive than men if the exigencies of social life were to call for women to assume aggressive gender roles and for men to be more passive.
>
> (266)

Indeed, Harris (1999) goes on to claim that male dominance (political, military, commercial, agricultural, etc.) is the result of "cultural rather than biological selection" (105).

If only it were all a matter of cultural selection and learning. The problem with Harris's conclusions is that his premises are basically false. For example, recent studies show that when it comes to sports competitions such as judo, pre-contest testosterone levels are a *good predictor* of offensive fighting, with levels being high after winning and low after losing (Archer 2006, 329). Moreover, it has been found in primate studies that the level of testosterone in *most* males is low before an aggressive encounter but actually high in *dominant* (alpha) males. What happens in dominant males is that their stress hormones tend to be kept in check, allowing them to assess the threat, but also allowing them to maintain high levels of testosterone (which would otherwise be suppressed by the stress hormones) so as to attack the source of the threat if required (Sapolsky 1990). Given an evolutionary history of sexual selection and sexual dimorphism in humans, the same effect by analogy would naturally apply to us, making it naive to think that men have ten to twenty times the amount of testosterone than women simply because of culture and learning. Moreover, the effect of testosterone on the sex and aggression centers in the old mammalian part of our brain, on sex differences in brain organization starting right in the womb (Kimura 2002), and on muscle mass and glucose supply to the muscles (Bhasin 2001), is all sufficiently well known, certainly enough to render obsolete the SSSM approach to testosterone. The same can be said for what is known about estrogen and its relation to emotions such as caring and behaviors such as talking and flirting, which I will not belabor here (see Brizendine 2006).

In all of this, the question remains whether the science of evolutionary biology itself is sexually biased. Many feminists reject much of the science of evolutionary biology because of what they sometimes call *malestream* epistemology (e.g., Code 1991), "knowledge" manufactured by men and for men, in this case white elitist men, much the same way as the stories of Adam and Eve and of Pandora in ancient mythology were manufactured to put women in their place. Indeed, for Code and many others, knowledge is fundamentally and irrevocably relative.

I shall return to the relativity part. For the moment, there can be no question that science, including evolutionary science, has a substantial history of male chauvinism. For example, in *The Descent of Man, and Selection in Relation to Sex* (1871 II), Darwin writes,

> With respect to differences of this nature ['mental powers'] between man and woman, it is probable that sexual selection has played a very important part . . . Woman seem to differ from man in mental disposition, chiefly in her greater tenderness and less selfishness . . . Man is the rival of other men; he delights in competition, and this leads to ambition which passes too easily into selfishness. These latter qualities seem to be his natural and unfortunate birthright. It is generally admitted that with woman the powers of intuition, or rapid perception, and perhaps of imitation, are more strongly marked than in man; but some, at least, of these faculties are characteristic of the lower races, and therefore of a past and lower state of civilisation. The chief distinction in the intellectual powers of the two sexes is shewn by man attaining to a higher eminence, in whatever he takes up, than woman can attain—whether requiring deep thought, reason, or imagination, or merely the use of the senses and hands . . . The half-human male progenitors of man, and men in a savage state, have struggled together during many generations for the possession of the females. But mere bodily strength and size would do little for victory, unless associated with courage, perseverance, and determined energy. They have, also . . . to defend their females, as well as their young, from enemies of all kinds, and to hunt for their joint subsistence . . . requires the aid of the higher mental faculties, namely, observation, reason, invention, or imagination . . . and as in both cases ['the contest of rival males' and 'the general struggle for life'] the struggle will have been during maturity, the characters thus gained will have been transmitted more fully to the male than to the female offspring. Thus man has ultimately become superior to woman. It is, indeed, fortunate that the law of the equal transmission of characters to both sexes has commonly prevailed throughout the whole class of mammals; otherwise it is probable that man would have become as superior in mental endowment to woman, as the peacock is in ornamental plumage to the peahen.

(326–329)

Darwin immediately goes on to suggest how the mental powers of women can be increased over the generations—vigorous higher education combined with having lots of children—but no matter, the damage is done. What Darwin says in the quotation above makes many feminists cringe, much like fingernails on a chalkboard. The feminist biologist Ruth Hubbard, for example, in her now classic "Have Only Men Evolved?" (1979), says that "The Victorian and androcentric biases are obvious." It is, what she calls, "The Victorian picture of the active male and the passive female" (158), "the wish-fulfillment dream of a proper Victorian gentleman" (160).

I have to admit that what Darwin wrote above makes me uneasy. But on the other hand, as someone who has spent a lot of time and effort delving into the mind and character of Darwin (Stamos 2007), I have to seriously doubt that Darwin would have resisted modern evidence that went against his conclusions expressed above, even evidence that went against his views but did not conclusively prove him wrong. Darwin was, after all, *the* pioneer in evolutionary psychology, but a pioneer nonetheless. His was the first word, not the last, and he well knew it. Built of a kind and gentle character on the one hand, and a patient theorizer and sifter of evidence on the other, he was not the sort of man to be dogmatic on this or other questions.

The same must be said for modern science as a whole, including the science of gender differences. On the one hand, there is today a considerable accumulation of evidence showing that there are differences in the brains of men and women. It has long been thought that men tend to be, for example, more left-brained, more analytical and better at math, whereas women have a slightly larger corpus callosum and tend to be more holistic and intuitive in their thinking (see Springer and Deutsch 1993; Kimura 2002). On the other hand, there is a long history of male chauvinism in science. Extending her analysis to modern sexual selection theory, Hubbard (1979) suggests that it was no accident that minimal parental investment theory arose around the time of the women's movement of the 1960s and '70s. Comparing that movement to the women's movement as it existed in Darwin's time, "We should remember," she says,

> that Darwin's theory of sexual selection was put forward in the midst of the first wave of feminism. It seems that when women threaten to enter as equals into the world of affairs, androcentric scientists rally to point out that our *natural* place is in the home.
>
> (164)

In short, for Hubbard and many feminists, modern sociobiology and evolutionary psychology are no better than Darwin.

But is this a fair assessment? For some, such as the don of philosophy of biology Michael Ruse (1998), Darwinism, original and modern, is sexist and it does not matter. In many examples, he says, one can find theories that are demeaning to females. But just as easily, he says, one can find theories that give preferential status to females, such as the work done on queen bees or the work done by the evolutionary anthropologist Sarah Blaffer Hrdy (1981) on why human females are alone in the animal world in not going into heat (in short, to fool a chosen male into helping with child rearing). I might add that the picture painted of men by Ghiglieri examined in the previous chapter makes men look far worse than women and that the picture painted by Buss makes neither sex look better than the other. For Ruse, whether you are a male chauvinist or a feminist, "Darwinism will accommodate you" (121). In the end it does not matter. "Science is not," he says, "simply a matter of fulfilling personal agendas. There are controls and guidelines—epistemic values—aimed at bringing the products of science into correspondence with external nature" (127).

It is the controls and guidelines, however, that make many feminists suspicious. Are those controls and guidelines not, after all, the domain of mostly male scientists? And even if those scientists are not explicitly motivated by a sexist agenda, do they not have a way of knowing that is not congenial to women? Indeed, for many feminists, and not just feminists, knowledge is indeed the problem, what might be called *the myth of objective knowledge*. Some have argued that women have their own "ways of knowing" distinct from that of men (intuitive, holistic, noninvasive, and whatever else) and that feminists should promote these ways (e.g., Belenky *et al.* 1986). Others, such as Code (1991) briefly discussed above, look at all knowledge as relative, not just sexually but culturally as well.

Either idea just does not wash. Even if is true that men and women think somewhat differently (albeit statistically) when it comes to matters of knowledge (whether the evidence from modern psychology and neuroscience is taken into account or not), different ways of *thinking* may generate different hypotheses and theories, but they do not thereby generate different *knowledge*. Thinking something is so (no matter how strongly) is not knowledge, and science is not simply hypotheses and theories. What philosophers of science call the *context of discovery* is not to be confused with what they call the *context of justification*. The latter involves the testing of hypotheses and theories against evidence publicly available to all, male and female alike. (The context of discovery is simply the context of hypothesis and theory generation.) Thus, genuine knowledge cannot be generated simply by different ways of thinking.

But what if the context of justification is not simply the testing of hypotheses and theories based on their predictions (a rather naive view, after all, given that observation presupposes theory and that any finite amount of data supports innumerable, even infinitely many theories) but rather, as the feminist philosopher of science Kathleen Okruhlik (1994) accepts, *inference to the best explanation*? If scientific explanation is essentially *comparative*, saying why one theory is better than another, then there is no clean divide between the context of discovery and the context of justification. More specifically, if all the competing theories in a given area are infected with male bias, then the best explanation, what passes as scientific knowledge, will be sexist as well.

But even so, even if science, including evolutionary biology, has so far been mainly "malestream," infected by male bias and prejudice and power politics, it does not follow that it should be scrapped altogether or that it should be replaced with female "ways of knowing."[4] As the logician, philosopher of science, and feminist Susan

---

4   Okruhlik (1994) does not prescribe this, but her rather apparent rejection of sociobiology—from her rejection of adaptive explanations of the female orgasm and of innate differences between male and female skeletons, as well as her agreement that "*Sex* as well as gender is socially constructed, at least in part" (198)—certainly lends itself to these radical approaches.

Haack (1993) puts it in her critique of feminist epistemology—"the new cynicism" is her descriptive label for it—

> it doesn't follow that it is proper to allow prejudice [in this case feminist prejudice] to determine theory choice. Even if it is not possible to make science perfect, it doesn't follow that we shouldn't try to make it better.
>
> (583)

For Haack, the greatest threat to knowledge is not men, or women for that matter, but the idea that knowledge should be politicized, that it should serve political ends. Unfortunately that is the problem with much of feminist critiques of science, as Haack ably reveals in her article. The danger with such a political program, indeed with any political program that masquerades as epistemology, is that it creates a breeding ground for what she calls "sham reasoners seeking only to make a case for some foregone conclusion" (584).

In all of this, we need to distinguish between what is *called* "knowledge" and *knowledge*. It is the essence of the modern relativism and skepticism known as *postmodernism* to deny this distinction, and it is the essence of science to affirm it. There can be no middle ground.

Before closing, let me give one example of what many feminists call bad science, male sexist science, corrected by good, nonsexist science. It is the story of fertilization, the story of the sperm and the egg, the often-told story that one might call the flagship of feminist critiques of science. Given its importance as such, and the controversy surrounding it, it is worth examining here in some detail. The story goes back to Aristotle, who in his biology develops the idea that the semen contributes the form (essence) of the species and the egg merely the matter, and he develops this idea using phrases such as "the active and the passive sex" (*Generation of Animals* I.20). According to many feminists, this stereotype of the active male and the passive female, so characteristic of Western patriarchy, extended by Aristotle to the fertilization process, carried through over two millennia to the discovery of sperm and egg. Indeed, the idea of the former actively swimming and penetrating the passively awaiting latter, thus beginning the process of fertilization, lasted well up until the mid-1980s! What was found at that time was that the egg does not passively receive the penetrating sperm cell, indeed that the sperm cell is incapable of penetrating the egg. Instead, the egg cell has thousands of little finger-like projections on its surface which pull the sperm cell through the outer barrier of the egg cell. (I almost said "lucky" sperm cell, which might indicate a sexist bias on my part, in the sense of "getting lucky," but the fact remains that only one sperm cell from an ejaculate of roughly 100 million sperm cells is taken in by the egg. In a sense, each and every one of us is a lottery winner.) What particularly troubles feminists is that these little arms, called *microvilli*, were discovered in 1895 using microscopes. And yet the *description* of their action suffered a sexist timelag until the mid-1980s, when new data combined with feminist influences finally started to have a salutary effect on this bad biology.

That is the official story, but the problem is in the telling of the story and in the facts. Beldecos *et al.* (1988), for example, collectively known as "The Biology and

Gender Study Group," have the good scientists as being almost all female scientists (three out of four), such that their work on fertilization "can be viewed as feminist-influenced critiques of cell and molecular biology" (196). The feminist anthropologist Emily Martin (1991), on the other hand, to her credit, has the relevant scientists being almost all male (five out of six), but adds that they still tended to use sexist metaphors such as "shoots out and harpoons the egg," or "key" (sperm) and "lock" (egg), or "captures and tethers" the sperm, the latter of which make the egg cell appear "dangerous and aggressive, the *femme fatale* who victimizes men" (112). Moreover, they all are guilty of what she calls "endowing cellular entities with personhood" (114).

For the molecular biologist Paul Gross (1998), however, either version of the story is pure nonsense (see also Gross and Levitt 1998, 117–122). First, the sexist quotations, he says, "are from ancient or secondary sources [and] have little to do with the state of professional understanding since the nineteenth century" (63). In other words, the real scientists doing the hardcore work just do not infect their work with sexist talk. Second, as early as 1919 the embryologist E. E. Just, based on his own observations and following a yet earlier source, wrote that "Penetration follows as an activity of the egg; the spermatozoon does not bore its way in—the egg pulls it in." Indeed, according to Gross, "After the late nineteenth century, the egg was not seen, in any scientifically meaningful way, as 'passive.'" Instead, he says, since that time the number of studies on eggs have greatly outnumbered the number of studies on sperm, "and with good reason, for the egg is vastly more important for development" (65). Third, again by the late 19th century, an increasing amount of observation and experiment was done on *parthenogenesis*, the development of an egg cell into an individual organism without fertilization (i.e., without combining genetic material from a sperm cell). The idea here, again, was that the egg is active, not passive, and the research was well known. In tangent with this, the role of the egg in development has long been studied, leading to cloning as early as the 1950s, but nobody, says Gross, "however brainwashed in masculinity, has tried to clone animals from spermatozoa" (67). Fourth and finally, says Gross, even though scientists routinely use metaphors, and some have been misdirected by them, scientists in the main know what each other is talking about and are not fooled by their metaphors. (Indeed, no biologist is misled by Dawkins's metaphor of the selfish gene, to give another example.) In the case of developmental biology, Gross claims, knowledge "depends on those metaphors not at all" (67).

Controversial examples like these abound in the feminist debate over modern evolutionary biology. But aside from the controversies (which do at times seem a bit childish, like an extension into adulthood of sibling rivalry), why should it matter if the sex cells come out looking pretty evenly active after all? If that is the way it really is in nature, then we should want to know. And if it is not, we should want to know that too. In either case, what is needed is an end to the obsession over which sex comes out better or worse looking from a Darwinian point of view. It is a bit like the obsession of students over grades. What should be at issue instead is *knowledge*, understanding the way nature really is inasmuch as we can, rather than whether we like the results.

In conclusion, one would expect a division of labor to have evolved between the sexes, given the reproductive asymmetries between males and females. As Csikszentmihalyi (1993) puts it, "During most of evolutionary history, gender specialization was simple: men had to produce, women reproduce" (49). This is certainly putting it too simply, but the point remains that females would be serially pregnant following puberty, would be heavily involved in nurturing behavior, and would also be involved in activities such as gathering. Males, on the other hand, would almost exclusively be involved in hunting, which is very dangerous and physically demanding when involving big game, in dominance battles within the group, and also in battles with other groups, resulting in males on average being bigger and stronger than females. In all of this, it is only to be expected that the hormones and neural structures that promote and facilitate these different behaviors and natures would have evolved in the sexes by selection pressures. As Alcock (2001) puts it against the blank slate (SSSM) theory,

> How likely is it that millions of years of natural selection [many consider sexual selection part of natural selection] on humans and their immediate ancestral species would produce a male psyche easily culturally conditioned to be indifferent to the cues associated with fertile women? The likelihood is exceedingly low.
>
> (137)

I would add "virtually zero," whether applied to the plasticity of the male psyche or of the male body.

With the advent of civilization, things changed in many ways, but in all human cultures one cannot help but see the evolutionary division playing out its themes, even in small cultural experiments that have attempted to eliminate them (such as the hippies of the 1960s). We carry our evolutionary past in our genes. Indeed, as Darwin (1871 II) put it, mankind carries "the indelible stamp of his lowly origin" (405). But civilization has been increasingly changing the rules. What was, no longer still necessarily ought to be. In particular, the evolutionary logic of the division of labor between the sexes has gradually become invalid with the progressive advance of civilization. With civilization came the concept of justice, and a robust concept of justice demands that women be treated with respect, as equal beings alongside men. (In fact it demands more, but that is something I shall save for Chapter 7 on ethics.) With civilization also came the scientific method, an enormously successful method (or family of methods) for acquiring knowledge, both of our surroundings and of ourselves. Historically this method has been compromised to a degree by the male chauvinism carried over from the evolutionary division of labor. But justice makes its demands on science too, and there is no reason why science cannot be gradually purged of its harmful elements, of which male chauvinism is but one. Modern feminism is absolutely essential for that process, but it is best accomplished from *within* the halls and laboratories of science. Haack's (1993) suggested remedy of "diversity within science" (584) and Okruhlik's (1994) call for "the inclusion of diverse standpoints" (205) must—and happily already does!—include many feminist scientists motivated by the ideal of objective knowledge. But in all of this change and in all of the theorizing it would be pure folly to ignore the inertia of our genes evolved over countless generations by selection. To arrive at where we want to be, we need to take full account of where we came from and of what we are. The challenge is daunting, but not necessarily impossible. Much the same, we shall now see, applies as well to the question of human races.

6

Evolution and Race

The question of human races and racial differences, as E. O. Wilson (1978) put it, is "the most emotionally explosive and politically dangerous of all subjects" (47–48). It is emotionally explosive because of the horrid history of racism, from slavery and Nazism to the kind of discrimination that fuels the civil rights movements. Certainly the topic has to be treated with great sensitivity. Indeed, I should think that, as with the topic of feminism, *empathy* is the operative concept here. What is needed is the ability to imagine oneself, as seriously and as fully as one can, in the shoes of one who lives with racism every day.

Keeping that in mind, we shall look in this chapter at some common misconceptions about race that are undermined by a basic understanding of evolutionary biology. We shall then look at the standard arguments by biologists for why the concept of race is illegitimate and should be discarded. We shall also, however, look at some recent attempts to revive the concept and give it legitimacy. Next, assuming for the sake of argument that human races are real in some objective sense, we shall examine a variety of arguments concerning the question of whether there are innate behavioral differences between human races, focusing mainly on the hot question of race and IQ. Finally, we shall examine the question of whether there evolved in humans a racism instinct, an innate propensity for racial discrimination. With all of these topics, the search for scientific knowledge and understanding must be tempered with sensitivity toward those who have suffered and continue to suffer the injustice and humiliation—the pain—of racism, and I shall try my best to maintain this balance.

As mentioned by Wilson above, the danger of debates about racial differences is largely political; it resides not only in the fact that the mere suggestion of innate racial differences, especially evolved ones, invites the label of *racist*, but in the fact that it easily fans the flames of racial hatred and prejudice. This, however, was not always quite the case. According to Richard Dawkins (2003), "Race, in Victorian times, was not the political and emotional minefield it is today, when one can give offence by so much as mentioning the word" (76).

It was still a minefield in Victorian times, however. This is because the debate over slavery, which was an enormous issue at that time, was largely a debate over race. On the one hand, there were the *polygenists*, those who believed that the human races were independently created, that each was a separate biological species. Typically this involved the idea of a hierarchy, with Caucasians at the top and the rest of the human races/species filling the gap between Caucasians and the great apes (chimpanzees, gorillas, and orangutans). On the other hand, there were the *monogenists*, those who believed that all human races were derived from a single stock (like many believed about the breeds of dogs).

The relation between these two views and slavery was not as direct as one might suppose. Many monogenists were against human slavery mainly because they believed that no members of the human species should be treated as animals. Other monogenists supported human slavery, however, because they believed that there are many degenerate human races, such that within the human species there are races of natural masters and races of natural slaves (the institution of slavery being best for both). While many polygenists also supported human slavery, because they believed that there were inferior human species and it was morally justifiable to treat them as animals, there were also many polygenists who were against slavery, mainly because they had an even lower opinion of the practice of slavery than they had of what they believed to be the inferior human races (see Gould 1981, 31–39, 69–72; Jackson and Weidman 2004, chs 2 and 3).

Although the idea of evolution, variously conceived, did not create the distinction between monogenism and polygenism, the introduction of evolutionary ideas had a lot to do with the debate. Darwin, for example, was a monogenist and weighed in heavily against slavery. He thought the evidence strongly supported the conclusion that all human races evolved from a "single primitive stock" in Africa. External racial differences, interestingly, he thought were probably not adaptations to different climates and habitats but instead were primarily the result of sexual selection, while on the other hand he thought that there were intellectual, moral, and social differences between the races that were primarily the result of natural selection (see Darwin 1871 I, chs vi, vii, xix, xx, and xxi).

That was Darwin. The real question is whether modern evolutionary biology has any relevance for modern race issues. The answer, quite simply, is a most definite *yes*, although many people have it stuck in their heads that evolution can only make matters worse. Evolution teaches that we came from monkeys, after all. In the Appendix I deal with this and other misconceptions. Here, in this chapter, it is important to show how evolution undermines many racist assumptions and is actually a powerful ally in the fight against racism. Or at least it is to a large extent, as we shall see when we get into some of the controversies.

For a start, we can quickly dispense with the view that some of the human races came out of the trees later than others, or any similar view. A now dead theory of human evolution is the *multiregional model*. According to this theory, advocated most famously by the anthropologist Carleton Coon (1962), our species, *Homo sapiens*, did not evolve once from our immediate ancestral species, *Homo erectus*, but five times in five different localities and not all at the same time. *Homo erectus* evolved in Africa from an earlier hominid species, *Homo habilis*. Part of *erectus* then eventually migrated out of Africa, spreading to various regions of the world and diverging into four geographic races or subspecies, resulting in five races or subspecies once one includes the original African population. One by one each of these five races of *erectus* evolved separately into *Homo sapiens*, with European *erectus* passing through the threshold first and African *erectus* last. As universally recognized today, this is poor evolutionary reasoning (never mind the apparent racism). While *Homo erectus* did spread out of Africa in a number of migrations, beginning possibly as far back as two million years ago, being well established in China a million years ago and in Europe 800,000 years ago, all the evidence accumulated by paleoanthropology points to the conclusion that anatomically modern humans, *Homo sapiens*, evolved in a single locality in Africa roughly 200,000 years ago and spread out of Africa roughly 100,000 years ago (Tattersall 2000). But more importantly, the probability that five races of the same species in five very different geographic localities would evolve separately into the *same* species (let alone at different times), with all of the resulting races perfectly reproductively compatible, is so vanishingly small that the theory is not worth the paper it is written on. In spite of the fact that evolutionary convergence is a widespread and well known phenomenon—eyes, for example, evolved at least 40 times in the animal kingdom—convergence *at the species level*, let alone five times over, is the poorest kind of evolutionary thinking. No professional biologist takes it seriously today.

In short, we are all African under the skin. This brings us to another common misconception about human races, namely, that they divide along color lines. This view, popular among the public, is completely undermined by evolutionary biology. Quite simply, skin color does not correlate with geography. One cannot say that blacks, for example, are from Africa, or that whites are from Europe. This is because dark skin is an *equatorial* trait (not an African one), an adaptation to high UV levels. Those who are genetically lucky to have more melanin (pigmentation) in their skin would be less likely to get skin cancer, thus increasing their chances of survival and reproduction. If a light-skinned population migrated to an area with high UV, natural selection would kick in and over many generations the population would gradually evolve into a dark-skinned one.[1] Light skin, on the other hand, is an adaptation to environments with low UV. This is because UV stimulates in us the production of vitamin D and vitamin D is necessary for proper bone growth. The principle of natural selection would apply just as equally here as well (see Ehrlich 2000, 291). In all of this, there is no higher or lower, no moral superiority or inferiority. Instead, what we have here are traits that evolved adaptively to different environments. Nothing more, nothing less. The same applies to other traits that are typically focused on as racial characteristics, such as hair types, nose types, and body structure types, the short stocky build of Eskimos, for example, being an adaptation for minimizing heat loss, while the tall lean build of Kenyans being an adaptation for promoting heat loss (see Feder and Park 1993, 337).

This brings us to yet another idea about race that is often found in the public but that makes absolutely no sense from a biological point of view, namely, the *one drop of blood* rule. According to this rule, you are black, for example, part of the black race, if you have at least one black ancestor, no matter how far back. One can

---

1   This is apparently what happened with the Lemba tribe in southern Africa. They are as African-looking as can be, and yet they practice many Jewish customs and claim to be descended from a tribe of Jews that left the holy land well over 2,000 years ago. Remarkably, analysis of their DNA has confirmed that they are of Jewish origin (see Hamer 2004, 192–194).

perhaps understand the need for the reverse racism here. It is the same need that promotes the myth that the ancient Greeks did not invent democracy, philosophy, and science, but instead stole their ideas from Africa, particularly Egypt. Just as this is history at its worst,[2] the *one drop of blood* rule is biology at its worst. From a scientific point of view, one does not inherit blood, one inherits genes. Moreover, the genetic inheritance is 50:50, half of one's genes coming from one's mother, the other half from one's father. One's racial identity, then, is going to be a 50:50 mix in cases where one's parents are racially pure but belong to different races (this is to assume, of course, purely for the sake of argument, that the concepts of race and of racial purity make good sense biologically). To be sure, there are what biologists call *dominant* and *recessive* genes (e.g., the gene for height in Mendel's famous pea plants), but as biologists will quickly tell us, *dominant* and *recessive* refer to versions (alleles) of the *same* gene (typically a mutation produces a recessive allele of a gene). It makes no sense to think one gets one's dominant genes from one parent and one's recessive genes from the other. One gets them from both. Even less so does it make sense to think that any one race has a monopoly on dominant genes. But finally, and most ironically of all, if the *one drop of blood* rule were extended back through evolutionary time, then to the shock and horror of those who subscribe to it they would have to conclude that not just they but all humans, all humans everywhere, blacks and whites and everyone else, are of the same race. For once again, we are all Africans under the skin. (And if that doesn't do it, blood transfusions will!)

This leads us to what many consider the ultimate implication of modern evolutionary biology for racial issues, namely, the elimination of the concept of race altogether. This, however, is where the real debate begins.

Let us go back briefly to Darwin. Surprisingly, like most everyone else in his day, Darwin never seriously questioned the concept of race itself. The most he did was to point out that polygenists could not agree on how many human species there are (the number ranged from two to sixty-three), so that whether we talk of human species or of human races the apparent fact is that "they graduate into each other, and . . . it is hardly possible to discover clear distinctive characters between them" (Darwin 1871 I, 226). This did not stop Darwin from talking about different human races, however, such as the Hottentot, the Fuegian, and the Esquimaux races.[3]

---

2 See Snell (1960), Cromer (1993), Johnson (2003), but especially Lefkowitz (1997).

3 One might naturally suppose that Darwin thought that human races are not objectively real but instead are arbitrary mental constructs, since in the *Origin* (1859) he defined *species* and *variety* as arbitrary concepts made for the sake of convenience (52). This is extremely odd, of course, given his central claim that varieties are what he called *incipient species* and given the full title of his book: *On the Origin of Species by Means of Natural Selection*. In spite of this title, and other evidence to the contrary, almost everyone has taken Darwin at his word, as believing that species and varieties are not objectively real. I have elsewhere argued in great detail (Stamos 2007) that this received view is seriously mistaken, that Darwin did actually think that species and varieties are objectively real and that he routinely applied his own evolutionary species concept to animals and plants. If my analysis is correct, then it might well be that Darwin also thought that what are called *races* are in many cases objectively real too, including human races. This is not to say, of course, that Darwin was right.

Whatever the history of the concept of race, and especially given the prominence of race in our popular culture (racial pride, racism, and the various civil rights movements), the idea that there really are human races certainly needs to be questioned and examined carefully. What is interesting is that within the field of evolutionary biology in recent decades it has not fared well at all.

Without question the now classic—in the sense of the most cited and most approved—argument from evolutionary biology against the objective reality of human races was provided by the paleontologist Stephen Jay Gould in an article titled, "Why We Should Not Name Human Races—A Biological View" (1974b). Gould's argument is based on acceptance of the so-called *biological species concept* and on what is known as *multivariate analysis*. The biological species concept is a popular species concept among zoologists and was forcefully developed and long defended since the early 1940s by the ornithologist Ernst Mayr (e.g., Mayr 1970), regarded by many as the greatest evolutionary biologist of the 20th century. According to this species concept (oversimplified here), two populations belong to the same species if they are capable of interbreeding. In line with this species concept, humans and chimpanzees, for example, are two species rather than one because they are genetically incapable of producing offspring. Chimpanzees constitute a single species because all populations of chimpanzees are fertile with each other. Similarly, humans constitute a single species because all populations of humans are fertile with each other. Or as many people like to put it, we are a single race.

This feature of reproductive compatibility between all human populations need not, however, preclude the concept of human races (plural), since if geographically separate populations each breed from within (*endogamy*) for enough generations they could then evolve distinguishable features from other human populations (potential interbreeding is not actual interbreeding, after all). And, of course, ever since our species spread out from Africa roughly 100,000 years ago via the Middle East, it has migrated and settled in virtually all parts of the world, reaching Australia, for example, at least 40,000 years ago and North America roughly 15,000 years ago (Feder and Park 1993, 301). Moreover, cultural separation has often served to keep human populations apart when geographic separation has not. Not surprisingly, then, many biologists have argued that the concept of different races does indeed apply to the human species.

For example, this was the consensus arrived at in a symposium held in 1966 by the American Association for the Advancement of Science, a symposium dedicated to provide "an inventory of what science has to say about race" so as to "help to dispel the evil myths that persist about race" (Light 1968, vii, viii), in particular "pseudoscientific statements which . . . have attempted to prove the innate biological inferiority of the group of Americans who are socially classified as Negro" (Mead 1968a, 3). As Mayr (1968) put the consensus, "Quite rightly all the speakers said that if you define races properly then, yes, there are races" (103). For Mayr and many others, that meant geographic races, local populations that are describable and distinguishable in statistical terms. Indeed, for Mayr, "when you look at animals (and a botanist finds the same thing with plants), there is hardly a species that does not also have geographic races." The problem is when to group local populations into races. As Mayr points out, in one anthropology textbook one finds that there are five human races, in another that there are sixty-five human races. For Mayr, "how to draw the line between them is not only difficult, it is impossible" (103). However, if the idea that there is at present an objective number of human races is erroneous, there is also, according to Mayr, "the equally erroneous extreme of thinking of the total identity of everybody" (105).

Mayr's view seems to be, then, that the reality for most species (really, for all but extremely small and local species—*endemic* species) is geographic variation and that the term used to classify this variation, *race* (often also *subspecies* in zoology, *variety* in botany), is vague and flexible but necessary. This view was (and remains, as we shall see later) hardly idiosyncratic. Dwight Ingle (1968), for example, another participant in the symposium, pointed out that the concept of cancer is equally difficult to define and yet it would be ridiculous to say that cancer "does not exist and should not be studied" (114). Theodosius Dobzhansky (1968a) added that if the definition of *race* were so clear as to permit "exact, nonoverlapping, discrete entities," then "we would not have races, we would have distinct species" (165). Yet for Dobzhansky, one of the key architects of the Modern Synthesis, "To deny the existence of racial differences within the human species is futile" (166). Indeed, for Dobzhansky (1968b), "If man has no races . . . why then are the inhabitants of different countries often recognizably different?" (78).

Remarkably, only a few years earlier, in an anthology edited by the British physical anthropologist Ashley Montagu (1964), Montagu and nine others concluded that the concept of race, including that of human races, is illegitimate from a scientific point of view, that the concept is a myth. As Montagu put it, the "unanimity of the views independently arrived at" (xviii) was that "Race is the phlogiston of our time" (xii).[4] For Montagu and his contributors, there are no "different populations of the same species which are distinguished one from another by the possession of certain distinctive hereditary traits" (xi), no "discrete unit characterized by a particular complex of physical attributes" (xviii). Such "typological thinking" is mistaken. The reality, instead, is continuous variation within a species, such that "whatever populations are considered they are always found to grade gradually into or incline toward others" (xvii).

What is interesting when comparing these two anthologies is that both represent a mixed bag of academics (Mead's anthology includes nine biologists, six anthropologists, and four psychologists, while Montagu's anthology includes seven anthropologists and three biologists), both are against racism and any idea of racial superiority or inferiority, both accept evolution and the biological species concept, and yet both come to diametrically opposite conclusions on the concept of race, one concluding that it is a legitimate biological concept, the other that it is a myth, a social construction. We cannot attribute this difference to the national backgrounds

---

4   In the 18th century, *phlogiston* was the name given to what was commonly thought to be the substance emitted when something was heated or burned, a theory abandoned by the late 1790s based mainly on the work of Lavoisier.

of the contributors, since for both anthologies they were mainly American. Of course, the period in which both anthologies appeared is surely significant. It was the highly turbulent 1960s, the decade in which the civil rights movements reached their peak, and as many outside of science have argued, scientists are not as immune to social, political, and cultural influences as they generally like to think. But I do not believe that is the fundamental reason for the profound difference between the two anthologies. Instead, I believe the difference trades on a longstanding ambiguity lurking within the concept of race. If one focuses on the concept of race as a primarily geographic concept, in some degree that of an incipient species, as those in the Mead anthology did, then one gets the view that races are in some sense real, whereas if one focuses on the concept of race as a primarily character-trait concept, as those in the Montagu anthology did, then one gets the view that races are arbitrary and hence not real.

This brings us back to Gould's article (1974b). Gould's second and main argument for why we should not name human races is based on character traits. As Gould and all biologists clearly recognize, wide-ranging species vary geographically in their traits. When mapping these traits, however, serious incongruities often appear. For a given species, a map based on one trait will quite often not match a map based on another trait. Feder and Park (1993, 334–335), for example, provide two maps, one for the distribution of humans with type A blood, the other for the distribution of human skin color. The two maps are highly discordant, no doubt because human blood types have nothing to do with UV and vitamin D (instead they have a rather complicated and not yet fully understood connection with disease resistance). Paul Ehrlich (2000, 50–51) provides four maps, for skin color, hair structure, average height, and average head shape, and all four are highly discordant. The ultimate point here, then, is that a race division, no matter how statistical, based on any one of these character traits would not converge with a race concept based on any one of the other character traits. But more than that, when one simultaneously superimposes *many* traits of a species onto a single map—a procedure known as *multivariate analysis*, possible only with the advent of electronic computers—the situation becomes only worse for a division into races. As Gould puts it, then, "Geographic variability, not race, is self-evident" (232). Indeed, for Gould, multivariate analysis sounds the death-knell of the biological concept of race.

That was in 1974, in a decade, compared to the 1960s, that saw a rapid rise in the development and use of electronic computers. Today, many more biologists would agree with Gould's assessment. Paul Ehrlich (2000), for example, himself an ecologist, employs basically the same reasons as Gould and concludes that "As is the case with other species, geographic variation in human beings does not allow *Homo sapiens* to be divided into natural evolutionary units" (291). Instead, for Ehrlich, the concept of race is merely "a basis for social stratification, dominance of one group over another," such that "Sadly, many scientists in such stratified societies have reinforced popular theories of purported racial superiority to support the views of those in the dominant culture" (292–293). For many of the same reasons, most anthropologists today would agree with Ehrlich and with Gould before him. According to the 1998 American Anthropological Association Statement on "Race" (reprinted in Jackson and Weidman 2004), racial divisions are "both arbitrary and subjective" and "physical variations in the human species have no meaning except the social ones that humans put on them" (367).

One problem with statements such as these is that any scientist who still wants to argue for the validity of the concept of race in biology, especially in human biology— even if he does not want to argue for any kind of racial superiority—is going to have whatever arguments he makes dismissed out of hand—before he even makes them— simply because of a supposed ulterior social motive, whether that motive is thought to be conscious or not. This temptation to *poison the well* has got to be resisted. The evidence against the existence of human races is admittedly powerful, and arguably it is morally praiseworthy, but its moral worth is irrelevant to what is at bottom a scientific and conceptual issue. What has to be examined is whether something has been overlooked in the rejection of the concept of race.

What is interesting, for a start, is that those who argue that human races do not objectively exist continue to refer in their arguments to human groups that are commonly thought of as races. Dobzhansky (1968b) noticed this years ago and commented that "I find it amusing that those who questioned the validity of racial classifications have themselves used the word 'race,' or the term 'so-called race,' many times" (166). The situation is actually worse, since they continue to identify *specific* races. For example, Washburn (1964) in the Montagu anthology, when arguing against the division of the human species into three races, raises "the problem of the aboriginal Australian" (246). Similarly, when discussing a study on human blood types, Gould (1974b) refers to "Australian Aborigines" (236). Similarly again, when discussing human intelligence tests, Ehrlich (2000) refers to "an average Australian Aborigine" (295). This is extremely odd. No sooner do they kick human races out the front door, they let them in again through the back door. It seems that even those who argue that human races do not exist cannot help talking about human races as if they do exist. As Dobzhansky (1968a) put it, "Indeed, how else could they speak about human variation at all!" (166). The problem becomes even more acute when we notice, keeping our focus on Australian aborigines, that given their evolutionary history they are a good example of what Dobzhansky and Mayr refer to as *geographic races* (as with other fairly isolated groups, they eventually would have evolved into a different species had their isolation been extended far enough into the future). Perhaps, then, as Dobzhansky and Mayr maintain, character distribution cannot be meaningfully communicated without the use of geographic group labels. This might indicate something only about scientific communication (that there resides a constant need to refer to entities that do not exist, as some claim is true of other concepts, e.g., species and niche).[5] Or perhaps anthropologists and biologists such as Washburn, Gould, and Ehrlich are missing something deeper that other biologists such as Mayr and Dobzhansky glimpsed but only partially grasped.

---

5 Perhaps this is what E. O. Wilson (1978) had in mind when he wrote, "Most biologists and anthropologists use the expression 'racial' only loosely, and they mean to imply nothing more than the observation that certain traits, such as average height or skin color, vary genetically from one locality to another . . . most scientists have long recognized that it is a futile exercise to try to define discrete human races. Such entities do not in fact exist" (48).

This leads us to two recent attempts to revive the concept of biological race. One involves what is known as *cladistic taxonomy*. This is a system of taxonomy that has taken over biological classification in recent decades. Here, division of groups is not based simply on shared character traits (that approach is not necessarily evolutionary). Nor does it involve traditional evolutionary taxonomy, which combines phylogeny (evolutionary lines or branches) with a degree of similarity and dissimilarity. On this traditional view, birds, for example, constitute a class of their own, on a par with mammals and reptiles, because of their degree of evolutionary change. According to cladism, however, classification is based solely on phylogeny (*clade* is the Greek word for branch), specifically on branching points. The criterion of similarity (for grouping) is here dismissed as not objective. Thus birds, according to cladism, are actually dinosaurs, because birds evolved from one or a few of the branches within the dinosaur clade (e.g., Gaffney *et al.* 1995).

Robin Andreasen (2004) provides an elucidation and defense of the cladistic race concept. According to this concept, there are human races in an objective, cladistic sense of the term, because human evolution has largely conformed to the standard picture of branching evolution. Accordingly each branching point in a breeding population represents the beginning of a cladistic race, defined by Andreasen as "when a species splits into two or more breeding populations that experience different evolutionary forces under a significant degree of reproductive isolation" (431). The first major branching point in human evolution occurred roughly 100,000 years ago when an undifferentiated population of modern humans migrated out of East Africa through the Middle East (426). *Homo sapiens* continued to branch as populations of humans migrated to different parts of the world. The result was five major cladistic races: Africans, Caucasians, North-East Asians, South-East Asians and Pacific Islanders (including New Guineans and Australians), and Native Americans (437), with each of these major cladistic races being composed of smaller cladistic races. Although the reconstruction of human phylogeny is still a work in progress, the cladistic evidence for the details of the branching history comes mainly from genetic analysis, such as the work of the geneticist Luigi Cavalli-Sforza (e.g., 1991), although that evidence is supplemented by other kinds of evidence, namely, evidence from biogeography, archaeology, paleontology, and, interestingly, historical linguistics. In all of this, any discordance in traits between cladistic races is entirely irrelevant, as a cladistic race is not defined by a suite of character traits but exclusively by a branching point in the history of a breeding population. As such, a clade can undergo numerous changes in character traits and yet remain numerically the same clade.

An entirely different issue is that of gene flow between the clades, in this case the cladistic races, but I want to hold off on that for a moment, for there is a preliminary problem. While cladism has dominated professional biological taxonomy in recent decades, it has only dominated the classification of species into higher categories; it has not dominated the classification of organisms into species. Some biologists have indeed attempted to produce a *cladistic species concept*, but the results have been so counterintuitive that most biologists have rejected it (see Stamos 2003, 256–269). For example, according to cladism if a species is a clade then a species that buds off a new species automatically goes extinct at the branching point and two new species are born, even if the parental species was in a state of stasis throughout the process (this is because similarity is not, for cladists, part of the ontology of a clade). Also, a species that undergoes "infinite evolution," as one of its main advocates puts it, remains one and the same species throughout the entire time period, simply because it did not undergo any branching.

In all of this we need not concern ourselves with the methodology of cladistics (or rather methodologies, since there are related but competing methodologies), with how cladists arrive at *cladograms* (branching tree diagrams). Instead, we need only concern ourselves with whether it can be maintained that there are clades within a species, in particular within the human species. Massimo Pigliucci and Jonathan Kaplan (2003), for example, do not raise the preliminary problems I raised above. Instead, they maintain that, given human migration patterns and human interbreeding, especially over the past few millennia, there has been too much gene flow between human populations to allow for the application of a cladistic race concept, that human evolution has been mainly reticulate (netlike) rather than cladistic (1163). Indeed, they doubt that there ever were cladistic human races. Andreasen would have to maintain, in reply, that although there has been gene flow, there has only been enough to blur human cladistic races, not to eliminate them entirely (blurred lines, after all, are still lines). Interestingly, Andreasen (2004) suggests that cladistic human races "may be on their way out" (431). Nevertheless, she still believes that human evolution has thus far been more cladistic than reticulate (I suggest the term *cladistic inertia* to capture her idea), so much so that we can still talk meaningfully about cladistic human races.

This is a topic for further debate and I shall leave it at that. What is interesting is that Pigliucci and Kaplan (2003) provide a very different kind of race concept, one that is based on ecological niches rather than on clades, so that it is not affected by their gene flow argument. Interestingly, they begin their discussion by pointing out that, for a concept that is supposedly no longer in use in biology, the term *race* has shown up quite a lot in recent years in the biological literature on nonhuman species. Focusing on actual *use* in that literature, then, they argue that a biological race is best understood as an *ecotype*, an ecological type within a species. Ecotypes have five basic features. First, they have a necessary connection between genetic differentiation and ecological adaptation. Second, they are not necessarily individual branches of evolution within a species but rather "functional-ecological entities" (1163). What this means is that the same ecotype can possibly be found in distant geographic locations. Third, there can be gene flow (actual interbreeding), even significant gene flow, between ecotypes within a species without destruction of those ecotypes, just so long as the ecological selective pressures are greater than the opposing gene flow. Fourth, different ecotypes within a species can have intermediate populations; indeed, an individual can belong to more than one ecotype. Fifth and finally, ecotypes can be differentiated based on many or on few ecological/genetic differences.

This concept of ecotype, Pigliucci and Kaplan point out, is routinely overlooked in the literature against the concept of race in humans. Typically the concept used is geographic, in the sense of a bud in the tree of evolution, in the sense of an incipient species. But given the history of human migration and interbreeding, it is true that there are no human races in that sense today (although there might have been in our remote evolutionary past). Equally there are no human races in the sense of groups of humans delimited by discrete groups of character traits. But for Pigliucci and Kaplan all of this is irrelevant to the existence of human races in the sense of ecotypes. Also irrelevant are public or folk conceptions of racial divisions. The division into blacks and whites, for example, does not correspond to human ecotypes, or any other color division. In fact, there are far more human ecotypes than folk racial divisions. So one cannot say that blacks, to give a specific example, are better runners. But one can say, to use their example, that the region of Kenya contains a human ecotype involving a high proportion of great marathon runners. Hence for Pigliucci and Kaplan (2003), "on an ecotypic conception of race, there would in fact be 'races'—and indeed, races associated with athletic ability" (1167).

The concept of ecotype, unfortunately, suffers from many of the difficulties that plague the related concept of *ecological niche*, which is vague and suffers from different definitions (e.g., whether a niche is defined in terms of one or more environmental features). The concept of niche becomes only more problematic when used to define the concept of *species* (see Stamos 2003, 143–165). For many of the same reasons, the concept of ecotype is going to be problematic when used to define the concept of *race*. And yet, along with clade, the concept of ecotype is not the kind of concept that is likely to disappear from the biological lexicon. Moreover, both concepts keep the concept of race alive in different but interesting ways. This is something to keep in mind when we turn to our next question, that of race and IQ. Obviously a large part of the problem with the literature on this topic is the concept of race that lurks behind the arguments. But keeping in mind the possibility that not all concepts of race, including those of human races, are biologically illegitimate, this gives us reason to take at least a fairly brief look at some of the arguments and counter-arguments on race and IQ from an evolutionary point of view.

I shall here look at two fairly recent and highly controversial works on human race and IQ that actually feed off of each other. The first is *The Bell Curve* (1994) by Richard Hernnstein, a Harvard psychologist, and Charles Murray, a political scientist at The American Enterprise Institute, a conservative thinktank in Washington DC. The second is *Race, Evolution, and Behavior* (1995, 2000) by Philippe Rushton, a Canadian psychologist. In *The Bell Curve*, Hernnstein and Murray bring together a lot of evidence, most significantly from twin studies, showing that IQ is largely heritable, and they also bring together a lot of evidence showing that IQ tests are correlated with race, tests which indicate that, on average, Asians (Mongoloids) score at around 106, whites (Caucasoids) score at around 100, and blacks (Negroids) score at around 85. Because race is inherited, they conclude from the above statistics that racial differences in IQ are largely inherited. But their ultimate conclusion involves social consequences. Part of their argument is that women with low IQ tend to have more illegitimate children than women with high IQ, which spells dire consequences for welfare programs. Another part of their argument is that whites do better than blacks in America because generally whites are smarter than blacks, so that social programs such as affirmative action will end up causing more harm in society than good. Rushton's *Race, Evolution, and Behavior* also involves the same basic approach to race and IQ, but he extends the analysis to other propensities and behaviors such as the propensity for violence and aggression. But what mainly distinguishes Rushton's book is his evolutionary explanation for the differences between "the three races." I shall return to that below.

The use of tests to measure IQ is not, of course, new. Nor are any problems they might have. One is that we should expect, from an evolutionary point of view, that there evolved many different kinds of intelligence, such that the concept of overall intelligence might not have any meaning (see Ehrlich 2000, 295). Another is that IQ tests themselves are culturally relative, so that people who belong to the same culture as those who made the tests will tend to score better than those from other cultures (*ibid.*). Another problem is that environment plays an enormous role in IQ scores, not only in terms of level of education but also in terms of socioeconomic status. Hamer (Hamer and Copeland 1998, 225) cites a study which found that 99 black children adopted from poor black families in Minneapolis into white middle-class families ended up with an average IQ score that was not only higher than the average black IQ but higher than the average white IQ. The researchers concluded that environment added roughly 16 IQ points. Hamer cites another study conducted at Stanford University where some of the students were told beforehand that they were taking an intelligence test while others were told that they were taking a problem-solving test that had nothing to do with intelligence. With white students, it made no difference to their IQ scores what type of test they thought they were taking. With black students, however, those who thought they were taking an intelligence test scored more than 25% lower than their fellow black students who did not think they were taking an intelligence test. Jackson and Weidman (2004, 229) cite the work of the Nigerian anthropologist John Ogbu, who studied the role of class and oppression and found, for example, that Koreans in Japan, where they are a subordinate minority, suffered significantly lower test scores, even though both are Asians. The same goes for North American Indians, who are closely related genetically to Asians and yet score significantly lower than North American whites (230).

On the other hand, Hamer (Hamer and Copeland 1998, 227–228) cites studies on several hundred pairs of twins which compared the IQ test scores of identical twins with fraternal twins over their lifetimes and which found that the genetic influence increased as they got older, such that by age 15 the influence of genes had increased from almost nothing to roughly 50% and had increased to as much as 80% by old age. The explanation for this, as Hamer points out, is that our genes are not all switched on at the same time but are switched on and off in a programmed manner throughout our lifetimes.[6] Thus, when it comes to genes and IQ, as Hamer puts it, "Environment is most important early on, while genes become most important as we mature" (227).

---

6   For the same reason many genetic diseases do not kick in until a certain age, such as Huntington's chorea, a lethal neuromuscular disease caused by a single dominant gene which turns on in middle age in those who possess it, such as the late singer Woody Guthrie (see Medawar and Medawar 1983, 145–146).

This is of obvious consequence for public education policies, but it also brings us back to the issue of race and IQ. It is possible that IQ tests, conducted especially on adults, reflect racial differences. According to Rushton (2000), "Hundreds of studies on millions of people" (P14) show the basic three-way pattern in IQ scores, with sub-Saharan Africans averaging around 70 (P15). For Rushton and others, environmental arguments do not negate these well-corroborated averages.

Nevertheless, returning to *The Bell Curve*, the mistake Hernnstein and Murray make is not so much their reliance on IQ tests, or even their division of *Homo sapiens* into three basic races—for arguably, these are broadly cladistic *and* broadly ecotypic. Rather, as many critics have pointed out, their fundamental mistake involves their use of the concept of *heritability*. The anthropologist Alexander Alland, in his book *Race in Mind* (2002), devotes an entire chapter to *The Bell Curve* and states that the book's "major flaw," its "key error," is in its use of the concept of heritability. As Alland puts it, "this measure tells us what contribution genetics makes to the observed variation of a trait *only* in a *specific* population living in a *specific* environment" (149). The problem with *The Bell Curve*, then, is that "heritability figures for one population can *never* be correctly applied to another population. Environmental effects may be responsible for *all* observed difference between populations even when the trait in question has the same genetic component in these same populations" (150). Clearly when it comes to blacks and whites in America, even though they live in the same country, they do not all (statistically) live in the same environment. The same point about heritability, elaborated by Hamer (Hamer and Copeland 1998), is that *The Bell Curve* "confuses individual differences, which is what twin studies and other genetic methods measure, and group differences, which can't be addressed by these methods" (224). Hamer gets the point across by getting us to imagine a bag of sunflower seeds where half the bag is planted in a sunny garden and regularly watered and the other half of the bag is planted in a shady garden and only infrequently watered. If the average height of the plants in the two gardens were compared, a clear cause of the difference would be the difference in the environments, but it would be impossible to say if any of the difference in height was due to genetic differences in the seeds. This would follow even though the height of sunflower seeds is, of course, partly genetic. In other words, you cannot do anything like twin studies on races, taking two populations of the same race and raising them in different environments to test their IQ scores. What works for individuals cannot work for races.

All of this cannot, however, be used to deny that the difference in average IQ scores between blacks and whites in America is partly genetic. It still might be. It might also be that blacks in America have a higher average genetic IQ than whites in America. But is that the end of it? Is there nothing else that weighs in on the question of race and IQ?

This is where Philippe Rushton's work comes in. I said earlier that Rushton's book and *The Bell Curve* feed off of each other. In *Race, Evolution, and Behavior*, Rushton often cites an earlier work by Herrnstein, agreeing with its data and conclusions. And in the Preface to the third edition of Rushton's book (2000), which serves as an abstract and an update, he does the same for *The Bell Curve*. In *The Bell Curve*, on the other hand, Herrnstein and Murray favorably cite earlier publications by Rushton in which he applies r- and K-selection theory to explain racial differences in IQ as well as other differences such as athletic ability, promiscuity, and crime rates. Briefly, r-selection and K-selection are the extremes of a continuum, where r-selection is a reproductive strategy characterized by large numbers and little if any parental care (e.g., oysters lay millions of eggs in a year but invest no parental care), while K-selection is a reproductive strategy characterized by low numbers and very high parental care. Humans clearly are a K-selection species, but according to Rushton the three basic human races did not involve the same degree of K-selection. Negroids, the oldest human race, evolved in Africa, where the warm climate, geographies with open savannas, lots of different kinds of predators and competitor tribes, and high rates of viral and bacterial diseases called for higher reproductive rates with lower levels of parental care, along with higher levels of sex hormones such as testosterone (which increases promiscuity, strength, and aggression). As human populations migrated out of Africa into the Middle East and Europe beginning roughly 100,000 years ago, they encountered cooler climates with geographies that offered less food, all of which made it harder to raise children, so they had to evolve larger, more intelligent brains for problem solving and had to evolve greater family stability, which meant less promiscuity and more parental care, which in turn meant differences in physical features such as lower testosterone (because of the need for less promiscuity and aggression) and wider hips in women (to accommodate the bigger brains). Populations of Caucasoids migrated to harsher climates and geographies still, beginning roughly 40,000 years ago, resulting in even greater K-selection and further development of the features above, resulting in the Mongoloids.

As Rushton (2000) is quick to point out, "these three-way differences are *averages*. The full range of behaviors, good and bad, is found in every race" (P11). This is the sort of thing one should expect for an evolutionary argument, since biology is, after all, statistical. What is interesting is the kind of replies Rushton's evolutionary argument has evoked. The biologist Joseph Graves, for example, in his book *The Race Myth* (2004), argues that Rushton's argument using r- and K-selection is "hopelessly compromised" (138), since it can easily be turned around against Rushton's conclusions. For example, he says that there does not have to be a tradeoff between greater intelligence and greater athletic ability. If the latter involves greater circulatory capacity, then that will also aid greater blood flow to the brain. (Of course, blood flow is one thing, brain circuitry and hormones quite another.) But his main criticism is that

> professional biologists now consider r- and K-selection theory as virtually useless. Biologists began to expose the fallacies in this concept in the late 1970s. Since that time, multiple experiments have failed to corroborate the core premises of r- and K-selection theory.
>
> (175)

The problem with this claim, however, is that it will come as a total surprise to many professional biologists. For example, in an anthology devoted to concepts in ecology

(Cherrett 1989) published over a decade after Graves's supposed demise of *r*- and *K*-selection theory, the theory continued to be applied to many examples in nature (e.g., 216, 296–297). Futuyma (1998) summarizes the situation best, pointing out that while *r*- and *K*-selection theory has its critics who have urged its abandonment, "many organisms do fit the predictions fairly well." His conclusion is that generally more than merely *r*- and *K*-selection theory is needed in order to "capture the full variety or causes of life histories" (573). The upshot of all of this is that while it is too strong a claim to say that *r*- and *K*-selection theory is dead, its application to human evolutionary history is at best controversial.

Even more serious is the main criticism of Rushton by Alland (2002), who devotes a chapter to Rushton in his book. According to Alland, Rushton and all the scientists who provide quoted support for his book in the inside cover of the third edition—his "coterie of admirers" (161), which includes famous race researchers such as Arthur Jensen, Hans Eysenck, and Charles Murray—received generous research grants from the Pioneer Fund. This organization, Alland is quick to point out, is "noted for its support of racist research," it had in its earliest years "praised aspects of Nazi Germany's racial policies" (8), and he quotes a newspaper journalist who reports that its original charter was devoted to "race betterment" in the form of financial support to needy white students in the original 13 states (9). Alland also points out that one of the supporters cited inside the front cover of Rushton's book, Glayde Whitney of Florida State University, "wrote the introduction to the autobiography of David Duke, former national director of the Ku Klux Klan" (161).

Whether these allegations are true or not, it does not really matter, since, from a logical point of view, they commit at least three fallacies when used, as they are by Alland, to reject an *argument*, in this case Rushton's evolutionary argument for racial differences. The first is the *genetic fallacy*, rejecting an argument, position, or institution because of its negative history. Chemistry evolved from alchemy, but that does not make chemistry a pseudoscience too. The next is *guilt by association*. Even if the Pioneer Fund is a racist organization (you can compare what Alland says with their website), the fact that it funded Rushton's work, whether in whole or in part, does not automatically make it, or him, racist too. Finally, there is *circumstantial ad hominem*. A person's motive in making an argument, whether that motive is conscious or not, has no bearing whatsoever on the logical worth of the argument. An argument is an autonomous entity, and it needs to be analyzed accordingly, in terms of whether the premises are true, relevant, and sufficient. The motive of the person who provided the argument is totally irrelevant. All of these points are readily available in books on what is known as *informal logic*, and students will readily agree with them when they examine them in theory and see them applied to mundane examples, but it all has a habit of going out the window in the context of hot topics such as race (others are feminism and religion). The ultimate point is that a person's motive, his associations, and the historical background to his argument might help to explain *why* a person puts forward the argument that he does, but they do nothing to help *assess* whether the argument itself has merit. In order to do that, one has to confine one's attention to the argument itself. (And if one is not willing to do that, then one cannot justifiably reject the argument.)

Indeed, the problem goes both ways. In other words, Alland's kind of attack can be used *just as easily* and *just as illegitimately* against those who *reject* arguments for racial differences in IQ and other traits. Rushton (2000), for example, says that "These attempts to deny race differences amount to a new form of creationism" (P27). (Can you identify the fallacy here? Clue: *The Wizard of Oz*.) One cannot possibly hope to produce any light with such tactics, only heat. But let me provide two better examples. They are instructive not because they involve highly loaded terms such as "creationism" but because they involve further issues and arguments about evolution and race that we have not yet dealt with but which are common enough and need to be addressed.

My first example concerns Stephen Jay Gould. In his article titled "Racist Arguments and IQ" (1974c), Gould does not deny that intelligence, however defined, has a genetic component. He does not hold that intelligence is purely a matter of the environment. No biologist would. But Gould thinks the genetic basis of intelligence is "trivially true, uninteresting, and unimportant" (246–247). This is odd. Given his claim in his article against naming human races (1974b) that geographic variation is a fact, why would geographic variation in West Indian land snails be interesting and important but not the question of geographic variation in heritable traits in humans such as IQ and aggression? I can only suggest that the motive or cause that Gould (1974c) throws at others interested in racial differences really applies to his own lack of interest in racial differences, namely, "The answer must be social and political" (247). Gould was a Marxist, dedicated like all good Marxists to the ideology of the plasticity of human nature. Hence he moves from his critique of craniometry (skull shape) and later the IQ tests as developed by Arthur Jensen, which were founded on the suspicious work of Sir Cyril Burt, to the implied claim that there can be "no valid data" (247) on the matter whatsoever. This claim does not at all follow. But it is not because he was a Marxist, even though he was.

The Harvard geneticist Richard Lewontin, a pioneer in the study of genetic variation within populations, provides an even better example. In the early 1970s, Lewontin (1972) argued that there is greater genetic variation within human populations than between them, no matter where in the world those populations are from. He concluded by saying,

> It is clear that our perception of relatively large differences between human races and subgroups, as compared to the variation within these groups, is indeed a biased perception and that, based on randomly chosen genetic differences, human races and populations are remarkably similar to each other.
>
> (397)

Later, Lewontin (1982) argued that if some catastrophe would kill humans from all over the world leaving only native Africans alive, the human species would retain roughly 93% of its total genetic variation (123). All of this has been used by Lewontin and others against arguments for racial differences in traits such as IQ, including Alland (2002, 162–163), Graves (2004, 9–11), and the authors of the 1998 American Anthropological Association Statement on "Race" (Jackson and Weidman 2004, 366–369). What it seriously ignores, however, are two basic facts about genes. First, all it takes is a single base change in a gene, a single change in a DNA letter, to make a significant change in the phenotypic expression of the gene. Hence a little genetic variation can go a long way. Second, not all genes are on a par. Most genes are *structural* genes, but some genes are *master* or *regulatory* genes, turning structural genes on and off during development. A small variation in a master or regulatory gene, then, can go an even longer way in making a difference between two phenotypes. Hence there should be nothing at all surprising in the often-heard claim that humans and chimpanzees are 98 or 99% genetically similar (see Ehrlich 2000, 354 n. 36). What this means for genetic differences between human races should be obvious. We can say that Lewontin never brings out these two fundamental points about genes because, like Gould, he is a Marxist, motivated by the Marxist ideology of the plasticity of human nature. (I suspect the others mentioned above are motivated by political correctness.) But this has got to be separated from the worth of his argument against racial differences. As it turns out, what he says about genetic variation in human populations is actually perfectly consistent with claims about racial differences such as Rushton's. The racial differences can reside in relatively few key differences in structural and in master or regulatory genes, some of them the result of genetic drift, but more importantly some of them favored by natural selection. The *averages* cited by Lewontin are just plain irrelevant, since they are only averages in genetic variation and nothing more; as such, they cannot help but obscure the above point. But it is not because Lewontin is a Marxist that his argument against racial differences is a bad one.

At this point my own motives have probably come under suspicion, but I really do not care. My only desire has been to argue that, from an evolutionary point of view, there is nothing inherently mistaken or wrongheaded, let alone evil, in supposing that there are racial (geographic, cladistic, or ecotypic) differences in IQ or in other character traits within wide-ranging species such as *Homo sapiens*. Any aversion to research in this area is basically socially and politically motivated, it is not biologically motivated. At the end of the day, when all is said and done, it remains *possible*—indeed, *quite* possible—that from a modern evolutionary point of view there are innate statistical differences, even significant differences, in aptitude and behavior between different human races. The problem is making one's way through the thousands upon thousands of pages of conflicting and contradictory research— something personally I really do not care to do. At the same time, there is the further problem, and it needs to be emphasized, that whatever the facts, they must contend with the enormous size and power of culture to bring out, or conversely to suppress, human potential. That potential is what just-minded people, along with our political and social institutions, need mostly to focus on. Equal opportunity combined with a positive environment is the moral imperative. But we should not let the value of this imperative fool us into believing that biology makes us equal. Like it or not, biology just does not work that way. We need to be realistic, and to remember that we are always dealing with statistical phenomena. At any rate, for my own part, I prefer to rest content with the wisdom of Martin Luther King, Jr., who said it is not the color of a man's skin that matters but the content of his character. Truer words were never spoken, possibly even if repeated from an evolutionary point of view.

That would bring us to our next chapter, on evolution and ethics. But before we get to that, I want to say something about whether there could have evolved in humans a racism instinct.

When we think of racism, we think of a continuum, ranging from racial jokes at one end to genocide at the other. The continuum is a descent into horror. Racial jokes, if not checked along the way, easily lead to scapegoating, which easily leads to hate, then to dehumanization (often involving slavery or war, and always words such as "subhuman" or "vermin"), ending finally in genocide. Racism has a twisted logic, but it seems a human universal: every culture studied seems to have it.[7] And it is not confined to the majority group in a particular society, or between societies. Minorities within societies amply display racism as well, and not only toward the majority but toward other minorities, a feature of human nature well brought out in the movie *Crash*.

Is this feature of human nature cultural or evolutionary/genetic? By now one should recognize this as a *false dichotomy*. There can be no doubt that culture plays a role, given that racism can be whipped up into a frenzy, taught in schools, and conversely proscribed. But does also evolution via genes play a role?

Surely, given the problems with the concept of race, as we have seen earlier in this chapter, it is highly unlikely that racism evolved in humans in the sense that we evolved an instinct to detect different human races and to be predisposed against them. Indeed, what is often called "racism" is evidently not racism at all, but more accurately *groupism, culturism*, or *differencism*, since the groups identified as races cannot possibly be races in any biologically meaningful sense of the term (e.g., Croats and Serbs, or Muslims). At any rate, how might evolution help explain, at least in part, the universal phenomenon of racism, real or so-called?

I suggest that in order to answer this question we need to analyze the various components of racism. Some parts might be adaptive, having been produced by natural selection, some parts might be mere byproducts of adaptations, and some parts, of course, might simply be cultural.

One obvious component of racism is *stereotyping*. Racism involves stereotyping to a high degree, beginning right at the start with racial jokes and continuing all the way to genocide. One often hears, of course, that it is wrong to stereotype, but one needs to think about this from an evolutionary point of view. Throughout our evolutionary past, and throughout that of every other animal species with mind, stereotyping would often lead to survival and hence reproduction. If a hominid failed to stereotype tigers, for instance, it may well have ended up a tiger's next meal. Those who stereotyped tended to survive and reproduce, those who did not stereotype or who did less of it tended to pass on their genes with far less frequency. We are all here today, in no small part, because our hominid and earlier animal ancestors routinely stereotyped.

---

7 Even genocide, defined as the attempt to extinguish a race or kind in one or more of a variety of ways—*genos* is the ancient Greek word for race or kind (Urmson 1990, 65–66)—would seem a human universal, given the many famous examples of it (such as in Germany, Rwanda, Yugoslavia, Armenia, and North America) and the far more numerous less famous examples of it studied by anthropologists, and given also the universality of the attitudes that lead to it, identified as xenophobia and ethnocentrism by Ghiglieri (2000, 211).

Of course, following Alcock (2001, 218–219), as we have seen in Chapter 4, an instinct can function maladaptively in an environment for which it was not evolved. Quite arguably that is what we see today when we stereotype each other racially. We did not evolve in cities, with minorities and multiculturalism. All it generally takes is to get to know someone from another race to see that our stereotyping of them was misplaced. Whatever our racial differences, they are minor compared to our similarities—we are not distinguishing ourselves from tigers, after all.

Another component of racism is *dichotomous thinking* in the sense of *us* and *them*, or in-group versus out-group thinking, with *us* being better than *them*. Since we evolved in hunting-gathering groups, it is quite plausible that this feature of human nature evolved in us as an adaptation. Hunting-gathering groups would be constantly competing for resources with other hunting-gathering groups, and us-and-them thinking would increase group cohesion.[8] The us-and-them thinking, moreover, would extend back not only through the evolutionary history of the human species, where we faced other groups of the same kind, but far back through the *Homo* genus, through even to the time of *Australopithecus* and beyond. At virtually every horizontal level in the time line, up until only roughly 30,000 to 25,000 years ago, our hunting-gathering forebears faced competition from groups of their fellow kind *as well as* from cousin subspecies and sibling species. Human evolution is the result of a bush pattern, not a single line (Tattersall 2000). Throughout our evolutionary past, then, not just the past 200,000 years, it would have been maladaptive to not instinctively think in terms of us and them.[9] Accordingly today, we can see this deeply embedded instinct everywhere, in local street gangs, in sports teams and their fans, in labor unions, in political parties, in nationalism and ethnocentrism, in religion, and, of course, in war.[10] What furthers the evolutionary/innateness argument is that there seems to be a poverty of stimulus for us-and-them thinking in little children, who all too easily divide into hostile tribes (Ghiglieri 2000, 211–212).

This brings us to the topic of *aggression*. What is typically involved with racism is aggression of some sort, if only in thought. Perhaps racism is intricately connected with aggression, as instincts co-evolved, or perhaps racism is a mere byproduct of aggression, aggression redirected in a manner for which it was not originally evolved.

---

8   This could have occurred either by selection *between* groups, where us-and-them thinking increased group cohesion against competing groups (selection favoring one group over another), or by selection of individuals *within* groups (where individuals who had more us-and-them thinking tended to reproduce more), or by a combination of both kinds of selection. I shall defer a fuller discussion of this to Chapter 8.

9   If ever any groups arose that were communal in spirit like the hippies of the Woodstock generation, with their purported love of everyone, they would have been quickly extinguished. Actually, even the Woodstock hippies exhibited strong us-and-them thinking, particularly in their attitudes toward capitalists and the police, the latter unkindly referred to as "pigs." The documentary *Woodstock* brings this out, but especially *Festival Express*.

10   There is much controversy over whether *xenophobia* (the fear or hatred of strangers) is innate in humans. For examples of the *yes* side, see Wilson (1975, 289–290) and Ghiglieri (2000, 211–212), for the *no* side, Lewontin (1991, 66) and Alland (2002, 182–183). Xenophobia, however, needs to be distinguished from us-and-them thinking, since with the latter the *them* need not be strangers at all.

In either case, one has to deal with the problem raised by E. O. Wilson. According to Wilson (1978, 101–102), a wide variety of research has shown that there cannot be an aggression instinct because aggression has different categories and these categories mix and match in various ways from species to species. Moreover, he says, these categories involve different control systems in the brain. Wilson (1975, 118–119) lists dominance aggression, sexual aggression, parental disciplinary aggression, weaning aggression, moralistic aggression, predatory aggression, antipredatory aggression, and territorial aggression.

If racism is linked to aggression, it would have to be only with the latter, *territorial aggression*. As Wilson (1975) argues, territorial aggression, or territoriality, defined as "any area occupied more or less exclusively by an animal or group of animals through overt defense or advertisement" (289), is a typical trait of hunting-gathering societies. Studies of extant primitive human groups show for each that the size of their territory and their population density follow "the rule of ecological efficiency" found in other animal species, where a meat diet requires roughly ten times as much territory as a vegetarian diet to get the same amount of energy. As Wilson puts it,

> Modern hunter-gatherer bands containing about 25 individuals commonly occupy between 1,000 and 3,000 square kilometers. This area is comparable to the home range of a wolf pack but as much as a hundred times greater than that of a troop of gorillas, which are exclusively vegetarian.
>
> (290)

If territoriality evolved as an instinct in wolves and in other animal species (one can even see one's own house cat mark its territory outside when it takes a whiz), in the absence of counter-evidence it would seem safe to extend the same conclusion to the human species. Interestingly, the anthropologist Michael Ghiglieri (2000), whose treatment of rape as an evolved reproductive strategy in males we have seen in Chapter 4, devotes further chapters in his book to war and genocide. In both cases, he argues not only that war and genocide are human universals, but that instincts for war and genocide reside in the DNA of human males as the evolutionary result of the competition between hunting-gathering groups for territory. Moreover, much like his argument about rape, Ghiglieri strengthens his case by focusing on our close cousins the chimpanzees, whose war parties and genocidal campaigns have now been well documented. In each case (humans and chimps), the battle is between groups of unequal size and power (equality being a deterrent to the would-be aggressor). In each case, the war parties consist of groups of males strengthened by male bonding. In each case, it is mainly males and infants that are killed in the opponents' territory. And in each case, territory acquisition is for territorial and genetic resources, i.e., for food and females necessary for survival and reproduction. Hence as Ghiglieri puts it, "The wars may change, but nearly all of them will be fought with genocide as a major, though unspoken, motivation" (210). Moreover, he says, "war is a male reproductive strategy" (165) such that "women use violence only to defend their reproductive interests; men use violence far beyond this to expand their reproductive interests" (197).

Ghiglieri (2000, 170, 209) seems to further suggest that racism is part of this overall male instinct for territoriality, for war/genocide. Most interestingly, he says, "The human male psyche seems compelled to categorize other men as 'us' or 'them' and to be biased toward 'us' and label 'them'—those with whom 'we' share the fewest genes and least culture—as enemies" (211). While I find this suggestion about racism (which apparently it is) quite attractive, it faces one glaring problem: it overlooks women. If, as Ghiglieri seems to suggest, racism is tied genetically to territoriality and war/genocide in men, then what we should find in women is not only far less or no racist behavior, but far less or no racist attitudes. But do we? It would be interesting corroboration for Ghiglieri's thesis if it were found that women were not as prone to racist attitudes as men. The problem, of course, is to separate cultural influence in some way. A good cross-cultural study might help. What would really help, however, is a test that involves a poverty of stimulus environment (an environment with little or no racism), or perhaps a study of identical twins raised in environments with different levels of racism. As far as I know, no such studies have been undertaken, or even conceived. Again, positive results would corroborate Ghiglieri's thesis, while negative results would support the SSSM. Interesting stuff either way.

7

Evolution and Ethics

Ethics is about right and wrong, good and bad conduct and motives, as well as virtues and vices, duties, obligations, rights, and justice. As an academic field, ethics is normally divided into three parts. *Descriptive ethics*, mainly the domain of anthropology, is about the study of people's actual ethical beliefs and behaviors. *Normative ethics*, mainly the domain of religion and philosophy, is about what we ought to believe and do. *Metaethics*, mainly the domain of philosophy, is about the nature of moral beliefs, statements, and reasoning. Evolution has implications for each of these areas and in this chapter we shall take a look at some representative theories of evolutionary ethics as well as some criticisms of evolutionary ethics.

The best place to begin is with what has come to be known as *Social Darwinism*. Some people today still equate evolutionary ethics with this doctrine. It is the view that, since we evolved by natural selection, we should continue to apply selection to our species in furtherance of our evolution, for the good of our species, that to not apply it, or to try to fight against it, will result in the degeneration and perhaps even the eventual extinction of our species. Although Social Darwinism is named after Darwin, it is well known that Darwin himself never really subscribed to it. In *The Descent of Man* (1871 I), however, he did indeed express a worry or concern that is behind it. "With savages," he wrote,

> the weak in body and mind are soon eliminated; and those that survive commonly exhibit a vigorous state of health. We civilised men, on the other hand, do our utmost to check the process of elimination . . . Thus the weak members of civilised societies propagate their kind. No one who has attended to the breeding of domestic animals will doubt that this must be highly injurious to the race of man.
>
> (168)

Darwin, as we shall see a little later in this chapter, promoted the evolution of conscience, and his own conscience would not allow him to support Social Darwinism. Instead, it was a contemporary and fellow countryman of his, a popularist of evolution named Herbert Spencer, who is most famously associated with Social Darwinism.

Spencer had become an evolutionist long before Darwin published *On the Origin of Species* in 1859, and he advocated evolution not just as a biological theory but as a theory applicable to many areas of life, such as morals, politics, and economics. His evolutionism, however, was mainly Lamarckian (Glossary). But with the revolution begun by Darwin, Spencer jumped on the bandwagon and promoted natural selection as well. Spencer's view was that evolutionary processes such as competition should be promoted in human society, not stifled, restrictions on individual freedom should be abolished, the poor are poor because they are failures in the struggle for existence, the weeding out of the unfit should be allowed because it encourages everyone to try to become fit and thus avoid the consequences of failure. In fact it was Spencer who suggested to Darwin that he replace his label "natural selection" with "survival of the fittest," since many critics did not like Darwin's label. Ever since then, Social Darwinism has had a checkered history, many advocating eugenics (the selective breeding and sterilization of humans), such as Darwin's cousin, Francis Galton, some advocating *laissez-faire* economics for the business world, such as John D. Rockefeller, and then there were the Nazis with their "master race" pseudoscience and their concentration camps, who even used the phrase "the law of natural selection" in their propaganda films.

It is often thought that Social Darwinism met its intellectual demise at the hands of G. E. Moore (1903), that Moore showed it to be guilty of what he called the *naturalistic fallacy*, a fallacy often equated with the *is-ought* fallacy attributed to David Hume. Indeed, most philosophers today (including Ruse and Woolcock discussed below) treat both as if they were fundamentally the same. But the tradition is a mistake, and it is extremely important to see why if we are to make our way clear in evolutionary ethics.

The is-ought fallacy, first brought to our attention by Hume (1740, 302) and often claimed to be the most important rediscovery in ethics of the 20th century, is the claim that one cannot logically infer a moral conclusion from non-moral premises, that there is, in short, a logical gap between "is" and "ought" (or facts and values). For example, if one argues against the rock star Ted Nugent that bow-hunting causes unnecessary pain and suffering so that bow-hunting is wrong, one is then committing the is-ought fallacy. To avoid the fallacy and make the argument work, one has to add the premise that whatever causes unnecessary pain and suffering is wrong.[1]

Moore's naturalistic fallacy is actually quite different, and bears only a superficial resemblance to the is-ought fallacy. The fundamental difference is that Moore's naturalistic fallacy is a *supervenience* thesis (Glossary). Indeed, although Moore did not use the term, he was arguably the first to apply the idea of supervenience to the field of ethics (Kim 1993, 54–61). Moore, accepting the fact-value dichotomy, thought

---

1 To prove that this is an operational premise in the original argument, or ought to be, one has only to insert its denial and see what happens. Part of the beauty of the is-ought fallacy is that it forces us to bring hidden premises out into the open, premises which may indeed be false or at least questionable. I should add, further, that there are some philosophers who doubt that the is-ought fallacy is indeed a fallacy (e.g., Putnam 2002), but I shall ignore them here. The consensus, with which I am in agreement, is that the is-ought fallacy is indeed a fallacy, which shall have to suffice for the present work given the lack of space to argue the point.

that there are moral facts in the world, that there really is, for example, *good* out there outside of our minds. But he did not think that moral facts are out there existing by themselves. Instead, he thought they exist as "non-natural" properties of natural things (including events, processes, and actions), with necessary connections between them. Hence for Moore, contra Hume, you can indeed logically derive an "ought" from an "is" (Baldwin 1993, xviii)! Supervenience comes in when Moore focuses on the relation between moral properties and natural things. Using his favorite example of *good*, he claimed that if something is in fact good, it is good because of certain natural properties that are intrinsic to that thing. On the other hand, he claimed that knowing that something is good does not allow us to infer those particular natural properties. This is because, he claimed, there are other natural properties, intrinsic to things, that are also good. Hence, knowing the natural properties that are intrinsic to a thing allows us to infer that the thing is good, while knowing that a thing is good does not allow us to infer what the natural properties are that are intrinsic to the thing. It is this asymmetry that makes the relation between the moral property and the natural property a supervenience relation. The naturalistic fallacy, accordingly, is committed when one attempts to reduce a moral property to something natural. Because the relation of moral properties to natural ones is supervenient, the reduction is fallacious. Hence for Moore, it is perfectly acceptable to say, for example, "pleasure is good" or "evolution is good." Although such statements might in fact be false, there is no fallacy if the "is" used is the "is" of predication. It is only a fallacy if the "is" used is the "is" of *numerical identity* (one-and-the-same identity, as in "Clark Kent is Superman"), which would be to say "good is pleasure" or "good is evolution."

There is none of this in Hume and there is none of it in the is-ought fallacy connected with his name. For Hume, as for a number of modern philosophers such as Gilbert Harman (1977), there are no extra-mental moral facts. Moral predicates, instead, are something we *project* onto the world. In a classic and haunting passage, Hume (1740) says,

> Take any action allow'd to be vicious: Wilful murder, for instance. Examine it in all lights, and see if you can find that matter of fact, or real existence, which you call *vice*. In which-ever way you take it, you find only certain passions, motives, volitions and thoughts. There is no other matter of fact in the case. The vice entirely escapes you, as long as you consider the object. You never can find it, till you turn your reflection into your own breast, and find a sentiment of disapprobation, which arises in you, towards this action. Here is a matter of fact; but 'tis the object of feeling, not of reason. It lies in yourself, not in the object. So that when you pronounce any action or character to be vicious, you mean nothing, but that from the constitution of your nature you have a feeling or sentiment of blame from the contemplation of it. Vice and virtue, therefore, may be compar'd to sounds, colours, heat and cold, which, according to modern philosophy, are not qualities in objects, but perceptions in the mind.
>
> (301)

Hume then immediately proceeds to give his famous exposition of the is-ought fallacy. Moore, on the other hand, as we have seen, thought that there are objective extra-mental moral facts. He was not at all against the is-ought fallacy itself but was only against reducing properties such as *good* to natural properties such as pleasure or evolution, where reducing is in the sense of numerical identity. For Moore, although moral properties are dependent on physical properties, they cannot be reduced to them. For Hume, on the other hand, there are no extra-mental moral facts in the first place, and what we think are moral perceptions are simply subjective projections onto the outside world. The two fallacies are as different as the philosophies behind them.

The question then becomes whether Spencer and other Social Darwinists are guilty of the naturalistic fallacy, the is-ought fallacy, both, or neither. Spencer clearly thought that evolution is good, but he was guilty of the naturalistic fallacy only if he used the "is" of identity, not if he used the "is" of predication. But either way, he was not guilty of the is-ought fallacy. This is because in his arguments for what we ought to do in terms of Social Darwinism he used not merely factual premises but the moral premise that evolution is good. Those who argue in this manner cannot be guilty of the is-ought fallacy, only those who argue merely from the facts of evolution.

The issue between Hume and Moore is an important one, not only for Social Darwinism but for what is generally regarded as the newer kind of evolutionary ethics, the kind equated with arguments for the evolution of moral instincts. Moore, like many modern philosophers who think of ethics as autonomous from evolution and other disciplines, took ethics to be a matter of intuition, but intuition mysteriously connected to the existence of objective extra-mental moral facts. For Hume, instead, there are no such facts, we only think so because we naturally project such judgments onto the world from instincts that nature has implanted in us, instincts such as sympathy, empathy, and pity. The view that morality requires religion so that society would dissolve without religion does not follow, claimed Hume, given these innate instincts augmented by reason. Hume is clearly the modern here from an evolutionary point of view, not Moore, in spite of Hume's earlier place in time. The problem that we shall find with the newer kind of evolutionary ethics, however, is a tendency to slide from the argument for the evolution of these instincts to the promotion of these instincts as a moral *ought*.

But first we need to deal with the question of whether evolution can give us any justification for the belief in the existence of extra-mental moral facts. Moore was silent on this topic, but others have tried to fill in the gap. I shall give only one example, a well-known argument in environmental ethics provided by Paul Taylor. According to Taylor (1981), all living organisms, whether conscious or not, have "inherent worth"—are "*intrinsically* valuable" as ends in themselves (519)—because they are "equally teleological centers of life in the sense that each is a unified system of goal-oriented activities directed toward their preservation and well-being" (525). "*Conceiving of it as a center of life,*" he says, "*one is able to look at the world from its perspective*" (525). Even though humans have extra features such as language, the highest degree of consciousness, and moral freedom, this does not mean, for Taylor, that humans have more inherent worth, since the above are merely features that have furthered the good of humans in their evolution. To give greater value to those features than to, say, echolocation in bats and dolphins, or to photosynthesis in plants, simply displays in us a subjective bias.

While Taylor's last point is arguably a good one, his argument for inherent worth fails because of a fundamental equivocation over "goal-oriented activities." In modern biology it is standard to make a distinction between *teleology proper*, genuine goal-oriented behavior found in conscious organisms such as humans, and *teleonomy*, goal-oriented behavior due to nothing more than a genetic program (see Mayr 1988, 44–48). To attribute a "perspective" to the latter (e.g., an amoeba or a tree) is simply mistaken, an anthropomorphic illusion. Hence, the "is" does not entail an "ought," the fact does not entail a value.

Related to the above is the question of *natural rights*. Everyone agrees that there are conventional rights, rights given and taken away by society, such as the right to drive a car. It would be silly to believe that this is a right that all humans are born with. Instead, it is a social construction. But many believe that in addition to conventional rights there are *natural rights*, often called *human rights* or *equal rights*, rights that do not depend on any society or culture but are innate and inborn in all human beings. In the American Constitution, for example, it is stated that "We hold these truths to be self-evident, that all men are created equal, that they are endowed by their Creator with certain unalienable Rights, that among these are Life, Liberty and the pursuit of Happiness." Similar statements can be found in many other constitutions and in the Universal Declaration of Human Rights drafted by the United Nations in 1948. Today with all the rights talk the impression is often given that natural rights are written into our very DNA itself.

But can evolution provide any justification for these common sentiments? The answer, I suggest, is *no*, a further example of the universal acid of evolution. Humans are merely one of roughly 30 million existing species and most species that have ever existed are now extinct (no matter what the species concept). To single out one particular species, humans, and say they have natural rights but others do not is pure *speciesism*, to use Peter Singer's (1975) term for the equivalent of sexism and racism. And if we are going to grant natural rights to other species, then to which ones and by what criteria? Any answer is bound to be arbitrary.

To see just how damaging evolution is here, we need to keep in mind that evolutionary history is gradual and proceeded (as it continues to do) in a branching fashion. Any theory of natural human rights, then, has got to take into consideration the fact that all the incremental intermediates between humans and our nearest living relatives, the chimpanzees, existed in the past. But if modern humans have natural rights, how far back do they go? Back to the beginning of anatomically modern humans some 200,000 years ago? But that was not a point in time, and even if it were, what about the generation of humans before them, and the generations of humans before them? And what about all the branches of hominids, indeed what about the species that branched off chimpanzees in one direction and ultimately humans in the other? Many people consider abortion murder but they have no problem with experiments conducted on living adult chimpanzees. From an evolutionary point of view, however, all of this is completely arbitrary. As Dawkins (1986) points out, "The only reason we can be comfortable with such a double standard is that the intermediates between humans and chimps are all dead" (263). But imagine if they were not, what would happen then to the idea of natural human rights, or of natural rights at all?

It would seem the social constructionists have the solid ground here. Moreover, in addition to the facts of natural history the facts of cultural history are on their side. We commonly think of the idea of natural, innate, equal human rights as a beacon of Western thinking. But it is not to be found in the two main pillars of Western thought, namely, Greek philosophy and the Bible. Plato, for example, believed in human-animal reincarnation, he divided humans in his *Republic* (III) in the manner of metals, and he believed in a created scale of living beings, with human males on top, females the first level of degeneration below, and animals successive levels of degeneration below that (*Timaeus*). Aristotle after him rejected reincarnation and creation but argued that there are both masters and slaves by nature and women at best are in-between the two (*Politics* I). Likewise the idea of natural, innate, equal human rights is not to be found in the Bible. In the Old Testament, for a start, not only are slavery and the inferiority of women taken for granted, but God sometimes commands the wholesale slaughter of innocent men, women, and children, as with the destruction of Jericho and Ai (*Joshua* 6–8). Nor is the New Testament much better. Not only does Apostle Paul, the founder of Christianity as a religion, send a runaway slave back to his master and urge slaves to obey their masters and wives their husbands (*Philemon* 10–17; *Ephesians* 6: 5–6; *Ephesians* 5: 22–24), in the latter case because women were made for men (*I Corinthians* 11: 9), but he also claims that God predestined some to salvation and others not (*Romans* 8: 29–31) and that just as a potter has the right to make out of a lump of clay "one vessel unto honour, and another unto dishonour" (think of a flowerpot compared with a piss-pot) so too do we not have the right to complain to God, "Why hast thou made me thus?" (*Romans* 9: 18–21).

The sad fact is that it is not until we turn to the writings of philosophers of the European Enlightenment, beginning in the 17th century, that we find any talk of natural, innate, equal human rights, for example in John Locke (whose political writings served as the foundation for the American Constitution). It is an idea that most of us take for granted today (even, ironically, captured terrorists). But all the evidence points to the idea being a social construction nonetheless. Certainly it is nothing but a fanciful stretch to think that evolution can provide justification for such a doctrine. If anything, one could make a better case for the view that evolution by natural selection is not only unfair but downright evil. In fact, that very position has been made by the biologist G. C. Williams, but I shall save his argument for the next chapter, when we look at evolution and the theological problem of evil.[2]

If evolution cannot give us extra-mental moral facts, then what are we left with? Some argue that evolution gave us moral instincts, some that all of morality is simply

---

2 None of this is to say that we should abolish belief in universal human rights. The view of classical utilitarians is of interest here. While Jeremy Bentham called the idea of natural rights "nonsense upon stilts," his view was basically the same as his main successor, John Stuart Mill, who argued that, although natural rights do not in fact exist, a society that recognizes and upholds universal human rights is going to be, all other things being equal, a happier society than one that does not (see Warnock 2003, 226–227; Singer 1975, 8).

a social construction, while others smile at the intellectual babble and sing the praises of religion.

Evolution and religion is a question reserved for the next chapter, the only connection with ethics being the discussion on what evolution means for the problem of evil. Restrictions on space prevent anything more. Suffice it to say that, for those who think only religion can provide the necessary support for morality, the paradox raised by Plato in his dialogue *Euthyphro* has seemed to many a decisive blow against any divine command theory of ethics. The horns of the dilemma are basically as follows. Does God (or the gods) say something is good because it is good, or is it good because God says it is good? If we accept the latter, then not only does morality become arbitrary, as it depends on nothing but God's will and God's will could have been otherwise, but morality also becomes a matter of power worship, for no other reason for accepting God's will becomes relevant except that God is all-powerful. (Indeed, one might add that this approach is a further example of the is-ought fallacy.) On the other hand, if we accept the former alternative, even if it is because we think God is all-knowing and knows what is best, then morality becomes separate from God's will, so that morality becomes autonomous from religion. Either way, we are left with the conclusion that religion cannot serve as a proper foundation for morality.

What then of social constructionism? This is a view that has appealed to many, and it is in fact far from new. It was common among the ancient Greek Sophists, for example, such as Protagoras, who claimed that "Man is the measure of all things." Even earlier, the Greek historian and traveler, Herodotus, noted the differences in moral customs between different societies and concluded that "Custom is the king over all." In our modern time the social construction of morality was given a famous voice by the cultural anthropologist Ruth Benedict. According to Benedict (1934), "normality is culturally defined" (72), "most individuals are plastic to the moulding force of the society into which they are born" (74), "the majority of mankind quite readily take any shape that is presented to them" (75), and "morality differs in every society, and is a convenient term for socially approved habits" (73), in other words, "The concept of the normal is properly a variant of the concept of the good" (73), so that in ethics "all our local conventions of moral behavior and of immoral are without absolute validity" (79).

Benedict, like many anthropologists of her time, was greatly impressed by perceived variations between human cultures, particularly cultures relatively isolated from the West. Some cultures valued trance phenomena, some paranoia, some megalomania, some permitted homosexuality, some cross-dressing and cross-role playing, some required killing someone from another tribe when someone from their own tribe died (no matter what the cause), some practiced cannibalism, some even ate their dead grandparents, and on and on. Interestingly, Benedict (1934) compares all of this variation to the variation in languages from culture to culture, which she considers to be as wide as the "fashions of local dress" (72).

The comparison with languages, however, should awaken a suspicion given what we have seen in Chapter 3. Indeed, many scholars in recent decades have argued that the variation in cultural moralities studied by anthropologists such as Benedict is superficial, that beneath the surface variation there are cross-cultural commonalities. Interestingly, what these scholars argue is that the variation is caused by different conditions of existence and different beliefs, not really by different moral values. James Rachels (2003, 24–25), for example, focuses on Eskimos and Hindus. Eskimos leave their old and feeble to die in the snow and also practice female infanticide, all without any social stigma. But for Rachels, it is not because these people do not value human life any less than we do. Instead, it is the harsh conditions of their existence that make the difference. They live in the Arctic, they are nomadic as they cannot farm, having enough food for the community is always a problem, so when the old and feeble can no longer keep up they are left behind. Females, moreover, can only carry one infant, and males have a higher mortality rate because of casualties from hunting, so female infanticide is necessary as a last resort not only to control the number of mouths to feed but in order to keep the sex ratio even. As for Hindus, their vegetarianism stems from their belief that animals have souls and hence are persons. Christians, for example, do not have this belief, so they eat animals while the Hindus watch in horror. The difference is in beliefs, not in values. For Rachels, then, "the raw data of the anthropologists can be misleading" (25). Similarly, Louis Pojman (2000, 641) focuses on a tribe in the Sudan that throws its deformed children into the river. But again, it is not because they have different moral values than us, but rather because they have different beliefs, namely, in the above case, the belief that deformed children belong as property to the hippopotami, the gods of the river. As with our own debate over abortion, the difference is in beliefs about the facts (specifically, whether a fetus is a person), not in differences in values. For Pojman, then, following E. O. Wilson and a number of other thinkers (we shall return to Wilson below), beneath the cultural diversity in moral practices there is a common core of moral values, such as mutual obligations between parents and children, proscription of rape and murder, and truth telling, all of which "are necessary to any satisfactory social order" (644). Rachels (2003, 25–26), interestingly, adds that a selection process would result in certain common values cross-culturally, because those cultures without them would lack social cohesion and eventually die out.

The idea of moral universals along with selection brings us to possible biological explanations of those universals, to what is widely regarded as the new kind of evolutionary ethics, the kind that argues for the evolution of one or another of moral instincts. To help settle the matter, we shall have to look at representative samples of this new kind of evolutionary ethics, at select arguments for the evolution of particular moral instincts in humans, to make a reasoned judgment between the SSSM and evolutionary models.

The new kind of evolutionary ethics is actually not all that new. Perhaps not surprisingly, it began with Darwin (as with so much else). In the third chapter of his book *The Descent of Man* (1871 I), Darwin focuses on "the moral sense or conscience . . . summed up in that imperious word *ought*," what he calls "the most noble of all the attributes of man" and, when it comes to the differences with other animals, "by far the most important" (70). In spite of this great difference, Darwin says he is going to be the first to attempt to explain this feature of humans "exclusively from the side of natural history" (71).

Darwin's argument is an interesting and important one for critical evaluation, and I shall summarize it here in some detail. He begins by pointing out that social behavior is common in the animal world, not only the sacrificing behavior of parents toward their young but warning signals in flocks of birds and troops of monkeys, buffalo circles against predators, and "little services" like monkeys grooming each other for parasites. All of these are evolved instincts, and Darwin argues that evolving along with these would be pleasure and pain, pleasure when an instinct is satisfied, pain when it is not (hence with social animals, being in the group produces pleasure, generally speaking, being away from it pain). Equally important, Darwin also argues that sympathetic feeling is common in social animals (Darwin always gives plenty of examples, such as a blind pelican being fed by its fellow pelicans), and that however it started it would be augmented by natural selection in the sense that "those communities, which included the greatest number of the most sympathetic members, would flourish best and rear the greatest number of offspring" (82).

Darwin also points out that not only are instincts a matter of degree but they can be opposed to other instincts in the very same animal, for example when a mother goose is pulled between the instinct to migrate and the maternal instinct to stay behind with her late brood of chicks, or when the maternal instinct is opposed to the instinct for self-preservation, or when the instinct of a dog to obey its master is opposed to the instinct to tend to its pups. What makes one instinct "more potent" over the other, says Darwin, is usually if not always natural selection, in that whichever of the conflicting instincts tends to result in more offspring when there is conflict is going to be favored by selection. This point about degrees of instincts and conflicting instincts is extremely important for Darwin's argument, so much so that on it "the whole question of the moral sense hinges" (87).

Turning to humans, Darwin makes the obvious point that we are social animals. But we are also self-domesticated, and domestication, he observes, sometimes results in diminished or even lost instincts. Darwin nevertheless sees no reason why we should not have retained most of our social instincts to some degree. By *social instincts* he means sympathetic feelings, obedience to a leader, faithfulness to group members, love of praise and hatred of blame from group members, and defending and aiding group members (85). All of these instincts evolved because of the need for group cohesion, not for the good of the individual or the species but for "the good of the community" (103). Any human who happens to be without social instincts is an "unnatural monster" (90). On the other hand, we also have older instincts, instincts that predate the evolution of our social instincts, what may be called *selfish instincts*, such as self-preservation, hunger, lust, and revenge. As humans we also have language and reason, and these combined with the selfish and social instincts result in conscience in the following way. The social instincts are always present and operate at a more or less constant level of intensity; they are "ever present and persistent" (89). The selfish instincts, on the other hand, are "temporary" (90) and operate at widely varying levels of intensity, often below the level of the social instincts, but sometimes, on certain occasions and then only briefly, at a level that greatly exceeds and overpowers the social instincts. When this brief episode has passed, when the selfish instincts are satisfied at the expense of the social instincts—that is, when we return to normal—unlike other animals we cannot help but reflect on our past actions, because of our greater mental faculties. The typical result, apart from the "bad man" (92), is that

> Man will then feel dissatisfied with himself, and will resolve with more or less force to act differently for the future. This is conscience; for conscience looks backwards and judges past actions, inducing that kind of dissatisfaction, which if weak we call regret, and if severe remorse.
>
> (91)

All of this is further augmented, though it is not necessary, says Darwin, by the belief in God or gods (93).

This is an interesting theory. On the one hand, as Darwin himself suggests (97–98), it undermines the theory of *psychological egoism* (which holds that we always act selfishly) by appealing to unselfish instincts, while on the other hand it undermines the theory of ethics known as *utilitarianism* (which holds that we are motivated at bottom by pleasure and pain) by making pleasure and pain secondary to the gratification or frustration of instincts, whether selfish or social. (I shall return to utilitarianism below.) But is Darwin's theory of conscience a theory of adaptation or a byproduct theory, a theory that views conscience as a byproduct of the evolution of something else? Darwin himself does not seem to be sure. On the one hand, conscience appears in his view to be simply a byproduct of the conflict between social and selfish instincts in a social species with a high degree of reason and reflection. On the other hand, Darwin says,

> The imperious word *ought* seems merely to imply the consciousness of the existence of a persistent instinct, either innate or partly acquired, serving him as a guide, though liable to be disobeyed.

Recall that for Darwin conscience is "summed up in that imperious word *ought*." Moreover, Darwin in the previous quotation immediately proceeds to make a comparison to instincts in dogs, when he says,

> We hardly use the word *ought* in a metaphorical sense, when we say hounds ought to hunt, pointers to point, and retrievers to retrieve their game. If they fail thus to act, they fail in their duty and act wrongly.
>
> (92)

If Darwin's view is indeed that conscience is an instinct rather than a mere byproduct, however, he needs to make a clear case, but he has not done so.

Darwin's comparison of conscience to instincts in dogs suggests an even deeper problem, namely, an equivocation concerning the "ought" of morality and the "ought" of expectation. "Hounds ought to hunt" seems more like looking at an overcast sky and saying, "It ought to rain today." We neither think it immoral if old Duke does not want to hunt or if the sky does not rain. It is simply a failed expectation. True, hounds and pointers were bred for the purposes of hunting and pointing, but if they each fail in their purpose it does not make them or their actions immoral, any more than when a car built for racing stalls at the starting line.

But perhaps the deepest problem concerns Darwin's concept of *moral progress*. Rather than take "the test of morality" to be the utilitarian one, where we should pursue the greatest happiness of the greatest number (the *utilitarian calculus*), Darwin takes the test to be "the general good or welfare of the community," where "the general good" is defined partly in terms of reproductive success: "the means by which the greatest possible number of individuals can be reared in full vigour and health, with all their faculties perfect, under the conditions to which they are exposed" (98). With this in mind, Darwin then distinguishes between "higher and lower moral rules" (100), the higher relating to the social instincts, the lower to the selfish instincts. Among the higher moral rules there are, of course, "many absurd rules of conduct" and "many absurd religious beliefs," but these are merely the result of "ignorance . . . and weak powers of reasoning" (99). As small tribes unite into larger communities and these advance in civilization and reasoning, Darwin sees the social instincts extending in an expanding circle "to all nations and races," even beyond our own species "to the lower animals" (101). As this circle expands, "so would the standard of his [man's] morality rise higher and higher" (103), and Darwin rests assured, looking to future generations, that "virtue will be triumphant" (104).

In all of this, Darwin faces a number of problems. Not only does he see the expanding circle in progress, but he adds that "the simplest reason would tell each individual that he *ought* to extend his social instincts and sympathies" (100) (italics mine). I suspect that many will find themselves in agreement with Darwin here. But then they share his guilt in committing the is-ought fallacy. To avoid the fallacy, one would have to add that "the rearing of the greatest number of individuals in full vigour and health *is good*." This, however, is a questionable premise. For ancient Greeks such as Aristotle, for example, it would not be true, since moderation is a virtue and more of something good is not necessarily better. But even more basically, why should the social instincts be elevated above the selfish? Is this not simply a matter of perspective and biased interest? Equally problematic is the concept of the evolution of ethics. Many have held this concept. Jesus argued for the expansion of ethical concern to all humans, friend and foe alike, Peter Singer (1975) to all sentient animals, Albert Schweitzer (1933) and Paul Taylor (1981) to all living things, and Aldo Leopold (1949) to inanimate things such as rocks. The problem then becomes extremely thorny indeed, namely, that of determining the proper circle of ethical concern and trying to be objective about it.

When we turn to contemporary theories of the evolution of moral instincts, we find similar problems. E. O. Wilson (1975), for example, the don of sociobiology, argues for what he calls *innate moral pluralism*. Wilson begins by pointing out that intuitions reside at the core of philosophical systems of normative ethics, such as those of Locke, Kant, and Rawls (clearly also those of Ross, Bentham, and Mill). Philosophers typically appeal to emotive intuitions among their premises, such as feelings of justice and obligation, or that pain is bad. But these emotions, says Wilson, have control centers in the brain, namely, the hypothalamus and the limbic system. To really understand these emotions, then, we need to understand the origin of these systems. Of course they evolved by natural selection, but ethical philosophers typically ignore this. They treat the brain, instead, as a "black box," and this, says Wilson, is their "Achilles heel." The problem is that they neither consider the fact that "the human genotype and the ecosystem in which it evolved were fashioned out of extreme unfairness," nor do they consider "the ultimate ecological or genetic consequences of the rigorous prosecution" of their ethical systems (287), of, say, treating justice as fairness (Rawls).

The time has come, says Wilson, for ethics to be "biologicized" (4). This is the sociobiological approach to ethics. What this means, in part, is that one must look at the level of the gene, not the individual, the group, or the species, to understand the phenomena studied by ethics. Altruism, for example, genuine self-sacrificing behavior, whether in the behavior of bees or in the behavior and feelings of humans, is not something mysterious or divine but is explained by *kin selection* (55–58), a theory we have seen in Chapter 4. "Charity begins at home," after all, as the saying goes, and "blood is thicker than water." Altruism toward non-kin is explained to the satisfaction of Wilson (58) by what Robert Trivers (1971) called *reciprocal altruism*. The idea here is that an instinct for the Good Samaritan kind of altruism (self-sacrificing behavior toward non-kin) ought to evolve in a species with a high degree of social organization as well as the ability to recognize individuals and remember how they behave. In such a population, where the probability is high enough that those who receive altruism but do not reciprocate it when the opportunity arises will be found out, cheaters would eventually be deprived of altruistic benefits, possibly even punished directly, so that cheating behavior would lower individual fitness and an instinct to give and to reciprocate non-kin altruism would evolve.

But there is much more to the sociobiological approach to ethics. Wilson argues for a system of ethics that takes into account what he calls the *genetic evolution of ethics* (287). This theory is in some ways similar to Darwin's. We have much older genes which evolved for individual selfishness, and we have newer genes that evolved throughout our hunting-gathering past, with both being maintained by natural selection "at best in a state of balanced polymorphism" (288). The term *polymorphism* refers to a species with more than one form (subspecies, sexual dimorphism, caste polymorphism, ontogenetic polymorphism). Wilson's point is that we should expect the same for the genetic basis of ethical intuitions in humans. Since the vast majority of the history of the human species evolved in small hunting-gathering groups (beginning roughly 200,000 years ago), even more so for the *Homo* genus of which we are a part (beginning roughly 4 million years ago), we should expect intuitions appropriate for primitive forms of tribal organization (e.g., alpha male behavior such as macho posturing). Slight additions may have occurred as we adapted to the farming life (beginning roughly 12,000 years ago) and then city life (beginning roughly 6,000 years ago). The result is what Wilson calls "moral ambivalence," which he says "will be further intensified by the circumstance that a schedule of sex- and age-dependent ethics can impart higher genetic fitness than a single moral code which is applied uniformly to all sex-age groups" (288). For example, Wilson says it would be a selective advantage for young children to be self-centered and disinclined to altruism. Oddly, however, he does not say why. Nevertheless, an interesting answer can be found in Frank Sulloway's *Born to Rebel* (1996), in which he argues that the family unit is a Darwinian field of competition, specifically sibling competition for limited parental resources such as nutrition and affection (the interesting long-term result is that firstborns tend to be conservative, from playing parents to their younger siblings, while laterborns tend to be liberal and rebellious). A further example for Wilson is the selective advantage for teens to be "tightly bound by age-peer bonds within their own sex and hence unusually sensitive to peer approval," the reason being that alliances and status are more important at this age than later when they are parents, where parental morality becomes the main determinant of fitness. Further differences, says Wilson, ought to result from different kinds of population situations, populations experiencing overpopulation (on an island, say) experiencing selection pressures for competitive/selfish genes, founder populations experiencing a new environment experiencing selection pressures for cooperative/altruistic genes. From all of this, Wilson concludes that

> no single set of moral standards can be applied to all human populations, let alone sex-age classes within each population. To impose a uniform code is therefore to create complex, intractable moral dilemmas—these, of course, are the current condition of mankind.
>
> (288)

There clearly does seem to be an "ought" here, coming collectively from each and every "is." But even if we put that thought aside for a moment, let us try to consider what Wilson's approach to ethics really means. It means that what evolved as an adaptation is good. But does this mean that rape is good? As we have seen in Chapter 4, there might indeed have evolved in males an adaptation for rape. Wilson might pass rape off as "outdated" or "a relic of . . . the most primitive form of social organization" (287). But then he is caught in the thorny position of advocating some adaptations and not others, ultimately of being biased if not arbitrary. Indeed, one could just as easily argue that evolution by natural selection is evil at the core and that the purpose of ethics is to fight against nature, as mentioned earlier a position advocated by the biologist G. C. Williams that we shall examine in the next chapter.

A further problem is whether the theory of reciprocal altruism is really needed for the sociobiological approach to ethics. The theory is not simple, after all, but rather complex in its details (just read Trivers 1971). Certainly a simpler theory would be desirable if it would work just as well. Interestingly, Alcock (2001, 181–182) attempts to defend the theory of reciprocal altruism by arguing that behavior which seems to refute it—behavior which seems highly unlikely to receive a reciprocal payback, such as charity donations, or the work of people such as Mother Theresa among the poor—is typically undertaken only when it is known that other people will know about it, in other words only when it is a matter of reputation. Since reputation is one of the proximate mechanisms of reciprocal altruism, we seem in those cases to have reciprocal altruism after all. I suspect that reputation might indeed be a major factor in many cases of non-kin altruism. Some might go further and say most or all. But we need not argue the matter, for the possibility remains that reciprocal altruism is not needed to explain altruism toward non-kin, whether the altruism is likely to be reciprocated or not. As Alcock himself points out (2001, 218–219), which I quoted briefly in Chapter 4, sociobiologists have the byproduct hypothesis and the novel environment hypothesis to explain maladaptive behavior. In short, a generally adaptive behavior might produce a maladaptive behavior simply as a byproduct (instincts do not always operate perfectly, after all) or as the result of operating in a new environment (as with our penchant for foods high in sugar, salt, and fat). Reciprocal altruism is arguably not maladaptive in itself, but perhaps altruistic behavior toward non-kin is simply only a byproduct of the instinct for altruistic behavior toward kin, an imperfect instinct redirected and nothing more. The question would seem to boil down to evidence for reciprocal altruism as an adaptation, in the sense of increasing reproductive success.

Michael Ruse (2002), who co-authored a paper with Wilson on the sociobiological approach to ethics (Ruse and Wilson 1985), adds some interesting twists of his own which are worth looking at since they seem designed to avoid problems that Wilson's approach invites. For a start, Ruse avoids the is-ought fallacy by claiming that normative claims are not really statements at all, in the sense of correspondence to reality. They are neither true nor false and hence not in need of justification. In ethical theory this is known as *noncognitivism*. But Ruse gives it an evolutionary spin. "Normative ethics," he says, "is a biological adaptation" (658). Since there is no move from "is" to "ought," but rather we begin with the "oughts" in us already, there can really be no is-ought fallacy for the sociobiologist.

But do we all have the same "oughts" in us? Some of us are vegetarians, after all, and think others ought to be the same, while others think not, and so on for many other topics. Ruse's reply here is to argue for an underlying universal normative morality in the human species. He says,

> it is the philosopher's stock-in-trade to look for counterexamples to established moral systems. But most of the time, the well-known and tried systems [a little earlier he specifically lists "Christianity, Kantianism, probably utilitarianism, and more"] agree on what one should do. Kantians, Christians, and everyone else agree that you should not hurt small children for fun and that if you are blessed with plenty then you should help the poor person at your door. Standard moral systems do not urge you to do crazy moral things.
>
> (657)

For myself, the overlap argument automatically makes me suspicious, but especially when used by Ruse. For a start, when it comes to the content of normative ethics, Ruse is not in sync with Wilson. As we have seen, Wilson (1975) thinks that John Rawls's theory of justice as fairness is fine for "disembodied spirits" but that the human genotype was fashioned out of "extreme unfairness" (287). Ruse, on the other hand, claims that Rawls's justice as fairness is "just the sort of system favored and expected by the evolutionist" (656). But even more, Ruse used the overlap argument to solve a problem in a completely different area, namely, the species problem, the problem of deciding between the many different species concepts available in modern biology. Applying William Whewell's concept of a *consilience of inductions* (Whewell was an older contemporary of Darwin's), Ruse (1987) argues that the different major species concepts pretty much pick out the same organisms. He says, "There are different ways of breaking organisms into groups, and they *coincide!* The genetic species is the morphological species is the reproductively isolated species is the group with common ancestors." But not only does Ruse leave out ecological, phylogenetic, and cladistic species concepts (see Stamos 2003), the fact is, as biologists readily agree (and the consequences are profound for conservation biology), there just is no such consilience! This alone makes me wonder if Ruse's *kind* of approach is any more accurate when it comes to different systems of normative ethics. The topic of the ethical treatment of animals, for a start, should cause doubt. And contra Ruse, this cannot simply be passed off as a matter of philosophers focusing too much on counterexamples. Instead, it cuts right to the bone of the matter.

Of course, one could appeal to the kind of reply we saw earlier in this chapter against relativists such as Ruth Benedict. Maybe the variation is surface variation caused mainly by different beliefs. Humans are, after all, the combined product of biology and culture, as Ruse often remind us. Interestingly, Ruse appeals to the language analogy, saying "Morality is like speech" (661) and explicitly referring to the universal grammar of Chomsky, to explain the surface variation in normative ethics. For Ruse, then, beneath the human variation in normative ethics there is a universal morality evolved by natural selection, much like Pinker argues for the UG. And like the rules of the UG, Ruse believes that "the claims of normative ethics are like the rules of a game" (658). Moreover still, like Chomsky on the UG and Darwin before him on hive-bees (1871 I, 73), Ruse argues that morality could be very different if it evolved in other species, what he calls "intergalactic relativism" (661). But that is where the comparison with language seems to end. Even if we grant that there are moral universals in the human species analogous to linguistic universals, there just does not seem to be any way to test the theory that they are innate, as with the poverty of stimulus or creoles in the UG case. Certainly universality does not entail innateness. The universal morality, then, if there is such a set of rules, just might not be innate.

What is more plausibly innate, instead, are not specific normative oughts but (following Hume) simple and general propensities such as sympathy, empathy, pity, and parental instincts. Moreover, there is nothing to say these instincts cannot be sex, age, and population dependent in terms of degrees and objects. Nor is there anything to say they could not be modified by the environment, such as by the inculcation of beliefs. A further candidate is the propensity to believe in the objectivity of morality, to believe that normative claims suckled from the breast of one's society really are true in the sense of truth as correspondence to reality. This, it seems to me, is the strongest and most plausible part of Ruse's thesis. According to Ruse, what has evolved in our species is "a collective illusion of the genes, bringing us all in (except for the morally blind)" (659). Hence, Ruse does not expect us to buy the sociobiological approach to ethics, not because of any reasoning on our part but because of our genes. "Your genes are a lot stronger than my words," he says. "The truth does not always set you free" (661).

It is with the collective illusion, I suggest, and not with the universal morality, that we could find the poverty of stimulus. Perhaps even someday the collective illusion will find its Dean Hamer, the geneticist who finds the genes or genetic markers for the collective illusion, as Hamer seemed to do for male homosexuality (Chapter 4). At any rate, the theory of the collective illusion would be simple to test. One could compare the way two groups respond to, say, the sight of a baby being thrown against a wall, one group never having heard of sociobiology, the other group complete converts to its species relativism and noncognitivism. (Of course I do not recommend using a real baby, a really good Hollywood fake would do.) I suspect the response in both cases would be identical, namely, complete horror and an overwhelming desire to kill the perpetrator. Indeed, I suspect the response would be identical with just the sociobiology group alone, compared before and after their conversion. I really do not think their change in metaethical beliefs would make even the slightest difference to their moral behavior.

All of the above, of course, tends to diminish if not eliminate the role of moral reasoning. And that, as one might suspect, is a major bone of contention with critics of the sociobiological approach to ethics. Below are two major examples.

In an article titled "Ethics Without Biology" (1979), Thomas Nagel argues that ethics is fundamentally a rational activity, a matter of "practical reason," a field of critical investigation with reasoning "internal to the subject," a field with "progress" and "development" and "discoveries" that are "somewhat analogous" to those in science (143), so that "it would be as foolish to seek a biological evolutionary explanation of ethics as it would be to seek such an explanation of the development of physics" (145). Peter Woolcock (1999) provides some further arguments that amount to basically the same view. He argues that even if the sociobiological approach to ethics establishes that we have moral instincts (he does not think it has), ethics and morality are more than just about instincts or feelings. Instead, they make claims about what we ought to do, claims that are both universal (they apply to everyone) and categorical (they are about what we ought to do irregardless of our feelings, desires, or beliefs). The sociobiological approach, on the other hand, he says, misses the obvious nature of morality. It treats altruism, for example, as a cause, even a compulsion cause, but not only is altruism not a compulsion cause (since people can be trained out of it), causes are not reasons. If we look at how we actually deal with moral issues, he says, "Justification does not collapse into explanation, nor do reasons collapse into causes" (291). We are moral agents after all. We do not justify our actions by explaining them, nor do we treat causes as reasons. In fact we treat causes, especially compulsion causes, as absolving us of moral responsibility. Instead, we appeal to our free will and justify ourselves morally only by the reasons we provide for our actions.

The claim that we are moral agents, of course, rests upon the claim that we have free will. But as we have seen in Chapter 2, if we take biology and physics seriously, free will looks to be an illusion. The point is that one cannot try to solve the problem of ethics by assuming something that is just as problematic. But that is exactly what Woolcock and so many others do.

Interestingly, Darwin (1871 I) provided a reply to the claim that moral justification requires reasons rather than causes, that it requires deliberation alone. He points out that quite often we do not know what people's motives are so we focus only on the actions, that the line between deliberation and impulse is not a clear one, that actions repeated enough times become habitual to the point that they can "hardly be distinguished from an instinct" (88), and that quite often we praise actions as moral when they are not done from deliberation but done impulsively, so that "in the case of man, who alone can with certainty be ranked as a moral being, actions of a certain class are called moral, whether performed deliberately after a struggle with opposing motives, or from the effects of slowly-gained habit, or impulsively through instinct" (89).[3]

Arguably, Darwin did not go far enough. Contrary to Nagel and Woolcock, and even the limited extent that Darwin allows, I am just not convinced that humans should be characterized as moral reasoners. It seems, instead, that *most people most of the time* do not reason at all when they behave morally. They simply follow the norms of their society, often combined with instincts such as sympathy. Nor does counter-argument seem to have much of an effect in changing their moral or immoral behaviors and attitudes. For example, Peter Singer (1975) argues that we should become vegetarians based on the fact that we live in a modern society and do not need to eat meat, that we eat meat primarily only to satisfy our palate, that the meat industry causes enormous pain and suffering to the animals, and that most of us believe that unnecessary pain, suffering, and killing is wrong. In spite of the fact that this is a good argument, possibly even a sound one, very few people who are exposed to it (e.g., in practical ethics classes in universities) become vegetarians. Instead, when asked for justification most people typically provide remarkably lame arguments in defense of the status quo, if they provide reasons at all. Given that this is so, and that the case of ethical vegetarianism is not unique but a representative example of a statistical tendency in humans, it would seem that Hume (1739) was more right than wrong when he wrote that "Reason is . . . the slave of the passions" (266).

At any rate, what the is-ought fallacy teaches us is that genuine moral reasoning (which must be granted), if it is not to be fallacious, requires at least one moral or value premise. And this is what fundamentally ruins Nagel's comparison of ethics with sciences such as mathematics and physics. Not only do arguments in the latter not involve moral or value premises, but discovery and progress in those fields are in the realm of facts, not values. Hence they are objective in a way that ethics can never hope to be.

---

3   Part of what Darwin says here has a famous precedent in the writings of Aristotle. In *Nicomachean Ethics* (II.1), Aristotle argues that moral excellence does not come from teaching (for then we could get it from studying a book), nor is it innate (because what is innate cannot be changed by habit), but comes instead from repetition developed into habit, so that "we become just by doing just acts, temperate by doing temperate acts, brave by doing brave acts." The disagreement between Aristotle and Darwin is not really deep here, as it hinges merely on the origin of habit. Aside from that difference, they both point to a major feature of our moral experience, namely, that a moral action does not necessarily require having reasons.

The question, of course, remains concerning where the moral and value premises in moral arguments come from and whether they can be justified. To justify them with further reasons only leads to an infinite regress. Ultimately, one has to stop with either self-evident truths, or with the cultural relativism of the SSSM, or with evolutionary biology. It hardly seems likely that morality boils down to necessary truths, in the manner of axioms in geometry, and appealing to cultural relativism is really no justification at all. What we are left with, then, in the very least, is a Darwinized Hume.

But what of Woolcock's claim that ethics gives us prescriptive, universal, and categorical oughts? I see no reason why these cannot be subsumed under Ruse's collective illusion of our genes. If we have an instinct to objectify morality, to make it appear to transcend our personal interests, then prescriptive, universal, and categorical oughts are exactly what we should expect. But they would be illusions nonetheless.

There still remains the point that sociobiology has not established that we do in fact have moral instincts. Woolcock (1999), for example, says not only must sociobiological ethics show that there are no other equally plausible explanations of the data of human morality, but it would also have to prove itself as a testable and confirmed theory in terms of reproductive success (and he knows of no such tests). Otherwise, he says, the sociobiological theories of ethics and morality are "just so" stories (279).

But this seriously misconstrues not only sociobiological theories but scientific theories in general. Gone are the days when it was thought that for theories to be scientific they must be testable, in the sense of either verification or falsification (of course these remain epistemic values). What has come out of both the history and philosophy of science is that testability is sufficient for a theory to be scientific but not necessary. String theory in physics, for example, is not testable, and yet it is nonetheless scientific (Galison 1995). Instead, as I have stressed a number of times in previous chapters, what is at the core of science is what has come to be known as *inference to the best explanation*. At the heart of this non-deductive form of inference is *contrastive explanation* (Lipton 1990, 1991), not simply explaining "Why this?" but "Why this rather than that?" To use an example from Lipton, to explain why Kate rather than Frank won the essay prize, it will not do to simply explain why Kate's essay is excellent. Instead, one has to explain why it is better than Frank's. Likewise, science is characterized by competing explanations and inference to the best explanation. Often this involves testing, but not always. Darwin's "one long argument" in the *Origin*, for example, was not a matter of proving evolution and disproving creationism, but rather a matter of gathering as much evidence from as many quarters as possible and showing that evolution by natural selection makes the most sense of the evidence, that it rather than creationism is the best explanation.

Woolcock entirely misses this and concludes that "It looks as if we shall have to resolve our moral differences through the hard grind of normative justification" (303). The problem is that normative theory as an autonomous enterprise, both past and present, the kind of moral reasoning basic in courses on ethical theory, leaves ethics with no foundation whatsoever but only conflict. As Hume (1779) put it in the case of conflicting religions, "all of them, on the whole, prepare a complete triumph for the sceptic" (186).

Hence the attraction of evolutionary ethics, since not only does it offer a unifying foundation for ethics, but in so doing it provides a powerful reply to moral skepticism. Indeed, from an evolutionary perspective, it is tempting to view the major normative theories in ethics—Aristotelian virtue ethics, the psychological egoism of Hobbes, the utilitarianism of Bentham and Mill, the deontologies of Kant and Ross, Rawls's veil of ignorance, and even more recently environmental ethics and feminist ethics—in somewhat the same way that John Hick views the world's major religions (of course, whether he is right is an entirely different matter). For Hick (e.g., 1985, 37), each of the world's major religions is like a blind Indian in the Buddhist parable of the blind men and the elephant, in which each of the blind Indians feels and describes a different part of the elephant and each confuses the part for the whole. Given that humans and human moral intuitions evolved over millions of years in the complex group dynamic of hunter-gatherers, it may well be that what each major normative theory gives us is a blind Indian, describing and attributing full reality to only a part of the whole of normative ethics as evolved in humans.[4]

Interestingly, when it comes to the "hard grind" of normative ethics, no one has contributed as much in the past few decades as the don of practical ethics himself, namely, Peter Singer, with his many books such as *Animal Liberation* (1975) and *Practical Ethics* (1993). Surprisingly, unlike almost everyone else in mainstream professional ethics, Singer takes sociobiology seriously, having contributed a book to sociobiology and ethics (Singer 1981) and having argued more recently that those of the political left, which typically embraces the SSSM, should embrace instead the Darwinian revolution, including both sociobiology and evolutionary psychology (Singer 1998, 1999). Given the liberal values that the left has, says Singer, they would do better to recognize human nature as the product of evolution by natural selection and give up their utopian dreams of a classless or genderless society, but they need not give up their other goals. Embracing Darwinism does not preclude embracing causes that fight for the poor, the oppressed, the ripped off. Instead, it results in a better understanding of the obstacles and costs and may lead to the most effective means of achieving those goals. The difference is between an unrealistic optimism that thinks the sky is the limit and a realistic optimism that recognizes down-to-earth limits and degrees of success. We might think of Ghiglieri's chapter on rape here. If Ghiglieri is right about males, then social progress in terms of egalitarianism could

---

4   What needs to be added to this is not only that the literature on evolutionary ethics is enormous, but that it includes its own conflicts and many more perspectives than I could possibly examine in this chapter, such as Owen Flanagan's (2003) argument that normative ethics is part of the science of ecology, that it involves the study of what is conducive to the flourishing of natural systems such as a wetland or a species such as *Homo sapiens*. Evolutionary ethics is enormous even if we focus only on Ruse. Starting with Ruse's main discussion on evolutionary ethics (1986, ch. 6), see, e.g., Campbell (1996) for an argument that moral beliefs can have truth value, and hence objective justification, all the while accepting Ruse's premises, see Lahti (2003) for a cultural rather than genetic explanation of Ruse's "collective illusion," and see Ryan (1997) for an attempt to slightly modify Ruse's position so that it may be successfully defended against the attack by Woolcock.

only hope to reduce rape, not eliminate it. Elimination could only be achieved at the genetic level, in terms of genetic engineering on an unimaginable scale.

Acid has positive applications, after all. But the universal acid of evolution cannot be so easily contained. In the case of Singer's argument for the ethical treatment of animals and for ethical vegetarianism (Singer 1975), the argument closest to his heart (and mine) and for which he is by now most famous, the acid threatens to leak out and corrode his Darwinian view of life. I suggest it does this in basically two ways.

A few pages earlier in this chapter I summarized Singer's argument for animals, and it is indeed an *argument*, albeit an unusual argument, as utilitarian arguments go, since it focuses not on the maximization of happiness but on the minimization of unhappiness (pain). It is an argument, moreover, that is a strong one, to my mind, as long as one adds the premise, as Singer indeed does, that unnecessary pain, suffering, and killing is wrong and something ought to be done about it. A problem, however, arises with the concept of pain. One of the basic premises of utilitarianism is that pain in itself is bad and ought to be avoided. Along with pleasure, pain is what Jeremy Bentham, the 18th-century father of utilitarianism, called one of nature's "two sovereign masters." Similarly for Singer (1975), "pain and suffering are in themselves bad" (17). Certainly our minds, our intuitions, tell us that pain is bad and ought to be avoided. We also pretty clearly have an instinct for avoiding pain. This is because pain is a sign of a problem in our bodies. From an evolutionary point of view, the capacity for pain and the instinct to avoid it evolved in animals because they increase survival and reproduction. Although not based on evolutionary theory, utilitarianism takes as one of its axioms that our intuitions about pain are correct. The problem is that, from an evolutionary point of view, pain gives us, in a sense, a false consciousness. Certainly in saying this I do not mean to deny that pain is a sign of a problem with our bodies. It certainly is. Instead, I mean to say that pain does not at all indicate to us that it is, in a very large sense, good. Or rather, to side with Hume against Moore, I should say not that pain is good but that it is creative. That is what a Darwinized Hume would say. Pain evolved in us as a proximate cause of survival and reproduction. The ultimate cause is natural selection, one of the agents of evolutionary change and the only agent of adaptive evolution. That agent, however, as Richard Dawkins (1986) so strikingly puts it, is none other than the "grim reaper" (62). By the invisible hand of this grim reaper, pain, suffering, and death become creative, not simply destructive. The very pain that we abhor is the pain that helps us survive and pass on our genes. It is the very same pain, moreover, that helped our hominid ancestors survive and their non-hominid ancestors before them. Indeed, the pain and death of untold numbers was and continues to be part of the creative aspect of evolution. That is what follows from evolutionary biology. This is not to advocate Social Darwinism. Far from it. But the evolutionary nature of pain does create a problem for one of the basic premises of utilitarianism. Pain is not so bad as utilitarianism makes it out to be, so that if we are going to fight against pain it would seem best to seek our principles from somewhere other than nature.

We might in fact go so far as to side with G. C. Williams and argue that nature is evil and morality ought to fight against it (an argument we shall examine in the next chapter). Or we might side with Hume and view nature as morally neutral.

Either way, the acid of evolution threatens to leak out in yet another direction against Singer's argument for animals. This is because, as many evolutionary biologists and anthropologists have argued, humans seem to have evolved in the direction they did, and to have out-competed all previous hominid species, because they gradually evolved a strong instinct for hunting and killing animals, as opposed to scavenging meat and bones or vegetarianism.[5] Even Albert Schweitzer himself was not a vegetarian, though one would naturally expect it given his famous teaching and lifestyle which he called *reverence for life* (Schweitzer 1933). It turns out that Schweitzer never practiced vegetarianism or imposed it on his missionary hospital in equatorial Africa (even though he helped both humans and animals at that hospital and was against hunting for sport) simply because he took it that evolution ingrained in humans such a deep instinct for meat that it was useless trying to deny it (see Berman 1989, 196–197). If one is going to take evolution seriously, then, if one is going to take a Darwinian view of life, one has got to take evolved instincts seriously too, something that Singer does not seem to do enough of, or he does so only selectively. Focusing on our instincts for altruism and cooperation is not enough. If it is true that humans, as a species, have a deep instinct for meat, then, sad to say, arguments and social conditions that foster ethical vegetarianism are going to have only marginal success at best. In short, not only are there always going to be far more Ted Nugents than Peter Singers, but more Albert Schweitzers than Peter Singers too.

The central idea explored in this chapter is that morality in humans is neither socially constructed, primarily rational, nor divinely revealed, but instead has a common core or denominator that is the product of our evolutionary past in hunting-gathering groups. As such, morality is not absolute (eternal and unchanging) and it is not personally subjective (a matter of what each individual thinks), but it is objective, in the sense that adaptive traits are objective in a species. For most people in this world, interestingly, morality (or at any rate "true" morality) cannot be separated from religion, either in the sense of divinely revealed commands and values or in the sense of a divinely implanted conscience (or both). In fact, for most people in this world, morality without religion is unimaginable. The question of ethics, then, naturally leads to our next big question, namely, religion.

---

5   See, e.g., Feder and Park (1993, 206–207, 235–238, 302–303). E. O. Wilson (1992, 244–253) is especially interesting for bringing together evidence and arguing that wherever human colonists arrived, mass extinction of species of large animals followed. This includes Europe, Asia, Australia, North and South America, and islands such as Madagascar and New Zealand. (Africa is the exception because humans evolved in Africa, which allowed animals to evolve an instinctive fear of humans.) In each case, argues Wilson, human hunting was the cause.

8
Evolution and Religion

If one were an alien anthropologist come to Earth from another planet to study humans, one could not help but be struck by a number of features that separate them from all other creatures on Earth. One is the ubiquity of religious beliefs and practices, both at present and throughout human history. Indeed, the human world teems with religion. Religions number in the many thousands, no matter how one defines the term "religion" within the normal usage of the term. This number becomes even greater if one includes cults, which one should, as the only real difference between a religion and a cult is in the number of adherents, and what are today called the "great religions" each began as a small cult. As such, almost every human subscribes to one or another religion (or cult). Moreover, almost every human believes they belong to the truest of all religions, while many believe that only their religion is the true religion and all other religions are false. Even those who belong to religions that are ecumenical in spirit believe in a hierarchy of religions, with their religion on top, a view expressed, for example, by the Dalai Lama with regard to Buddhism (Cabezon 1988). Moreover still, humans typically invest substantial, even enormous, time, energy, and material resources into the practice of their religion and are often willing to sacrifice their very lives for it.

All of this cries out for an explanation. One problem with trying to provide an explanation, especially from an evolutionary point of view, is that as most humans are religious to some significant degree, they will have a problem stepping out of their own shoes and viewing the matter objectively. But this is required for the topic. The fact is, from a logical point of view, these thousands of religions cannot all be true, as they contradict each other on a great many factual matters. Hence, one has to be open to the possibility that most of the central doctrines of religions are false, perhaps even all of them. And it will not help any to appeal to strength of belief. Strength of belief is a psychological matter, and the fact remains that "true believers" are found in every religion. Moreover, every true believer thinks that they have sufficient evidence for the truth of their beliefs and that true believers in other religions are simply misguided or deluded. But what true believers do not realize is just how much they have in common with true believers from other religions. Each tends to cite their own scriptures, but the scriptures of one are no more logically compelling than the scriptures of another. Often the supposed sufficient evidence is simply strength of belief itself. But again, strength of belief provides no evidence for the truth of what is believed; instead, it only tells us something about the nature of the mind that believes. Since that nature is so common, it raises the question whether evolution can help explain it.

In this chapter we shall look at three main questions. First, we shall examine whether it is useful to view religious beliefs and practices as *memes* subject to evolutionary processes. Second, we shall examine whether a religion instinct evolved in the human species, in similar fashion to the language and moral instincts examined in previous chapters. Finally, we shall examine whether evolution and religion can be legitimately combined into what is known as *theistic evolution*.

The concept of *meme* was first published in the eleventh chapter of Richard Dawkins's book *The Selfish Gene* (1976, 1989). As part of his thesis of Universal Darwinism (Dawkins 1983), Dawkins argues that evolution by natural selection is not domain specific, that it applies not only to biological evolution but to cultural evolution as well. As genes are the units of inheritance and selection in the biological domain, Dawkins coined the term *meme* for the units of transmission and selection in the cultural domain (*meme* is taken from the Greek word *mimeme*, not only because *mimeme* means "that which is imitated," but also because *meme* sounds like *gene*). Examples of memes, given by Dawkins (1976), are "tunes, ideas, catch-phrases, clothes fashions, ways of making pots or of building arches" (192). The study of memes is, quite appropriately, now known as *memetics* (the analog of genetics). Memes exist in *meme pools* (the analog of gene pools), the study of changes in meme frequencies in a meme pool is *population memetics* (the analog of population genetics), memes can band together into what are known as *meme complexes* (the analog of gene complexes), and they are subject to evolutionary forces such as mutation (as copies of memes are passed from brain to brain), drift (*memetic drift* is the analog of genetic drift), and natural selection.

In memetics, two dominant views of memes are to be found, namely, memes as replicators (the units of replication, the analog of genes) and memes as viruses of the mind (the analog of DNA and RNA viruses). Like genes, memes behave as if they were selfish. This is not to say, of course, that memes, or genes, have minds and are literally selfish. Dawkins makes this abundantly clear in his book. Rather, the sense of selfishness is a metaphor and purely behavioral. Memes, like genes, are selfish in the sense that they behave *as if* their only goal is to make more and more copies of themselves. Given that both genes and memes are really only units of information carried in very different mediums (Williams 1992, 11–16), one might go so far as to say that the ontology or nature of information is to disseminate itself. As Dawkins (1976) puts it with regard to memes, "What we have not previously considered is that a cultural trait may have evolved in the way that it has, simply because it is *advantageous to itself*" (200).

Memetics has indeed become a thriving field.[1] The perceptual and explanatory shift effected by memetics is to not only look at memes as real and as the constituents of culture, but more importantly to look at people not so much as having or acquiring memes as *memes having or acquiring people*. Lynch (1996) puts it as "Memetics . . . measures an idea's success by how much population it accumulates" (18). Dennett (1995) puts it more humorously as "invasion of the body snatchers" (342).

Memes want to do nothing but make more copies of themselves, but unfortunately for them, their hosts (our minds) exist in a limited quantity and each has a limited capacity to host memes. Hence memes exist in a field of competition. Accordingly, they are subject to evolutionary forces and can be expected to evolve strategies that improve their fitness (where "fitness" is defined in the good old way as the ability to survive and reproduce, i.e., reproductive success). Their evolution might improve the biological success of their hosts, or it might harm them, but in many cases it is going to be neutral. Indeed, as Dennett (1995) puts it,

> The most important point Dawkins makes . . . is that there is no *necessary* connection between a meme's replicative power, its fitness from *its* point of view, and its contribution to *our* fitness (by whatever standard we judge that).
>
> (363)

It needs to be added that the success of a meme or meme complex in terms of spreading to more and more minds need have nothing to do with its truth value (where truth is taken in the sense of correspondence to reality). In fact, many memes do not even have a truth value, such as the first four notes of Beethoven's Fifth Symphony (to use Dennett's example), a very successful meme whose spread does not even require the rest of its meme complex (the Fifth Symphony). For whatever reasons, some memes just naturally are "catchy" and easily spread from mind to mind, while other memes spread only with great difficulty, such as the idea of evolution by natural selection (something anyone who has ever tried to teach knows only too well). As Dawkins (1986) puts it, "It is almost as if the human brain were specifically designed to misunderstand Darwinism, and to find it hard to believe" (xi).

When applied to the phenomenon of religious beliefs and practices, memetics has become quite an attractive explanatory paradigm. Often one finds a comparison to viruses and their spread. Dawkins (1976), for example, focuses on the "God" meme, the "threat of hell fire" meme, and the "religious faith" meme. Why have these memes spread so easily in human history, seemingly like epidemics? It is not obviously because

---

1   Must reads are Dawkins (1976, ch. 11; 1993), Csikszentmihalyi (1993, ch. 5), Dennett (1995, ch. 12; 2006, app. A), Lynch (1996), Brodie (1996), Blackmore (1999), Aunger (2000, 2002), and Distin (2005). There are, of course, profound differences between genetic and memetic evolution, such as that memetic evolution operates generally much faster, memes are units of analog rather than of digital information, memetic evolution is in part Lamarckian, memetic mutation is often directed mutation, and there is no clear genotype-phenotype distinction for memes. These differences form part of the debate over memetics, along with whether it is a genuine science, all of which I cannot enter into here but the problems are amply dealt with in these readings.

they are true, because they are not obviously true. Indeed, the evidence for them is quite poor, as anybody who takes a competent course in philosophy of religion will find out. With regard to the "God" meme, Dawkins (1976) says its "survival value . . . in the meme pool," or its "infective power," resides in its "great psychological appeal" (193), mainly the answers it provides for the troubling question of the meaning of life, the comfort it provides for the pain and tragedy in life, and the hope it provides for rectifying the many injustices suffered in life.

What needs to be added, however, is that the "God" meme is really a *meme complex*, a complex that evolved over millennia in competition with polytheistic meme complexes. Anthropologists have long agreed that human religion began as a combination of ancestor worship and nature spirit worship (given the evidence from modern hunting-gathering tribes), from which evolved polytheism (given the records of the oldest religions). Monotheism came later on the scene. From the viewpoint of memetics, then, monotheism evolved to supplant polytheism not because of any truth value, but simply because it was more successful at spreading from one human mind to another. Why this would be so may be for the kind of reasons given by Dawkins. But those reasons speak to the nature of human minds, and that is where the infectious agent analogy runs into trouble. Epidemics, whether viruses or bacteria, harm their hosts. This might be said of memes such as the "God" meme. Indeed, many philosophers and scientists have held that religion does more harm than good, and Dawkins is one of them (see also Russell 1957, v–vii; Hitchens 2007). But the religion memes arguably do some good, at least from an evolutionary point of view (unless, of course, one takes survival and reproduction as bad). This is a point that I shall turn to shortly. But for the present we are not finished with the memetic point of view.

Dawkins (1976) also focuses on the "threat of hell fire" meme. Its function, of course, is to scare the hell out of believers into following the teachings of their leaders. Dawkins doubts that these leaders had the deep psychoanalytic insights into human nature that allowed them to see what a useful idea hell fire is, so that "Much more probably, unconscious memes have ensured their own survival by virtue of those same qualities of pseudo-ruthlessness that successful genes display" (198). The "threat of hell fire" meme spread "because of its deep psychological impact," which Dawkins says was linked to the "God" meme because "the two reinforce each other, and assist each other's survival in the meme pool" (198). This is an interesting idea. Of further interest and what needs to be added to it, however, is the fact that the "threat of hell fire" meme is not to be found in either the Old Testament or in ancient Greek religion and philosophy, the two background sources of Christianity. Instead, it is apparently a mutation, caused by the mixture of ancient Greek and Jewish ideas of the afterworld, its earliest known appearance being in *I Enoch* (22: 9–13) (Russell 1963, 154n), an apocalyptic book written around 100 BC while Jews in their homeland were fighting Macedonian oppression. The mutant meme spread in that environment and again when Judea was under Roman rule. When Christianity emerged in the same region, the meme quickly became part of the Christian meme complex and spread throughout the Roman empire via the spread of Christianity, reaching its greatest host population in the Middle Ages. The subsequent Age of

Enlightenment originated and spread memes that tended to confer immunity to the "threat of hell fire" meme, although we can still see it around today, something like malaria.

The "religious faith" meme is just as interesting. As usual, Dawkins pulls no punches. Religious faith, for Dawkins (1976), means "blind trust, in the absence of evidence, even in the teeth of evidence." Moreover, he says that "Blind faith can justify anything" (198). In the second edition of *The Selfish Gene* (1989) he is even more forceful, adding that "Faith is such a successful brainwasher in its own favour, especially a brainwasher of children, that it is hard to break its hold," so much so that "it is capable of driving people to such dangerous folly that faith seems to me to qualify as a kind of mental illness" (330). Dawkins (1993) adds that the "religious faith" meme evolved in ways to protect itself and further its propagation, specifically by combining with the "lack of evidence is a virtue" meme and the "mystery, per se, is a good thing" meme (138). Dawkins's three examples of the latter are the Catholic "Mystery of the Transubstantiation" meme (that the bread and wine served at Mass literally become the flesh and blood of Jesus), the "Mystery of the Trinity" meme (that God is three persons in one), and Tertullian's "*Certum est quia impossibile est*" ("It is certain because it is impossible"), to which one might add Martin Luther's condemnation of human reason as a "whore" and "the devil's bride" (Kaufmann 1958, 305).

But even if all of this is true (I'm inclined to think it is), memetics alone does not explain why human minds-brains are so susceptible to religious faith. What is needed, it would seem, is an evolutionary answer that goes much deeper than memetics, while at the same time not precluding it (given its attractiveness). In a number of previous chapters we have seen Alcock's (2001) point that when it comes to maladaptive instinctual behavior sociobiologists routinely employ as an explanation either the byproduct hypothesis or the novel environment hypothesis. Religious faith might well be at bottom an evolved instinct (I shall argue below that it is), and if it is we can then explain the maladaptive nature of it (when it is maladaptive) either as a byproduct (instincts are not perfect after all), as when it drives some cults and religions into self-destructive behaviors, or as the result of operating in a novel environment, an environment in which it did not originally evolve, namely, modern civilization (again, humans evolved in a very different environment, in small hunting-gathering groups or tribes).

But we are still not done with memetics. Other authors have attempted to add more substance to what Dawkins says about religion memes, and what they have come up with is quite interesting. Dennett (1995), for example, claims that many memes have evolved in such a way that they create an environment in their hosts that tends to prevent intrusion from competing memes. The religious faith meme, he says, does just that by discouraging critical reasoning. And indeed we can see this not only in the major religions but also in cults, where critical reasoning directed at the faith is often claimed to be a great sin, sometimes even an unforgivable sin, a claim that is sometimes accompanied with threat of violence. Indeed, Dennett predicts that religious faith memes will tend to flourish and "secure their own survival" in environments with "rationalistic memes" and will tend to lie dormant in "a skeptic-poor world" because in such a world they do not "attract much attention" (349). But again, I should think this would only follow if there is an evolved component to human nature that allows for the behavior of such memes, otherwise the prediction seems without a basis. Simply whether they attract attention is not a basis; indeed, it is no explanation at all.

Aaron Lynch (1996) provides more detail to Dawkins's framework by looking closely at the particulars of various religious doctrines and practices. With regard to Judaism, the commandment in Genesis to "be fruitful and multiply" ensures the proliferation of the Judaism memes by means of the parental transmission mode, which is especially important since Judaism evolved in an environment not conducive to horizontal transmission of memes by proselytizing or by force (99). For Lynch, it is all a matter of "how ideas program for their own propagation" (ix). In the Islamic religion, right from its start, the adversarial mode predominated, such that "God rewards those who fight and kill for Islam" (7). As Lynch points out, this stratagem succeeded either by killing those who refused to convert to Islam, thereby eliminating unreceptive hosts for the Islam memes, or by frightening non-Muslim hosts so they do not proselytize their own religion memes. Either way, Islam might in fact be the fastest growing religion in the world, but that need have nothing to do with truth value.

Indeed, it is attractive to look at the whole world of religion and cults as a highly competitive field of memes, each meme combining with other memes, recombination and mutation and natural selection resulting in the rough and tumble of ever-evolving propagation strategies, some strategies working better in some environments and not as well in others. Lynch (1996) has a fascinating chapter on the three big religions of Judaism, Christianity, and Islam, in which he finishes on a number of small religions or cults, namely, the Mormons, the Hutterites, the Jehovah's Witnesses, and the Shakers. Returning to Judaism, the first three of the Ten Commandments, he says (102), act directly to spread the Judaism memes, namely, the commandment to put the Jewish god above all others gods (hence God), the commandment to worship no other gods, and the commandment to not use God's name in vain. Similarly, the injunctions against marrying outside the faith, against divorce, and against adultery all serve to propagate the parental transmission of the Judaism memes (especially when the injunctions are said to be from God), and not surprisingly most other religions and cults have incorporated these memes as well. The same goes, argues Lynch, with regard to moral codes such as not killing and not stealing, as well as dietary laws. The first two, especially when said to come from God, increase the feelings of safety and trust, and hence aid in transmission of the religion memes, while the latter, again especially when said to come from God, makes conversion to another religion/culture more difficult and hence less likely.

In the case of Christianity, which bears much resemblance to its parental religion, Judaism, one sees some unique memes, argues Lynch, that make perfect sense from the perspective of memetics. Christianity added the idea of preaching the "gospel" (good news) to all, of proselytizing, which allowed it to survive and spread in spite of, or even more so during times of, great persecution. But that is not all. The idea of a savior, the son of God, who came and died to save all, not just a chosen few, aided the spread of the Christianity meme complex, as well the memes for eternal reward in heaven for belief and eternal punishment in hell for unbelief (unlike Christianity, Judaism has always focused on life here on earth). The memes for Resurrection and Armageddon, as well as the memes for the Lord's Supper and official forgiveness of sin (Holy Communion, Eucharist, Mass) serve to reinforce the heaven and hell memes. Similarly, the Christian meme for loving your neighbor as yourself aids the proselytizing program, both in connection with the eternal reward meme and also by making non-Christians more receptive to conversion. It also increases cooperation within Christian communities, providing socioeconomic benefits which in turn further help its propagation. All in all, as Lynch puts it, "'Love your neighbor' does double duty in memetic natural selection" (109). This served particularly well in the early centuries of Christianity, when Christianity was under Roman rule, but once Christianity gained political power, "evil apostasy," "infidel," and "heresy" memes were either added or gained prominence, which accounted for numerous "holy horrors" throughout the Middle Ages and beyond, toward both Christians and non-Christians alike (see Haught 1990; Hitchens 2007).

Islam, the most recent main addition to Western monotheism, borrowed much from its parent, Judaism, and from its older sibling, Christianity, but it involves some memetic strategies of its own which have made it rival Christianity in terms of number of hosts. Claiming to be the latest version of "Truth" certainly helps. But as Lynch points out, the requirement of public prayer five times a day especially helps preserve the Islam meme complex in populations with low literacy levels. Moreover, since it did not have any great empire to deal with in the early days of its rise, Islam benefited from what Lynch calls "religious warfare memes" (127), encapsulated in the concept of *jihad*, which allowed it to take over whole societies. Islamic law, moreover, provides the death penalty for Muslims who convert to another religion, while it forces pagans with the choice of either converting or dying. Christians and Jews are not faced with this choice, but are forced instead to pay special taxes (unless they convert to Islam, or, of course, emigrate), which lowers their capacity to support large families and pass on their faith through parental transmission. And naturally the promise of paradise filled with 72 young female virgins for each martyr for the faith inspires bravery in warfare as well as suicide/homicide terrorist acts in the name of Islam. The allowance of up to four wives at a time in this life, says Lynch, is a further meme that helps spread the Islam meme complex, since it redresses the imbalance in sex ratio caused by the death of males in jihad, allowing for more females to have children than in a monogamous system.

Just about every book on memetics has a chapter on religion. The phenomenon of religion seems to lend itself particularly well to memetic analysis. Unlike science memes, such as the "Earth is a planet" meme, or the "evolution by natural selection" meme, which spread mainly because of the force of evidence, religion memes do not spread because of evidence. If they did, there would not be thousands of different religions in existence today. Science as a whole, aside from its history, has a unity to it because there is only one reality and the main driving force of science is evidence of that reality. Biology and chemistry, for example, do not compete against each other like competing religions but instead complement each other, intersecting at the level of DNA. Religion is so utterly different, and the reason has to be because it is not grounded in evidence. But then we have to ask, why are religion memes so ubiquitous in the human species? Why are there so many religion meme complexes competing with one other? Why so much religion at all? The answer, I suggest, has to be rooted in human nature, in the evolved nature of the host minds-brains of religion memes.

Granted, memeticists have not ignored gene-meme interactions. Susan Blackmore's book, titled *The Meme Machine* (1999), is especially important here (and for memetics in general). The title, interestingly, is taken from the end of Dawkins's (1976) chapter on memes, in which he says, "We are built as gene machines and cultured as meme machines, but we have the power to turn against out creators. We, alone on earth, can rebel against the tyranny of the selfish replicators" (201). With regard to meme–gene interactions, Blackmore distinguishes between two situations, the one where the evolution of genes drives memes, the other where the evolution of memes drives genes. The former is the domain of sociobiology and evolutionary psychology, but not the latter, because the latter need not have anything to do with biological advantage. Instead, the latter is strictly about memetic advantage, hence it is the domain of memetics. Blackmore has some interesting hypotheses here, such as that meme evolution drove humans to evolve bigger brains (all the better for the proliferation of memes).[2] When it comes to religion, however, she sees it almost exclusively in terms of "meme–meme interactions" (110). Almost, but not quite. Blackmore speculates that memetic evolution could have caused the evolution of "genes for religious behavior" (197). Her main suggestion is that memes affected genes through *group selection*. Group selection requires very low inter-group migration rates, very high group extinction rates, as well as heritable group traits. For Blackmore, "Religions are a good example of a mechanism that decreases within-group differences, while increasing between-group differences and rates of group extinction" (200). Religions employ all sorts of what she calls "memetic tricks" to gain conformity within the group and hence spread the memes, such as the altruism trick, rules about sex and marriage, and dietary laws. These memetic tricks also serve to greatly reduce migration of individuals between groups.

Although we can see how memes might increase group cohesion, it is difficult to see how group selection driven by religion memes would operate in our religious past. But perhaps the greatest problem with Blackmore's thesis is the causal power she attributes to memes. Genes have causal power because their mediums, DNA and RNA, are physical molecules with physical properties. But if memes are to have a comparable causal power, it would be difficult to say it comes from their abstract nature alone. It would seem far more plausible that their causal power comes from something physical as well, in this case the neural structures in the brain. I do not

---

2 Apparently overlooked by all memeticists is the implication of memetics for *neoteny*, the evolution of juvenile characteristics into adulthood. In Chapter 1, we have seen Gould use this concept to explain why human nature is so variable. To extend Blackmore's thesis, it is possible that memes are responsible for neoteny in human evolution, for the fact that adult humans resemble juvenile chimpanzees far more than they do adult chimpanzees. Memes might have driven neoteny in early hominid evolution by exploiting the fact that juveniles are much more imitative than adults.

want to say that information, as information, has no causal power of its own. The issue is an enormous metaphysical and conceptual one. But the clear case of physical causation should lead us toward a more substantial evolutionary explanation of religion in terms of sociobiology and evolutionary psychology rather than the kind of explanation afforded by memetics. It is simply not enough to say, as Blackmore does, that "human minds and brains have been moulded to be especially receptive to religious ideas" (201). When it comes to religion, what is needed is an explanation for the proliferation and nature of religious thought and behavior rooted in an evolutionary view of human nature. After all, the world also teems with, say, fashion memes, but we do not see people swearing that they have the fashion "Truth," or willing to martyr themselves for fashion, or willing to kill innocent people in the name of fashion. To explain religion memes, then, we need to dig much deeper into human nature, far below the level of memetics.[3]

What happened in linguistics is very important here. Language is part of culture, and prior to the Chomskyan revolution (as we have seen in Chapter 3), language change was thought to be unconstrained. Since the Chomskyan revolution, however, the idea of a numerator–denominator distinction has prevailed. Comparative linguistics, sociolinguistics, and historical linguistics constitute the numerator, while psycholinguistics constitutes the denominator. What is studied in the denominator, from an evolutionary point of view, has been aptly called by Pinker "the language instinct." The language instinct both constrains and helps explain the nature of various natural languages. We have seen in the previous chapter an analogous numerator-denominator distinction for human morality. When it comes to human religion, the same idea, I want to suggest, applies as well. Memetics (and not just memetics) belongs to the numerator, while what might be called *psychotheology*, the study of the religion instinct, belongs to the denominator. But, of course, a good case for the existence of such an evolved instinct needs to be made.

In what follows I provide only a brief sketch, bringing together a lot of different material. What, we might ask, would an evolved religion instinct be like? From viewing the phenomenon of religion itself, the safest suggestion would be that it is not a single instinct but a *suite* of instincts (although for simplicity I shall continue to call it "the religion instinct"). The fundamental fact that we have to keep in mind

---

3   There is much more, it should be noted, to the view that religion is not a direct product of human evolution but instead a *byproduct* of the evolution of one or more adaptive products of human evolution. Dawkins (2006, ch. 5) himself defends this view, with his focus on the evolution of the mimetic capacity of the human brain. Dawkins also entertains further possibilities, such as the evolution of believability in children and the evolution of our capacity for falling in love. Interestingly, he compares religion to the phenomenon of moths flying into candle flames, which is not an adaptive behavior but instead a very harmful byproduct, a misfiring, of their light compass adaptation (172–173). But Dawkins's main approach remains memetics, with religion as a virus of the mind (188–201). The anthropologists Boyer (2001) and Atran (2002) focus on religion as a byproduct as well (though not so negatively), a byproduct of cognitive functions such as our emotional wiring, our social nature, our moral instincts, and our agent-detection systems. Much of the debate, of course, is empirical: it is a matter of bringing together evidence and weighing it against theory. But much of the debate is also conceptual, in that it hinges precisely on what one means by "religion," as we shall see in my treatment of religion below.

is that humans evolved in small hunting-gathering groups. One line of approach, then, is to think of the religion instinct evolving because it increased group cohesion. Group cohesion would be essential for group activities such as hunting, gathering, avoiding predators, raising infants, and the nomadic way of life in general. It would also be important for direct or indirect competition with other hominid groups. This approach, however, smacks of *group selection*, an idea that has had a checkered history over the past fifty years. Critics of group selection (e.g., Williams 1966) have argued that groups are too ephemeral to have group-related adaptations and therefore to function as units of natural selection, while explanation at the level of the individual or gene is simpler and more powerful. The criticisms against group selection are not so strong today, such that group selection has taken its place, albeit as a minor force, in modern evolutionary theory (e.g., Williams 1992, 45–49; Sober and Wilson 1998, ch. 1). But it does not matter. The dichotomy is, after all, false. The religion instinct might have evolved by group selection, simply because it increased group cohesion and hence fitness, or it might have evolved by individual selection, simply because those individuals within a group who tended to be more religious tended to reproduce more and hence pass on their genes with greater frequency, or it might have evolved by a combination of both, involving various mixtures of individual and group selection. The latter scenario will become more plausible as we look at candidates for instincts that together make up the religion instinct.

E. O. Wilson (1978), the don of human sociobiology, provides a chapter on religion in which he argues for the evolution of a religion instinct. This is no simple matter. Indeed, according to Wilson, "Religion constitutes the greatest challenge to human sociobiology" (175), first because religion is unique to the human species, and second because its genetic mechanisms are hidden from the conscious mind. Important for Wilson's analysis is his distinction between religion and theology. *Theology* is an intellectual enterprise, and Wilson sees it doomed to extinction from the encroachment of science. As science has progressed, theology has lost ground, increasingly giving up miracles, for example, and placing God's action either outside of the universe or somewhere inside subatomic particles, even reconfiguring God as an imperfect entity evolving along with the universe (process theology). But *religion* is something quite different. As Wilson puts it, "The predisposition to religious belief is the most complex and powerful force in the human mind and in all probability an ineradicable part of human nature" (169). This is not only because religion is fundamentally social and down to earth, but because it is visceral, not intellectual. Religion, quite apparently, is ultimately about struggles in *this* life. One has only to look around the world and see what people typically pray about. It is about getting enough food, about income, about marriage and children, about safety, about health and sickness, about surviving the harsh elements, about victory over enemies, and about dealing with death. In short, religion is about the mundane and everyday. This basic truth has been rediscovered again and again by every anthropologist and sociologist who has taken the time to study religion. If a religion instinct evolved in humans, then, it has to be looked at from this perspective.

And that is exactly what Wilson does. His basic idea is that religion increases group fitness, and it does this by actually being a suite of co-related instincts. Among these are obedience to a leader (especially susceptibility to a charismatic leader), hierarchical dominance systems, internal altruism, which he defines as "subordination of immediate self-interest to the interests of the group" (176), xenophobia (the fear or hatred of strangers), trophyism, the dichotomization of objects into sacred and profane, which he calls "binary classification" (181), trance induction, and the predisposition to indoctrination. With regard to the last of these, Wilson has stated elsewhere (1975) that "Men would rather believe than know" (285) and "Human beings are absurdly easy to indoctrinate—they *seek* it" (186). We can get our backs up and take exception to these statements, as Richard Lewontin (1991, 65) does, for example. But the fact remains that if we look at our species objectively we cannot help but notice how quickly and easily people, especially at the childhood stage, become indoctrinated into an ideology, whether religious or political and usually the ideology of their parents, how powerfully they take up the banner of righteousness, often to the point of killing and being killed, and how utterly difficult it is for them to change. Viewed objectively, it is hard to see indoctrinability as anything other than, as Wilson (1978) calls it, "a neurologically based learning rule that evolved through the selection of clans competing one against the other" (184).

But how might this have occurred? Wilson (1978) sees "three successive levels" of natural selection for religion. At the surface is what he calls "ecclesiastic selection," selection of varying beliefs and practices depending on their "emotional impact" (176). Selection here does not immediately involve genes. The next level is "ecological." As Wilson puts it, "If religions weaken their societies during warfare, encourage the destruction of the environment, shorten lives, or interfere with procreation they will, regardless of their short-term emotional benefits, initiate their own decline" (177). The third level is "genetic," the effect on gene frequencies that the first two levels have over long periods of time (i.e., in terms of many generations). As genes affect not only physical traits but also behavioral traits, selection forces will favor some behavioral traits, depending on the environment, and disfavor others. The result is a variety of instincts in the human species as a whole having to do with religion. Even the "emotional impact" that Wilson writes about has a possible genetic basis. According to Wilson, religions "congeal identity," they "classify him," they give one "unquestioned membership in a group claiming great powers," one's "strength is the strength of the group" (188).

But how might this "emotional impact" and other religious instincts have evolved in the sense of conferring biological, genetic advantage? Wilson sees both individual and group selection at work. Group selection would operate when conformity becomes strong in some groups and weak in others, such that groups with a higher frequency of religion genes will tend to successfully compete against those with a lower frequency, where personal conformity and sacrifice increases the fitness of the group as a whole even though it might lower the fitness of the individual. Individual selection would operate within the group, such that those individuals in the group that are relatively better than other group members at conforming would tend to reproduce more and hence pass on their genes with greater frequency, while those not as good at conforming would have a more difficult time at reproducing and might even suffer ostracism. Group selection and individual selection can operate against each other, but for Wilson

> These two possibilities need not be mutually exclusive; group and individual selection can be reinforcing. If success of the group requires spartan virtues and self-denying religiosity, victory can more than recompense the surviving faithful in land, power, and the opportunity to reproduce. The average individual will win this Darwinian game, and his gamble will be profitable, because the summed efforts of the participants give the average member a more than compensatory edge.
>
> (187)[4]

In all of this, it needs to be emphasized that religion, politics, and morality evolved together as a package, so that the division of them into different fields is mainly conceptual, not genetic. Science was never part of this package, no matter how often adherents of both sides use the phrase "love of truth." My main concern with Wilson's analysis is his suggestion at the very end that "The spiritual weakness of scientific naturalism is due to the fact that it has no such source of primal power" (192). Science, he says, in particular evolutionary biology,

> denies immortality to the individual and divine privilege to the society, and it suggests only an existential meaning for the human species. Humanists will never enjoy the hot pleasures of spiritual conversion and self-surrender; scientists cannot in all honesty serve as priests.

And so he asks, "Does a way exist to divert the power of religion into the services of the great new enterprise that lays bare the sources of that power?" (193). A more important question, it seems to me, is *should* the power of religion be injected into science? Given the nature of the religion instinct, I should have to say the answer is a loud and resounding *No*, that the nature of religious thought and practice is inimical to science and its underlying power would only serve to destroy it. Like Plato who argued that reason should rule the emotions and the visceral appetites below them, any other arrangement would seem to be a recipe for disaster, whether for the individual scientist or for science as a whole. Not all hot pleasures should be indulged in, and priesthood is not necessarily a good thing.

That science as science is spiritually impoverished and suggests only an existential meaning of life is a topic I shall reserve for the next and final chapter. That theology and evolution are incompatible is a topic for later in this chapter. But that the gonadic hormones of religion should not be injected into the brain of science becomes only more apparent the closer we look at the religion instinct. And indeed the idea of the evolution of a religion instinct has become quite an interesting research program, and I want to explore some of the developments in this program before we move on to the final section of this chapter, the section on theistic evolution.

---

4   The most ambitious explanation of religion in terms of group selection theory is provided by D. S. Wilson (2002). For criticisms, see, e.g., Dennett (2006, 179–188) and Dawkins (2006, 169–172).

In Chapter 1 we have seen experimental evidence suggesting that conformity in the sense of herd mentality or groupthink is an innate part of human nature. The experiments of Asch (1955) and Bogdanoff et al. (1961) are famous in psychology. Even more famous is the work of the social psychologist Stanley Milgram (1974), who in the early 1960s conducted a series of experiments on obedience to authority. In the basic experiment, each individual subject was told that the study was about the effects of punishment on memory and learning. After watching the "learner" being strapped into an electric chair, the subject was taken to another room where Milgram had a scientist in a lab coat instruct the subject, the "teacher," to administer increasingly painful electric shocks to the learner. The control panel of the "shock generator" had 30 switches, with labels ranging from 15 volts to 450 volts and "Slight Shock" to "Danger—Severe Shock." The learner, of course, was a plant, an actor, and no electric shocks were administered. The teacher, however, could hear the faked responses, which ranged from grunts to wild screams mixed with pathetic pleadings. The purpose of the study was to see how far people would go against their conscience and sense of sympathy and inflict pain on others because they were told to do so by someone they considered an authority. Milgram's results were shocking (excuse the pun). Conducted on more than a thousand subjects, he found that almost two-thirds continued to the last shock on the generator. Conscience and sympathy gave way to obedience to authority, to whom the sense of responsibility was transferred. Subjects cared more about whether they were living up to the expectations that the authority figure had for them, and they rationalized their actions as justifiable because they contributed to the greater good of society.

Milgram did not draw any implications of his work for religion (he was more concerned with the Nazis and with the war in Vietnam), but he did place his findings within an evolutionary framework. Beginning with the observation that members of our species "are not solitary" (123), Milgram argues that tribes with a division of labor (warriors, hunters, those who take care of children, etc.) would greatly enhance their chances of survival over tribes without a division of labor, and they would also have less friction and greater stability within them, the more those roles are defined and adhered to. What evolved in humans, then, was not a simple instinct for obedience, but what Milgram calls a "capacity for obedience" (125). This capacity he explicitly likens to Chomsky's universal grammar (though he does not mention Chomsky), such that "we are born with a *potential* for obedience, which then interacts with the influence of society to produce the obedient man" (125). Like Pinker's "language instinct," of course, I should think there is no problem calling the capacity for obedience an "obedience instinct." At any rate, for Milgram there might have been a time when people gave fully human responses to any situation, but "as soon as there was a division of labor among men, things changed" (11). Using cybernetic (control systems) theory, he argues that an inhibitory mechanism would have had to evolve in humans in order to allow for social organization, a mechanism that would inhibit innate impulses that would be destructive to fellow individuals in the organization. That inhibitory mechanism would be conscience. But for the next stage in social organization, hierarchical organization, a further inhibitory mechanism would have had to evolve, namely, obedience to authority. His ultimate point is that when the individual is working *on his own*, conscience is brought into play. But when he functions in an organizational mode, directions that come from the higher-level component [authority figures] are not assessed against the internal standards of moral judgment.

(129–130)

All of this, Milgram points out, involves statistical variation, both in the existence and operation of conscience and in the critical shift in attitude from what Milgram calls the "self-directed mode" to the "systemic mode," where "the person entering an authority system no longer views himself as acting out of his own purposes but rather comes to see himself as an agent for executing the wishes of another person" (133).

Interestingly, Milgram regards obedience to authority as "a fatal flaw nature has designed in us" (188). Of course, it was hardly a fatal flaw in the environment in which it evolved, but arguably in the modern environment known as "civilization," with our modern technological advances and hundreds of nations driven by competing ideologies, Milgram might be right in supposing that the obedience instinct "in the long run gives our species only a modest chance of survival" (188). Naturally, only time will tell.

At any rate, the implications for religion of Milgram's obedience instinct are easy to see. From stark examples such as the Inquisition, the burning of over a million women as witches, the Crusades, and terrorist suicide/homicide bombings, to more banal examples such as marriage ceremonies, ritual prayer, and dietary laws, the history of religion is replete with obedience to authority. And there should be no wonder why. Religion is enormously group oriented, with strongly pronounced in-group and out-group mentality. (In my own past, we were the "children of God," all others the "children of Satan.") Given that humans evolved in small hunting-gathering groups, groups with a division of labor including leaders, if there evolved a religion instinct, group mentality would have to be an inextricable part of it, along with obedience to a leader.

But when we think of religion, we think of much more, in particular we think of belief in supernatural agencies and places and in transcendental feelings. Is there any evidence that there is an innate predisposition to believe in the supernatural and to have subjectively transcendental experiences?

Certainly one can make consistency arguments. The biologist Donald Broom (2003), for example, provides the kind of argument we have seen in the previous chapter for the evolution by natural selection of a moral instinct in humans. But he argues further for the evolution of a religion instinct because, first, every religion has a morality at its core, and, second,

The religious framework makes it easier for the average person, or perhaps more importantly the likely transgressors of moral codes, to understand what should and should not be done. Those societies which formed such a framework were more likely to remain stable because anti-social, disruptive actions would have been less likely to occur.

(176)

Broom takes a very wide definition of religion: "a religion is a system of beliefs and rules which individuals revere and respond to in their lives and which is seen as emanating directly or indirectly from some intangible power" (164). This definition might be too wide, as we shall see below. But whatever the case, surely more is wanted by way of argument when it comes to the evolution by natural selection of a religion instinct. Every human society has involved walking at its core, shoes of some sort make it easier for humans to do their walking, but surely no one wants to argue for the evolution of a shoe-making instinct in humans. What is needed is much more than a "just-so" story.

In an essay review of the naturalistic study of religion, the evolutionary psychologist Joseph Bulbulia (2004) examines the arguments for an evolved religion instinct as opposed to arguments for religion as a byproduct of other evolved instincts (such as agent-detection systems, which are hyperactive and can make mistakes). Bulbulia comes down on the side of adaptationism, the side that views a trait as evolved specifically because it increases reproductive success. In line with Broom, he says it makes sense to think of an instinct for belief in all-seeing gods evolving in humans since "all-seeing gods impinge on our lives to hold us morally accountable" (667). The gods see what other humans cannot see, know what other humans cannot know, and they execute supernatural causation, rewarding the good and punishing the bad. Societies with humans who have this instinct would have an advantage over societies with humans who have it to a lesser degree or not at all, since "policing costs are substantially reduced in communities of prudent individuals who believe their transactions are perfectly policed by supernatural beings" (668). And all the better for the evolutionary account of this instinct if religiosity increases individual health and longevity. "Very basically," says Bulbulia, "if believing in supernatural causation helps us to recover from illness or meet the terrors of life, then tendencies to fall into such deceptions will be conserved and more intricately articulated" (680).

Of course, again, this could all be said to be a "just-so" story. What is interesting in Bulbulia's article is his main category of evidence, namely, recent psychological evidence of religiosity in small children. What we seem to have here is a growing body of "poverty of stimulus" evidence, evidence for the existence of a trait that cannot be explained by external stimulus alone. A number of recent studies have found that children below the age of five readily reason about the world in teleological terms, in terms of purposive entities with godlike powers which explain natural phenomena such as the pointiness of rocks or the existence of the sun. What makes this evidence the poverty-of-stimulus kind is that it does not matter what cultural background these children come from or whether they were raised in religious or non-religious homes. According to Bulbulia, then, "It may well be that a child's default theory of the world includes an 'intuitive theism.' Speculating further, it may be, as with language, that children are endowed with all possible religions, acquiring their religious idiolect largely by forgetting" (677). In other words, in religion acquisition the cultural environment does not really provide but rather triggers the specifics of the religion they will have, much like it triggers what language they will have. The disanalogy, of course, is that some people end up with no religion at all. But this could simply be a matter of them rebelling against the tyranny of their genes. With religion we can do that, with language we cannot. Either way, it does not weigh against them being evolved instincts.

There is further evidence to support the instinct view of religion, namely, evidence from identical twins, from neurophysiology, and from genetics. Waller *et al.* (1990), for example, studied religious interests, attitudes, and values in 53 pairs of identical twins and 31 pairs of fraternal twins, members of each pair having been reared apart, and compared these with the religious interests, attitudes, and values in 458 pairs of identical twins and 363 pairs of fraternal twins, members of each pair having been reared together, and came to the conclusion that roughly 50% of the observed measures for religiosity were genetically influenced.

Of equal interest is the evidence from neurophysiology. The Canadian neuroscientist Michael Persinger (1993), for example, has shown that seizures in the temporal and parietal lobe regions of the brain (roughly, the left and right middle and upper middle-rear regions of the outer brain, the cerebral cortex), as well as electrical and magnetic stimulation of these regions, can produce religious/mystical experiences, such as visions and out-of-body experiences, but generally a sense of presence which people interpret as God or some other mystical being. Interestingly, and this is relevant to the work of Milgram, Persinger found in a later study (1997) that, compared to the 7% of 1,480 university men and women who answered *yes* to the question whether they would kill someone if God told them to, 44% of the men who reported a religious experience, attended church weekly, and displayed at least partial epileptic signs stated that they would kill if God told them to. This is clear evidence of a link between Milgram's obedience instinct and a religion instinct.

Of further interest is the work of the radiologist Andrew Newberg and the psychiatrist Eugene d'Aquili, summarized in their book *Why God Won't Go Away* (2002). During the meditation sessions of a number of Tibetan meditators and the prayer sessions of a number of Franciscan nuns, they subjected their brains to a special kind of brain scan (single photon emission computed tomography) which detects the location of an injected radioactive tracer carried in the blood. Because the tracer is carried by blood and heightened brain activity is accompanied by greater blood flow, they could detect which parts of the brains of their subjects had heightened as well as reduced activity during the peak moments of meditation and prayer. What they found was greatly reduced brain activity in the posterior superior parietal lobe (the right upper middle-rear section of the cerebral cortex). The function of this section of the brain is to orient the individual in physical space. To do this, it must first generate a clear distinction between the self and everything else. Greatly reduced brain activity here means that this distinction is lost. Accordingly, the Tibetan meditators during their peak experiences reported a sense of union with the universe combined with a sense of timelessness and infinity, while the nuns reported a sense of union with God.

Newberg and d'Aquili suspect that this feature of the human brain evolved incrementally by natural selection because it increased survival and reproduction, but they do not believe it evolved directly as an adaptation. Instead, like the evolution in birds from wings with feathers that merely conferred temperature regulation and then were *exapted* for gliding and eventually flight, they conjecture that "the neurology of transcendence" in the human brain evolved first for mating and sexual experience. During orgasm, one experiences both a state of union and ecstasy, followed by a state of bliss. Interestingly, as Newberg and d'Aquili point out, the language of sexual pleasure (think of "Oh God! Oh God!") and the language of religious mysticism are basically the same:

> Mystics of all times and cultures have used the same expressive terms to describe their ineffable experiences: *bliss, rapture, ecstasy,* and *exaltation.* They speak of losing themselves in a sublime sense of union, of melting into elation, and of the total satisfaction of desires.
>
> (125)[5]

Moreover, mystical experience is often generated by repetitive and rhythmic stimulation—such as mantras in Eastern religions or prayers and songs in Western religions—akin to foreplay in sex. Mystical experience, then, on their view, originated possibly as "an accidental by-product" (126) of the evolution of sexual pleasure, which was then co-opted into the evolutionary direction of religious transcendence. Presumably (and they only vaguely fill in the necessarily link) this was accomplished because the beginnings of mystical experience opened the door to the sense of a higher being and a higher purpose, which over human evolutionary history aided hunting-gathering groups in their struggle for survival and hence in their reproductive success. As they put it,

> By lifting us out of fear and futility, and giving us the sense that wise and capable hands are steering the cosmic bus, religion has served as a powerful force of confidence and motivation that has not only shaped much of human history, but may also have been a crucial reason the human race has managed to survive.
>
> (132)

For Newberg and d'Aquili, religion *per se*, religions with their many beliefs and rituals, arose from the *interpretations* of mystical experiences, and they find it hardly a coincidence that the founders of religions and cults tended to be people prone to mystical, visionary experiences (e.g., Moses, Buddha, Jesus, Paul, Mohammed,

---

5   I will give two famous examples. Plato's mysticism consisted of intellectual union with the Forms, abstract essences (e.g., Triangle, Beauty, Good) existing outside of space and time. In the *Republic* (490b), he describes this intellectual union as an "erotic love" for truth, such that the true philosopher "neither loses nor lessens his erotic love until he grasps the being of each nature itself with the part of his soul that is fitted to grasp it, because of its kinship with it, and that, once getting near what really is and having intercourse with it and having begotten understanding and truth, he knows, truly lives, is nourished" (Grube and Reeve 1992, 163–164). Consider next the orgasmic vision of an angel by the 16th-century Catholic mystic Saint Teresa of Avila: "In his hands I saw a great golden spear, and at the iron tip there appeared to be a point of fire. This he plunged into my heart several times so that it penetrated my entrails. When he pulled it out, I felt that he took them with it, and left me utterly consumed by the great love of God. The pain was so severe that it made me utter several moans. The sweetness caused by this intense pain is so extreme that one cannot possibly wish it to cease, nor is one's soul then content with anything but God. This is not a physical, but a spiritual pain—though the body has some share in it—even a considerable share" (Cohen 1957, 210).

Emanuel Swedenborg, Joseph Smith, and Jim Jones). Via the evolution of mystical transcendence, then, they find it "very likely that natural selection would favor a brain equipped with the neurological machinery that makes religious behavior more likely" (139).

Finally, building on twin studies and the kind of work provided by Persinger and also Newberg and d'Aquili, the behavior geneticist Dean Hamer, whose work on homosexuality we examined in Chapter 4, has put forward evidence for a specific genetic component to religion in his book *The God Gene* (2004). By "the God gene" Hamer does not claim that he has found *the* specific gene responsible for religious beliefs and behaviors, for what he calls "spirituality" (indeed no one gene could possibly do all that), but rather *a* gene (no doubt one of many) that plays a specific role in what he does indeed call "an instinct" (6). The gene is actually an allele of the VMAT2 gene, named A33050C for its specific location on chromosome 10 and for the single DNA letter change (C rather than A) that distinguishes it from the common form of the gene. What Hamer found is that

> There was a clear association between the VMAT2 polymorphism and self-transcendence. Somehow, this single-base change was affecting every facet of self-transcendence, from loving nature to loving God, from feeling at one with the universe to being willing to sacrifice for its improvement.
>
> (73)

This startling conclusion is the result of a number of pieces of evidence. First, the VMAT2 gene codes for a protein responsible for the packaging of monoamines (namely, dopamine, adrenaline, noradrenaline, and serotonin), which are "the biochemical mediators of emotions and values" (103). Second, monoamines are linked to no human personality trait other than the propensity for mysticism (77, 82–89). Third and finally, Hamer and his team examined 1,001 subjects—men and women from a variety of age, education, and racial backgrounds—each of whom had been subjected to personality testing and placed on a personality scale, which measured specifically for self-transcendence, i.e., "people's capacity to reach out beyond themselves—to see everything in the world as part of one great totality" (10). These subjects then had their genotypes sequenced, specifically for the VMAT2 gene. What Hamer and his team found was that, in Hamer's words, "There was a clear association between the VMAT2 polymorphism and self-transcendence. Individuals with a C in their DNA—on either one chromosome or both—scored significantly higher than those with an A" (73). The difference was statistically significant, but not so significant as to be a matter of either-or. Twin studies have shown that the heritability of religiosity is roughly 50%, meaning that the contribution of genes is roughly 50% and the contribution of the environment is roughly 50%. But more than that, the statistical difference in the VMAT2 study was significant only to the point of showing that VMAT2 was one of many genes that influence the religion instinct (in the sense of self-transcendence). Hamer supposes that "There might be another 50 genes or more of similar strength" (77). Nevertheless, he concludes that the A33050C allele "was enough to tip the spiritual scales and predispose one toward spirituality" (74).

Such is the nature of genetics. When geneticists say that a particular gene is "for" something or other, all they mean is that that gene contributes to that effect. Or as Dawkins (1989) puts it, "to speak of a gene 'for' something only ever means that a *change* in the gene causes a *change* in the something" (281). In spite of this, there is nothing wrong with asking what the selective advantage of a particular gene might be, recalling that the selective advantage of a gene is always going to cash out ultimately in terms of reproductive success. Hamer provides a number of possible selective advantages for the VMAT2 gene in an attempt to explain why it would be favored throughout human evolutionary history. He points out that dopamine makes people "happy, confident, and optimistic," so that someone with more dopamine "may be more likely to get up and hunt for food, build a shelter, and—above all—want to have children" (159). He adds that dopamine also plays a role in novelty seeking, such that those with more dopamine would be disposed toward having more sexual partners, something that is of interest given the analysis by Buss that we have seen in Chapter 4. As for serotonin, Hamer points out that a change in serotonin levels (it is not sure whether it is an increase or a decrease) can result in a person feeling lonely, depressed, and worried about the future, but the same shift is often associated with a higher libido, which again is relevant to reproductive rates.[6]

Hamer admits that the evolutionary role of these chemicals is speculative at best and for proof "we would need to replay the tape of human evolution with and without these 'God genes'" (160). Barring that, however, something more still seems to be needed. Since it is really A33050C that is Hamer's "God gene" (or really "God allele"), the one that tips "the spiritual scales," he needs to provide an argument for why natural selection would favor this allele. He has not done that. Perhaps arguments provided by Wilson or by Newberg and d'Aquili would fill this glaring lacuna. The problem would then be that of testing the theory. The only suggestion I can think of is a massive study that compares the reproductive rates of people with

---

6   One has to be very careful here. Just because something statistically increases libido, that does not automatically mean that it would statistically increase reproductive success. The evolutionary psychiatrists Stevens and Price (1996, 63–71) provide the standard arguments for the evolution of the biochemical mechanisms of depression. According to *attachment theory*, depression evolved because it helps one deal with loss and bereavement by ultimately achieving detachment, at which point the individual can adjust to new circumstances and get on with life. More relevant is defeat depression. According to *rank theory*, depression evolved because it increases social cohesion. When one loses rank and is relegated to a lower social status, depression would inhibit attempts at regaining rank (which would typically be violent), and it would also indicate acceptance of the lower rank to those of higher rank. The lower friction within the group would result in greater social cohesion, so that genes for depression would be favored over evolutionary time in the struggle for survival against predators, other hominid groups, or the environment in general. Evolution of defeat depression would not increase individual reproductive success directly, because it involves lower rank in the social hierarchy, but it would be favored by selection because it increases the reproductive success of the group as a whole. Hence even if it is true that changes in serotonin bring on depression and at the same time increase libido, it by no means follows that mechanisms for a change in serotonin evolved because they increased *individual* reproductive success.

the A33050C allele with the reproductive rates of people who do not have it, a study that would have to cross many cultural and racial boundaries.[7]

Interestingly, according to Newberg and d'Aquili (2002) the central message of what they call *neurotheology* is that "science and religion do not have to be incompatible: One need not be wrong for the other to be right" (173). In fact they go further, stating that they

> believe that we saw evidence of a neurological process that has evolved to allow us humans to transcend material existence and acknowledge and connect with a deeper, more spiritual part of ourselves perceived of as an absolute, universal reality that connects us to all that is.
>
> (9)

Further still, they believe that the "miraculous organ" they studied, the posterior superior parietal lobe, is "the link between mind and spirit" (10). Even Hamer (2004) does not want to draw anti-religious conclusions from his research on "the God gene." "If God does exist," he says, "he would need a way for us to recognize his presence" (211).

These suggestions bring us to our final topic in this chapter, namely, whether theology and evolutionary biology are genuinely compatible. The combination usually falls under the umbrella term *theistic evolution*. There are a number of ways that one can look at this approach. At a level we have already examined, theistic evolution might simply be a memetic strategy evolved to help propagate religion memes in environments that are no longer suitable for the spread of pure religion memes and pure science memes, something like the evolutionary symbiosis of lichens from algae and fungi (see Stamos 2003, 334–335). This is a mere suggestion and I will not pursue this sense of compatibility any further. What I want to pursue is something deeper. Whether theology is really compatible with science all depends, of course, on what one means by "compatible." If the sense is that of logical consistency (where two statements are consistent simply if both can be true), then of course many theologies can be made consistent with modern science. But that does not mean they are compatible in a deeper, more relevant sense, the sense akin to compatibility in marriage. (Just because two people live together in marriage does not mean they are compatible.) Certainly the view that theology and science are compatible does not follow from the mere fact that some people have combined them, even some very intelligent people. One has to get over the temptation to think that because someone very intelligent believes something it is therefore legitimate to believe it. This follows from the fact that very intelligent people typically hold beliefs that contradict one another. At any rate, what I shall argue below is that the *spirit* of theology is deeply

---

7   In defense of Hamer, it is surely significant that he found that while only 28% of his 1,001 samples had the A33050C allele, the allele is apparently dominant (meaning that it needs to be found on only one of a homologous pair of chromosomes in order to be expressed), since fully all of the 28% was among the 47% who scored in the "higher spirituality group" (73–74). This is a remarkable finding in itself, and it opens the door to the possibility that natural selection has been playing a role in spreading the allele. But still, further evidence is needed to raise the possibility to the level of a respectable theory, and Hamer has not provided this—which is not to say that none can be found.

incompatible with the *spirit* of science. To accomplish this task, we need to take a tour of some representative examples of theistic evolution. But preceding that, we need to take a look at what biologists typically conclude about directed versus undirected (contingent) evolution.

We begin with some famous and characteristic examples, all focusing on the contingency of the evolution of humans (which is typically the bone of contention). The paleontologist George Gaylord Simpson, one of the architects of the Modern Synthesis, stated (1949) that

> Man is the result of a purposeless and materialistic process that did not have him in mind. He was not planned. He is a state of matter, a form of life . . . akin . . . to all of life and indeed to all that is material.
>
> (344)

The molecular biologist and Nobel laureate Jacques Monod (1971) stated that "Pure chance, absolutely free but blind, at the very root of the stupendous edifice of evolution: this central concept of modern biology is no longer one among other possible or even conceivable hypotheses. It is today the *sole* conceivable hypothesis, the only one that squares with observed and tested fact" (112–113). E. O. Wilson began his book on human nature (1978) by stating, "If humankind evolved by Darwinian natural selection, genetic chance and environmental necessity, not God, made the species" (1). The paleontologist Stephen Jay Gould, in spite of his thesis that science and religion comprise "two nonoverlapping magisteria" (a thesis we shall return to below), concluded (1989b) that

> if you wish to ask the question of the ages—why do humans exist?—a major part of the answer . . . must be: because *Pikaia* survived the Burgess decimation . . . I do not think that any 'higher' answer can be given.
>
> (323)

And then of course there is the Oxford zoologist Richard Dawkins, who has stated (1986) that "although atheism might have been *logically* tenable before Darwin, Darwin made it possible to be an intellectually fulfilled atheist" (6), since "slow, gradual, cumulative natural selection is the ultimate explanation for our existence" (318). For Dawkins and most professional biologists, the contingencies of environmental change, of mutation, and of processes such as natural selection and genetic drift mean that (1995) "The universe we observe has precisely the properties we should expect if there is, at bottom, no design, no purpose, no evil and no good, nothing but blind, pitiless indifference" (133).[8]

---

8   I say "most professional biologists" based on a number of recent surveys. While roughly 40% of American scientists believe in a God influencing evolution (Scott 1997, 405)—the number is going to be lower in many other countries, such as England or Germany—where "scientist" is taken widely to include those in applied sciences, such as physicians and engineers, the rate drops significantly with increased levels of education and accomplishment, such that at the elite level, represented by the roughly 1,800 members of the National Academy of Sciences, the rate of belief is a little under 10%, with biologists possessing the lowest rate (5.5%) and mathematicians the highest (14.3%) (Larson and Witham 1998, 1999).

All of this is sometimes called *atheistic evolution*, and there is no doubt that atheism is compatible with evolutionary biology (no matter what sense of "compatible" one chooses). Ever since Darwin, gone are the days when gaps in biology needed to be filled by appealing to God. Nevertheless, many outside of biology have taken exception to the atheistic interpretation of biology proffered by most biologists. David Young (1992), for example, in the conclusion of his textbook on the history of evolutionary thought, complains about Simpson's "jumping from scientific results to philosophical conclusions" (234). Similarly, Eugenie Scott (1997), Executive Director of the anticreationist National Center for Science Education, claims, with Simpson primarily in mind, that "Some prominent scientists do indeed confuse personal philosophy and science" and that "We cannot say that there is no absolute or ultimate 'plan or purpose' to life without stepping outside of what the empirical data can show us" (404). Against Dawkins, Michael Ruse (1999), the don of philosophy of biology, accuses Dawkins of "militant atheism" and spouting a "secular religion" (132), such that "In his animus toward religion, real nonepistemic values surely are showing through" (134). For Ruse (1997), instead, although himself an avowed skeptic on both God and immortality, "If you want evolution plus souls, that is your option, and if you want evolution less souls, that is also your option. Either way, evolution is untouched" (394). What he calls "tolerant disagreement" is his position. Much less harshly against Dawkins, Ian Barbour (1997), a physicist and theologian, states that Dawkins (1986) "makes some rather dogmatic philosophical and religious statements without careful discussion" (243) and that with his "atheistic philosophy" he has "failed to respect the proper boundaries of science" (244).

The first thing that strikes me in all of this is how readily critics of atheistic evolution dismiss it with the label "philosophical," as if philosophy is the intellectual equivalent of finger painting (i.e., simply expressing how you feel). This approach might appease the sentiments of our religious and politically correct time, but nonetheless it violates scientific and philosophic duty, where "duty" is taken to mean *getting to the truth of the matter as closely as one can.*

This brings me to the second thing that strikes me in all of this, namely, the narrow and constricted concept of science shared by those who criticize the expressions of atheistic evolution that we have seen above. Something definitely needs to be said about "the proper boundaries of science" before we continue with the issue of atheistic versus theistic evolution. One cannot attempt to settle an empirical matter simply by defining one's terms in a way that settles the matter in one's favor. In logic this is known as the fallacy of *question-begging definition.*

The nature of science, as opposed to pseudo-science and non-science, is, as one might expect, a central question in philosophy of science. The literature is enormous, and I can do little justice to it here except to brush in broad strokes. But the picture is no less accurate for it. In brief, long gone are the days when it was thought that science is *inductive*, proceeding from a build-up of particular observations and particular facts to general theories and general facts. Part of the problem with this picture is that observation requires theories in the first place. As Popper (1963) put it, "Observation is always selective. It needs a chosen subject, a definite task, an interest, a point of view, a problem" (46). Another problem is that scientists often do not proceed inductively anyway, but generate theories instead as a way of problem solving. Moreover, they often introduce concepts into their theories that are nowhere to be found among their premises.

What followed the strictly inductive view of science was a conception of science where *testability* was at the core. It did not matter where a theory came from. What mattered instead was whether it made predictions and was testable, either in the sense of *confirmation* (e.g., Hempel 1965, 1966) or *falsification* (Popper 1959, 1963). What typically went hand in hand with these views was the idea that a genuine scientific theory is a proposed law of nature. Laws figure in both explanations and predictions, so that in real science a proposed law, a scientific theory, was thought to be one that was repeatedly brought before the court of testability. More recently some have favored *experimentationism* (e.g., Hacking 1983), the view that genuine science is characterized by the use of controlled experiments. The main problem with all of these approaches is that they are based on physics and chemistry, not on science as a whole. Physics and chemistry are law-based sciences, so the above views remain attractive so long as one focuses only on those sciences. But once biology is taken seriously in philosophy of science, as it has in recent decades, especially evolutionary biology (which is not law-based but instead history-based), then the conception of science based on physics and chemistry is no longer acceptable.

What has happened is the expansion of the concept of science so as to accommodate what are clearly genuine sciences such as biology. Two approaches have converged here. One is the view of science as *inference to the best explanation*. In contrast to models of scientific explanation briefly discussed above, inference to the best explanation not only requires competing theories but employs *contrastive explanation* to explain "why this and not that" (Lipton 1990, 1991), as discussed in the previous chapter.

Contrastive explanation is at the very core of the conception of science as inference to the best explanation and we can see a clear example of it in Darwin's methodology. In arguing *for* evolution by natural selection, Darwin was also arguing *against* creationism (as well as Lamarckian evolutionism), showing why the latter was not a good explanation of the evidence. Inductivist, confirmationist, falsificationist, and other models of science fail to capture this.

The case of Darwin also highlights the other modern approach to science that converges with inference to the best explanation, namely, the *epistemic values* approach. This is a cluster class approach, such that for something to be science it need not involve any one or a particular set of epistemic values. Instead, inclusion depends only on how many epistemic values are involved and to what degree they are involved. Hence, inclusion in genuine science is not a matter of yes or no but of degree. The values distinguished by this approach are not always the same from advocate to advocate, but in general they include simplicity, testability, objectivity, consistency, consilience, and fruitfulness.[9]

---

9   See, e.g., Kuhn (1973, 321–322; 1983, 568), Quine and Ullian (1978, 66–79, 98), Kitcher (1982, 44–49), McMullin (1983, 15–20), Dupré (1993, 243).

In the case of Darwin, we can see that he employed them all: simplicity in the sense that he kept his explanations down to earth and his explanatory principles to a bare minimum; testability in the sense that he often made predictions that would either confirm or falsify his theory; objectivity in the sense that he was not motivated primarily by what he wanted to believe; consistency in the sense that his theory was not only internally consistent but was consistent with, for example, the latest developments in geology; consilience in the sense that his theory drew on and in turn explained a wide variety of facts, such as those found in geology, paleontology, biogeography, comparative anatomy, embryology, ethology, domestic breeding, and taxonomy (indeed, it is remarkable how much Darwin explained with so few principles); and fruitfulness in the sense that his theory opened up other fields of research such as evolutionary psychology.

Inference to the best explanation and an epistemic values approach to science work hand in hand, in the sense that it is the epistemic values that determine which of the competing explanations is the best one overall. With regard to consilience, for example, a creationist can, with some plausibility, raise his cavils to each one of Darwin's fields of evidence taken individually, but that plausibility vanishes once those fields are viewed collectively. If consilience is an epistemic value, then evolution is clearly the best explanation. And so on for simplicity and the rest of the epistemic values. (Indeed, when it comes to epistemic values, creationism scores remarkably low.) It was Darwin's great achievement to bring it all together and to make the first genuinely scientific case for evolution as the best explanation of the diversity of life. Modern professional biology is simply the expansion of this program on a massive, worldwide scale.

Keeping all of this in mind, when we turn to atheistic evolution we can see that it is no mere philosophy but that it fits perfectly with this expanded, thoroughly modern concept of genuine science. Atheistic evolution is simple (it is completely down to earth and involves nothing supernatural), it is objective (it is not what one wants to believe), it is internally consistent and externally consistent with what is known in the other sciences (such as physics and chemistry, which also share methodological naturalism), it is consilient (this is what makes it a universal acid), and it is fruitful (it has led to memetics, for example). But is it testable? One thing to remember is that not every genuinely scientific theory makes predictions, so that not every genuinely scientific theory is testable, string theory in physics being a common example (see Newton-Smith 1995, 26; Galison 1995). But even so, atheistic evolution *is* testable. At the genetic level alone one can conceive of observations that would refute it, such as a string of sequences in DNA that would be ridiculous to explain by chance or by natural selection.

Just as importantly, atheistic evolution is inference to the best explanation at its best. In order to see this, however, we need to turn to theistic evolution. What we shall see is just how utterly different the two are.

As mentioned before, theistic evolution is an umbrella term for the combination of theology and religion. Perhaps the best entrée to this field is the view of Stephen Jay Gould. For Gould (1997), science and religion comprise what he calls "two nonoverlapping magisteria," a principle he dubs NOMA, where the domain of science is "what is the universe made of (fact) and why does it work this way (theory)" and the domain of religion is "questions of moral meaning and value" (274).

In reply to this view, Dawkins (1997) points out that Gould's NOMA principle is untenable since "Religions make existence claims, and this means scientific claims" (399). In the case of Christianity, think of the claim that God exists, that Jesus existed and was born of a virgin, that he walked on water and raised the dead, and so on. But there are two problems with Dawkins's reply. First, existence claims are not necessarily scientific claims. Plato, for example, claimed that this world is only partly real and that the fully real world consists of abstract essences existing outside of space and time, the world of the Forms. Under no respectable concept of science is this a scientific claim. But even more seriously, Dawkins has missed Gould's central point. Gould was quite aware that religions often make existence claims. Over the years he wrote numerous articles against Christian creationism, for example. What comes out clearly in his book in which he develops his NOMA principle (Gould 1999) is that NOMA is not a *descriptive* thesis but rather a *prescriptive* thesis. NOMA is the way science and religion *should* be. Imagine the worlds of science and religion enjoying "respectful noninterference—accompanied by intense dialogue between the two distinct subjects, each covering a central facet of human experience" (5).

You know what they say (never mind who the proverbial "they" are), "If it sounds too good to be true, it probably is." And that, I believe, is the case with NOMA. It is simply too good to be true. First, *should* implies *can*. But *can* religions confine themselves only to moral matters and the meaning of life? Perhaps certain theologies can, but not *religions* in the down-to-earth sense (recall Wilson's distinction). Their claims about morality and meaning *depend* on their truth claims about matters of fact. Christianity, for example, is a historical religion, it is largely based on claims of historical fact, and St. Paul was surely right when he claimed that "if there is no resurrection of the dead, then Christ has not been raised; and if Christ has not been raised, then our preaching is in vain and your faith is in vain" (*I Corinthians* 14: 13–14).

Second, for Gould (1999) NOMA "cuts both ways" (9, 164). Scientists should no more cross the line into matters of morals and meaning than religionists should cross the line into matters of fact and theory. But this is hardly desirable. The field of morality, for one, is a hodgepodge mess until it is made sense of in some way by theory. And as we have seen in the previous chapter, evolution is highly relevant to the topic of morality, so much so that, arguably, it would be an enormous mistake to keep the two separate. The same, as we shall see in the next chapter, is true for the topic of the meaning of life. To return to morality, Dawkins (1997) is surely right when he says that even religious people do not use religion as the bedrock of their moral beliefs. "In practice," he says, "no civilized person uses scripture as ultimate authority for moral reasoning. Instead, we pick and choose the nice bits of scripture (like the Sermon on the Mount) and blithely ignore the nasty bits (like the obligation to stone adulteresses, execute apostates and punish the grandchildren of offenders" (397). The same is true, says Dawkins, when it comes to picking and choosing our religious leaders. All in all, then, it would appear that religion cannot be the foundation of morality.

202 | EVOLUTION AND RELIGION

Indeed, Gould's (1997) "respectful, even loving, concordat" is a pipe dream, no matter how good were his heart and intentions. It is as if Gould viewed science and religion as two separate and very different biological species, occupying the same habitat but different niches. This analogy might help to understand his further claim (1999) that "I do not see how science and religion could be unified, or even synthesized, under any common scheme of explanation or analysis; but I also do not understand why the two enterprises should experience any conflict" (4).

While many will agree with the latter claim, many of these same people will disagree with the former claim. This brings us to *theistic evolution* proper, to the view that God is in some way involved with evolution. There are two basic approaches here. One is to suppose that God set up the conditions for evolution but once evolution started he did not play a guiding role. The other is to suppose that God continued and is still continuing to play a guiding role. Let us look at some prominent examples, beginning with the former approach.

Arthur Peacocke (1985), a professional biochemist and Anglican priest, sees "God as 'exploring' in creation" (125), having constituted the universe with chance and necessity (law) so that "exploration of all the potential organized forms of matter (both living and non-living) were to occur" (117), such that "In the actual processes of the world, and supremely in human self-consciousness, God is involving himself and expressing himself as creator" (124–125). Paul Davies (1992), the Australian physicist famous for his argument for the probable existence of God mainly from the apparent "fine tuning" of the basic laws and constants of physics,[10] proceeds to "the clear implication that God has designed the universe so as to permit . . . life and consciousness to emerge. It would mean that our own existence in the universe formed a central part of God's plan" (213). Ian Barbour (1997), a physics professor and accomplished theologian, acknowledges that "It would be grossly anthropocentric to assume that we are the goal or the only purpose of creation" (220). Nevertheless, he believes that "There can be purpose without an exact predetermined plan" (216), that "The emergence of intelligent persons would be a plausible goal of an intelligent, personal God" (219), that in human intelligence "we can see here the work of a purposeful Creator" (220), and that there may be "beings superior to us" (215).

It will be noticed that none of these thinkers are evolutionary biologists. Unlike other natural scientists, evolutionary biologists are much more sensitive to the nature of contingency in evolution (thus vastly reducing, contra Peacocke, the ratio of actual to potential forms). For them, any of innumerable contingencies would have been sufficient to prevent humans, or anything like them, from evolving on Earth. They even extend this thesis to the improbability of life, especially intelligent life, occurring elsewhere in the universe.[11] Raup (1991), for example, argues that it was not bad genes but bad luck in the form of a large meteor impact that ended the 100 million year rule of the dinosaurs, allowing for the subsequent adaptive radiation

---

10   I will have more on the fine-tuning argument in note 1 of the next chapter.
11   See the papers in Regis (1985), especially for the recurring theme that physicists tend to be optimistic and biologists pessimistic.

of mammals, without which we humans would not be here. Gould (1989b) argues that not only could the dinosaurs have ruled for another 100 million years or more, but that they were not evolving larger and more complex brains, so that "we must assume that [human-like] consciousness would not have evolved on our planet if a cosmic catastrophe had not claimed the dinosaurs as victims" (318). Gould adds other considerations, such as a mutated virus that could have wiped out early *Homo erectus* or even earlier a change in climate that could have converted the African savanna (which brought our distant ancestors out of the trees) back into an inhospitable forest. Thus, says Gould, "*Homo sapiens* is an entity, not a tendency" (320).

Theologians will quickly reply that they have long claimed that God alone is necessary and that everything else is contingent, including mankind. But this confuses contingency with goal. The belief that God has loaded the dice in our favor, whether indirectly by setting the initial conditions and laws of nature or more directly by influencing mutations and meteors, is a far cry from the nature of contingency found in modern evolutionary theory. To think otherwise is to share the illusion typical of lottery winners.

Not all theistic evolutionists ignore the emphasis of contingency in evolutionary theory. A good example is the cell biologist Kenneth Miller. In his book titled *Finding Darwin's God* (1999), Miller—following excellent chapters rebutting in detail the twisted claims and rhetorical tricks of creationists such as Duane Gish and Henry Morris and neo-creationists such as Phillip Johnson and Michael Behe—proceeds to argue that God chose to create by setting up a system of evolution by chance variation and natural selection so as ultimately to procure beings with genuine free will: "free beings in a world of authentic and meaningful moral and spiritual choices" (291), beings that "could know Him and love Him, could perceive the heavens and dream of the stars," and that "would eventually discover the extraordinary process of evolution that filled His earth with so much life" (238–239). Preordained creatures could not be free creatures (273). Thus, in Miller's view, the evolutionary system of nature, with all the pain, suffering, and killing that is built into it, is "the only way" (279, 291) that God could get the happy accidents he wanted, namely, creatures like us with genuine free will. Never mind the horrendous callousness that such a view ultimately attributes to God (I shall return to the problem of evil below). And never mind that the concept of free will is plagued with enormous difficulties, not the least of which is that quantum indeterminism does not entail free will (see Searle 1984, 86–87). Miller, like so many others, ignores all of this and evidently believes only because he *wants* to believe.

Indeed, theistic evolutionary biologists tend to find themselves caught in what my late biology mentor Robert Haynes (1993) called *intellectual schizophrenia* (150), something akin to Orwell's *doublethink*. It is a schizophrenia (of course not in the technical sense) characterized by the superimposition of opposing belief systems. Perhaps the best example of this is that of the geneticist Theodosius Dobzhansky, one of the architects of the Modern Synthesis. As both a scientist and an avowed Christian, he believed that "Evolution is God's . . . method of Creation" (Dobzhansky 1973, 382) and that "Evolution (cosmic + biological + human) is going towards something, we hope some City of God" (Greene and Ruse 1996, 462). On the other hand, strictly as a scientist he could not believe this:

> To direct evolution . . . God must induce mutations, shuffle the nucleotides in DNA, and give from time to time little pushes to natural selection at critical moments. All of this makes no sense to me. . . . I cannot believe that God becomes from time to time a particularly powerful enzyme.
>
> (462)

As an evolutionary biologist he was constrained by the overwhelming evidence from modern biology to hold that "Natural selection is at one and the same time a blind and a creative process" and that "natural selection does not work according to a foreordained plan" (Dobzhansky 1973, 381). The only recourse he saw to resolve these apparently irreconcilable differences was to take the Augustinian position that "the Alpha and Omega of evolution are simultaneously present in God's eyes" (Greene and Ruse 1996, 462). But then comes the confession:

> I do not doubt that at some level evolution, like everything in the world, is a manifestation of God's activity. All I can say is that *as a scientist* I do not observe anything that would prove this. In short, as scientists Laplace and myself "have no need of this hypothesis," but as a human being I do need this hypothesis!
>
> (464)

In all of this it is difficult to see anything else but wishful thinking. Evolutionary biology directs his conclusions in one direction, his needs as a human being direct him in another. The problem, as I pointed out in Chapter 2, is that one should not let what one thinks *ought* to be the case cloud one's judgment about what one thinks *is* the case. We reject this way of thinking in courts of law and in other areas of practical concern. Why should it all of a sudden become acceptable when we turn to bigger questions such as God and evolution?

Bad reasoning is the same no matter what we call it and no matter who provides it. A good example of bad reasoning about evolution and God is to be found in the endorsement given to evolution by the late Pope John Paul II. In his *ex cathedra* "Message to the Pontifical Academy of Sciences (1997)—the "Academy" is a committee of scientific advisors to the Vatican—the Pope calls evolution an "unfolding of living matter" (383) and the science of it a "theory" enjoying a "convergence" from "various fields of knowledge" (presumably from within science), such that the convergence "is in itself a significant argument in favour of this theory" (382). Hence the Pope accepts the evolution of the human body from earlier forms. But he cannot accept the evolution of the human mind-soul "as emerging from the forces of living matter, or as a mere epiphenomenon of this matter" (383). One problem with this view, he says, is that it is unable to "ground the dignity of the person." Another is that with the human mind-soul, with its "experience of metaphysical knowledge, of self-awareness and self-reflection, of moral conscience, freedom, or again, of aesthetic and religious experience," we find "an ontological difference, an ontological leap," such that in the unfolding of the tree of evolution "The moment of transition to the spiritual cannot be the object of this kind of observation ['the sciences of observation']" (383).

There are a number of serious problems with this view, and given the influence of this Pope in particular and the papacy in general it is important to go over them. First, to call the history of evolution an "unfolding" indicates a poor understanding of evolution. Evolution is not like the development of the adult from the embryo. That was a view of evolution that preceded Darwin and it is not taken seriously by any modern biologist. Species do not have an internal genetic program, and the processes influencing evolution (mutation, natural selection, genetic drift, changing environments, and so on) are contingent, not deterministic.

Second, Darwin gave the perfect reply to the claim that evolution undermines the dignity of the person. At the end of the *Origin* (1859) he wrote: "When I view all beings not as special creations, but as the lineal descendants of some few beings which lived long before the first bed of the Silurian system was deposited, they seem to me to become ennobled" (488–489). Rather than lower the dignity of man, evolution raises the dignity of animals (as well as of plants, by the way). This, it seems to me, is an elevated worldview, without which no modern person should be considered fully civilized. To think otherwise is *speciesist*, a prejudice that has always resided at the core of the creationist theology of the West.

Further problems involve the Pope's claim of an "ontological divide." As Dawkins (1997) puts it,

> In plain language, there came a moment in the evolution of hominids when God intervened and injected a human soul into a previously animal lineage (When? A million years ago? Two million years ago? Between *Homo erectus* and *Homo sapiens*?).

For Dawkins, "The sudden injection of an immortal soul in the time-line is an antievolutionary intrusion into the domain of science" (398).

Dawkins is surely right that it is antievolutionary, but it is important to see why. One reason is that there is nothing sudden about evolution. Complicated features such as eyes do not just pop into existence. Instead, they are the product of gradual increments over vast stretches of time, of cumulative natural selection.

Another reason is that the modern science of evolution, like modern science in general, does not use what Dennett (1995) calls *skyhooks* (74)—"miraculous lifters, unsupported and unsupportable" (75)—i.e., explanations that appeal to the supernatural. If anything characterizes modern science, it is that its explanations come from below, not from above. This is what has made it so successful. Explanations from below are testable, explanations from above are not. Nor have explanations from above proven even necessary.

The Pope, on the other hand, employs the ultimate skyhook, God. Quite possibly this is why he refuses to call evolution a "fact," calling it, instead, not a "mere hypothesis" but a "theory." He is dead on, of course, when he says the modern case for evolution is the result of a convergence of research from different fields. As mentioned earlier, Darwin painstakingly drew his evidence for evolution by natural selection from a wide variety of fields, all of it added to enormously by modern science and clinched by the field of molecular biology (the study of DNA and RNA).

But to call the sum of all of this a "theory" is simply mistaken, even antievolutionary. On the contrary, modern evolutionary biologists routinely call evolution a "fact," not a "theory" (at least when they are not teaching undergraduates, who tend to be easily upset), and for the best of reasons: the evidence for evolution is so overwhelming that it is in the same category with the fact that the earth is spherical and orbits the sun, which began as a hypothesis, graduated to a theory, and is now a fact (see the Appendix).

This leads us to the final and perhaps biggest problem with the Pope's message. The interposition by God in the evolutionary "time line," the "moment of transition to the spiritual," the "ontological leap" (383), all of this is not only *not* suggested by science, but it goes flatly against it, especially evolutionary science. A fundamental principle in evolutionary biology, ever since Darwin, is that any trait that exhibits heritable variation is a candidate for evolution by natural selection. Since mental traits exhibit heritable variation, they are subject to evolutionary explanation no less than are physical traits. Not surprisingly, it so happens that there is an enormous amount of evidence converging on the conclusion that all the qualities of human mind have both a material and an evolutionary basis. The evidence from similarities between humans and animals concerning the expression of emotions (Darwin 1872), from the modular nature of the mind-brain (Pinker 1997), from the functional topology of the brain (Penfield 1975), from brain damage (Crick 1994), from split-brain research in particular (Springer and Deutsch 1993) and neuroscience in general (Churchland 1995), and from numerous aspects of the adapted nature of mind (Barkow *et al.* 1992), all converge on this one explanation. To accept convergence of evidence for physical evolution but not for mental evolution is inconsistent in the extreme. If, as the Pope (1997) says, "truth cannot contradict truth" (381), then clearly something has to go.

Maybe, then, God did not intervene in the evolutionary tree in the manner of a sudden injection, but rather directed evolution, both physical and mental, in a gradualistic, incremental manner. This view still has many problems along with its own fair share of bad reasoning, as specific examples make clear.

One rather striking example is provided by Sir John Eccles, the Nobel laureate in brain science. Near the end of his book on the evolution of the brain, Eccles (1989) states that "we have to recognize that we are spiritual beings with souls existing in a spiritual world as well as material beings with bodies and brains existing in a material world" (241) and that "In some mysterious way, God is the Creator of all the living forms in the evolutionary process," in particular humans with their "immortal souls" (243). Earlier in his book, Eccles employs Eldredge and Gould's theory of punctuated equilibria to take away some of the mystery, claiming that the ascent from earlier apes to *Homo sapiens*—via *Australopithecus africanus*, *Homo habilis*, *Homo erectus*, and *Homo sapiens neandertalis*—was accomplished at each stage by a "peripheral isolate . . . saltatory genetic change" in a "transcendent hominid relay" (12–34). Never mind that the theory of punctuated equilibria reduces to gradualistic evolution of the modern Darwinian variety (see Dawkins 1986, ch. 9; Stamos 2003, 220–231), or that human evolution has long been known to have been a bush pattern and not a direct line (Tattersall 2000). Eccles clearly believes because he *wants* to believe.

Not all advocates of God-directed evolution are as vague about how God operates as are Eccles or the former Pope. A number of philosophers of religion provide more sophisticated accounts, views that place God's operation within the laws of nature (hence no need for miracles), specifically at the interface of quantum physics and genetics. Nancey Murphy (1997), for example, takes the position that "The apparently random events at the quantum level all involve (but are not exhausted by) specific, intentional acts of God" (339), such that in quantum physics "God is the hidden variable" (342). Although in her view "God is free to bring about occasional extraordinary events at the macro-level" (343), by implication God is involved with all mutations. Thomas Tracy (1998) combines chaos theory with the position that at the quantum level God "determines some, but not all, of the events that are left under-determined by their finite causal conditions" (518), thus guiding evolution by "triggering" certain mutations at the right time and place. Robert Russell (1998) takes the position that via quantum indeterminism God guides evolution by generating the required variability wherever there is a single hydrogen bond, thus not only in point mutations but also in chromosome mutations and in genetic recombination.

Tracy (1998) repeatedly stresses that his view is a "theological interpretation of evolution" (513); he does not think there is any specific evidence that favors his view over a completely naturalistic account of evolution. I suspect that Murphy and Russell would say the same of their own views. Murphy's view, however, does not take modern physics seriously enough, and in fact is incompatible with it. Her view of quantum physics, physics at the subatomic level, is deterministic. The problem is that a deterministic view of quantum physics is not the consensus view in physics. From a long accumulation of evidence, experimentation, and penetrating thought, the conclusion forced upon physicists starting at least two-and-a-half decades ago is that there are no hidden variables hiding determinism behind the statistical laws of quantum physics. Instead, the strong consensus is that the quantum world does indeed involve genuine chance events (Rohrlich 1983; McMullin and Cushing 1989). Russell and Tracy, therefore, have a more compatible view with modern science when they have God influencing evolution through the statistical laws of chance at the interface between the quantum world of atoms and the genetic world of DNA. In an earlier publication (Stamos 2001) I brought together the evidence from quantum physics and molecular biology to show how a quantum event, an event of pure chance, could trigger a mutation in DNA (I found basically three different ways). For most molecular biologists, there is no God there. As one molecular biologist (Alan Weiner) put it to me, "chemistry constrains but chance rules." With all of this one cannot simply and easily superimpose a theological interpretation.

One problem that theorists such as Tracy and Russell face—and it is a problem that plagues all of theistic evolution—is that they do not make falsifiable claims. The problem is not that they do not make falsifiable claims in the sense of Popper's falsificationist criterion of science. They do not claim to be doing science, after all. The problem, instead, is that they do not even seem to be making genuine assertions. Antony Flew (1984) hit the nail on the head when he argued that a genuine assertion logically implies a class of other assertions, assertions contradictory to the original assertion. For example, if I state that my cat is at this moment lying on my couch, then I am also asserting that he is not lying on the floor, etc. Thus, Flew claims not only that "if you are going to say something which is both relevant and substantial, then you have to pay the corresponding price" (77), but that "if there is nothing which a putative assertion denies then there is nothing which it asserts either: and so it is not really an assertion" (73). Accordingly, Flew issues to theology what he calls the *Falsificationist Challenge*: "What would have to occur or to have occurred to constitute for you a disproof of the love of, or of the existence of, God?" (74). Applied to theistic evolution, we might weaken it as follows: "What would have to occur or to have occurred to constitute for you a disproof of, or at least strong evidence against, evolution influenced by God?" At this point theistic evolution falls completely silent (a silence not shared by atheistic evolution). What follows is not that their positive claims are meaningless (Flew does not take that position either), but that they are not really assertions after all.[12]

Any marriage between theistic evolution and science, then, has got to be an uneasy one for this reason alone. Other reasons we have examined make it hardly a marriage made in heaven. But no such problems face a marriage between philosophy (Anglo-American) and science. Indeed, arguably, they belong to a single continuum (Quine 1969, 126–127), with no "proper boundary" between them. Jealous at this thought, theistic evolutionists have resorted to a much too narrow concept of science.

But enough of that. A further reason that believers in God-directed evolution cannot simply rest with a theological interpretation of evolution is that there are facts about evolution that powerfully resist their interpretation.

One problem comes from DNA repair mechanisms (excision, direct, and mismatch). Found within the cells of virtually all organisms, these "proof-reading" mechanisms attempt to repair damaged DNA and correct micromutations (point and frameshift). Mismatch repair operates as follows: when a nucleotide (DNA letter) is inserted by DNA polymerase during the process of DNA replication, its $3' \rightarrow 5'$ exonuclease site removes the wrong nucleotide (based on the complementary nucleotide in the

---

12   Flew's Falsificationist Challenge has evoked a wide variety of responses. Hick (1990, 103–105), for example, agrees that falsifiability is sufficient for genuine assertions but disagrees that it is necessary. Using the example of the assertion that there are three consecutive 7s in the decimal of pi, he claims that such an assertion is not in principle falsifiable (since the decimal is an infinite series) but is in principle verifiable, so that verifiability is sufficient for genuine assertions. For Hick, theology similarly makes many assertions that are not falsifiable but are nevertheless verifiable, such as the belief in life after death (capable of what he calls *eschatological verification* but not falsification), and hence many genuine assertions. Gaskin (1984, 107), using the example "On Hallowe'en, if and only if no observation is being made, the face of the gorgon from the temple of Sulis Minerva in the museum at Bath smiles," argues that neither falsifiability nor verifiability are necessary or sufficient for the making of genuine assertions. Both of these examples, however, do not seem to be legitimate counterexamples to Flew's claim. Hick's example assumes a Platonic realism about mathematics, an assumption that is very much in need of argument, since many hold mathematics to be fundamentally a matter of definition, a matter of tautologies rather than of empirical fact. Similarly, Gaskin's example is by definition unempirical. If it were truly empirical, then it would be logically possible for the gorgon to be fooled and to have its smile observed on Hallowe'en. (This reply, incidentally, was inspired by my reading of Anselm's reply to Guanilo.) Thus, if Gaskin's example is modified to be a genuinely empirical claim about the world, it does indeed entail a class of contradictory empirical statements, so that it fails as a counterexample to Flew's claim about genuine assertions.

template strand) and replaces it with the right one. This is truly mind-boggling. Nevertheless, our imaginations are limited, and the sober reality is that all the repair mechanisms exhibit heritable variation, which entails differential fitness and hence adaptive evolution by blind natural selection (Freemann and Herron 1998, 109). What makes all of this especially difficult for God-directed evolution is the discovery that, as Barry Glickman (1987) puts it, "a significant fraction of spontaneous mutation arises as the consequence of attempted dna repair" (48). If DNA repair is directed by God, then we shall have to conclude that in correcting his initial mistakes he sometimes makes further mistakes!

But even more basically, theistic evolutionists have to wonder why there are any mutations at all. The problem is that the vast majority of mutations are in no way creative. Most mutations are deleterious, some are neutral, and fewer still are advantageous. Surely if an efficient and beneficent God were directing mutations, he would produce, if he wanted to create by evolution, only the advantageous mutations. Variation, and hence individuality, would still be maintained by genetic recombination. All the uncreative mutations would thus lack a sufficient reason.

This is a problem for theistic evolution on two fronts. On one front, there is the theology of the German philosopher Gottfried Wilhelm Leibniz, who flourished around the end of the 17th century. Premised on the perfect rationality of God, Leibniz required what he called a *sufficient reason* for any action of God: among contingent states of affairs, there must be a sufficient reason for why God would make something one way and not another. Hence Leibniz argued that there cannot be two things perfectly identical in all their attributes, since God could have no reason for why one would be in one place and the other in another place rather than *vice versa*. In the case of uncreative mutations, sufficient reason is lacking. Sufficient reason is also lacking in the case of the genetic code itself, which is arbitrary rather than chemically necessary and is not optimal with regard to minimizing the effects of errors (Freeland and Hurst 1998; Graur and Li 2000, 22–29).

The case of negative, deleterious mutations brings us to the other front. If all mutations are totally random and blind with respect to the environment, and hence with respect to fitness levels, we should then expect most of them to be deleterious. As Dawkins (1995) puts it, that

> most mutational effects are bad . . . is to be expected in principle: if you start with a complicated working mechanism—like a radio, say—there are many more ways of making it worse than of making it better.
>
> (130)

In reality, many micromutations are rendered harmless by the adaptive redundancy of the genetic code (the synonymy among codons) and by the DNA repair mechanisms. Nevertheless, many micromutations get through to make a difference, which, along with macromutations (inversions, deletions, translocations, etc.), range in negative effect from slight to painfully lethal, accounting not only for many spontaneous abortions but also for numerous cancers and genetic diseases such as hemophilia (abnormal bleeding), Huntington's chorea (gradual deterioration of the brain and voluntary movement resulting in death by middle age), and Tay-Sach's disease (a fat storage disorder resulting in nervous system damage and death by age three). Given all of these horrid facts and countless others like them, one has to wonder why a genetic system, if it was evolved by a rational and benevolent mind, would be so negatively mutational when a much more rational and benevolent system could have been produced.

All of this raises, of course, the infamous *problem of evil*. The problem can be put in the form of a *modus tollens*: If God exists (where "God" means all-knowing, all-powerful, all-loving creator of the universe), then there should be no evil. Evil exists. Therefore, there is no God (or at least one of the "all" attributes has to go, which for some might mean no God just the same). The problem becomes particularly poignant once it is noticed that in theology, traditionally, God's purpose for creation was not thought simply to be the production of happy beings, or a surplus of happiness over unhappiness, but something much higher, namely, beings of a high moral order, beings that could share in his glory. Evolutionary biology certainly did not create the problem of evil (it has been around for as long as there has been theology), but if evolution is thought to be God's mode of creation, then the problem of evil becomes accentuated and theistic evolution needs to provide a special answer to it.

Darwin faced this problem right from the start. For example, his main advocate in the United States, the Harvard botanist Asa Gray, was a deeply religious man who could not help but superimpose a benevolent God directing evolution. According to Gray (1860), "variation has been led along certain beneficial lines," and to think otherwise was "philosophically untenable" (414). Darwin and Gray often corresponded on this subject. For Darwin, given his knowledge of random variation among breeding stocks, Gray's view made no sense whatsoever, since, in the case of pigeons, it meant that "the Fan-tail was led to vary in the number and direction of its feathers in order to gratify the caprice of a few men" (Burkhardt *et al.* 1993, 496). What was true of domestic animals and plants had to be no less true in the wild. Moreover, Gray's view for Darwin had no empirical content:

> If you say that God *foreordained* that at some time & place a dozen slight variations should arise, & that one of them should be preserved in the struggle for life, & that the other eleven should perish in the first, or first few generations; then the saying seems to me verbiage.—It comes to merely saying that everything that is, is ordained.
>
> (238)

Moreover still, Gray's view seemed to Darwin "to make natural selection entirely superfluous, & indeed takes whole case of appearance of new species out of the range of science" (226). But it was over the problem of evil that Darwin felt the strongest and deepest objection to Gray's view, a moral objection. For Darwin, evolution directed by God did not square with the enormous pain, suffering, and killing that is necessarily part of the very fabric of nature from an evolutionary point of view. As he put it in a letter to Gray,

> I had no intention to write atheistically. But I own that I cannot see, as plainly as others do, & as I sh^d wish to do, evidence of design and beneficence on all sides of us. There seems to me too much misery in the world. I cannot persuade myself that a beneficent & omnipotent God would have designedly created the Ichneumonidæ with the express

intention of feeding within the living bodies of caterpillars, or that a cat should play with mice.

(224)

The problem of evil for theistic evolution only becomes stronger once evolutionary biology of recent decades is taken into account. This is because not only have so many more examples of the horrid cruelty of nature been documented, but because even apparent kindness and cooperation are now explained in terms of selfish genes. The biologist G. C. Williams makes the strongest and most famous case here. For Williams (1989), the romantic image of the beauty of nature is an illusion. Behind the veil is the ugly reality of enormous suffering caused by mutation, disease, starvation, and temperature extremes. But more than that, Williams discusses in detail numerous documented cases in nature of incest, cuckoldry, rape, cannibalism, infanticide, and wasteful predation. Even the institution of motherhood does not remain unscathed, given, for example, that in the langur monkeys of northern India, when a male usurps a harem and manages to kill a mother's nursing infant, she immediately goes into estrus and accepts the sexual advances of her baby's killer. It all boils down to selfish genes, so that, unlike Darwin's bulldog, T. H. Huxley, nature in Williams's view is not merely morally indifferent but is guilty of "*gross immorality*" (180). Indeed, for Williams, nature is immoral in much the same sense, but only worse, as war is said by pacifists to be immoral.

Before Williams is accused of making a rank mistake here, of attributing evil to something non-personal, it needs only to be remembered that theologians have long made a distinction between what they call *natural evil* and *moral evil*, where only the latter is manmade. Moreover, the idea that something non-personal can be immoral or evil presupposes a moral theory that connects blame with motive. From a *utilitarian* point of view, on the other hand, a theory of morality based not on motives but purely on consequences in terms of happiness, nature can indeed be evil. It is not so obvious, then, that Williams is mistaken when he characterizes nature, including evolution by natural selection, as evil.

Williams is also fond of calling nature "abysmally stupid" (209) and a "mindless fool" (210), given the many examples of bad design in nature such as the vertebrate eye and the tubing of the testes in male mammals. But he also views nature as a "persistent and powerful enemy" (210). The purpose of morality, then, for Williams, is not to fly from nature, let alone to imitate it, but instead, like Huxley before him, to rebel and fight against it (214).

But what has this powerful indictment against nature, based on modern evolutionary biology, have to do with God and theistic evolution? In a later book, Williams (1996) defines "God" as "Whatever entity or complex of entities is responsible for the universe being as we find it, rather than some other way or not there at all" (211–212).[13] He then provides a summary of the detailed kinds of evidence and argument

---

13   This definition is not new. In his *Dialogues Concerning Natural Religion*, Hume (1779) has his character Philo say: "the original cause of this universe (whatever it be) we call God" (142) and "By supposing it [the material world] to contain the principle of its order within itself, we really assert it to be God" (162).

that we find in his earlier work and draws his conclusion in the form of a striking question: "Do you still think God is good?" (217).

This is the question, posed most forcefully by Williams, that theistic evolution needs to answer. To my mind, the most heroic answer was given by Thomas Tracy, and we shall finish this chapter by examining it.

Tracy (1998) did not frame his answer specifically in reply to Williams, but it is easy to see how it would apply. For Tracy, as we have seen above, God operates within the laws of quantum physics by occasionally influencing quantum events that trigger key mutations in biological evolution. In this way God operates through chance to produce not a predetermined goal, but rather "to elaborate the possibilities of creation" (518), one of which is "soul making," possibly including kinds of creatures more evolved than humans. Tracy recognizes, however, that this view leaves us with an even more powerful version of the problem of evil, in particular with regard to the place of non-human animals. As he puts it,

> lives are generated and destroyed in a vast impersonal lottery of genetic trial and error. This process may appear too accidental, too wasteful, and too cruel to be the work of a God who possesses perfect knowledge, power, and goodness.
>
> (512–513)

As a possible solution to this problem, Tracy provides a modernized version of Irenaean theodicy. Irenaeus was a 2nd-century Church Father whose theological solution to the problem of evil, his theodicy, was that evil is necessary for the production of good, namely, souls with moral perfection and a personal relationship with God. Applied to evolution, this means for Tracy that

> Evils may arise as a 'by-product' or 'side-effect' of meeting various necessary conditions for the possibility of a good . . . as a collateral cost . . . If these goods cannot be realized without permitting or producing these evils, then a strong case can be made that God is justified in creating a world that includes such evils.
>
> (524)

In order to answer the critic who tries to point to specific examples of gratuitous suffering, Tracy appeals to what he calls our *epistemic finitude* (520). For each particular case of apparent gratuitous suffering, he says, we "must show how the world would be better off if God prevented instances of evil to which the objector points" (528–529). In other words, for any particular example we cannot say for sure if it is truly gratuitous, if it is collateral, or if it is meant for a purpose, because we can never know the bigger picture. In each particular case, he says, we would need to tell a "comprehensive story," but we cannot "*in principle* know enough to tell such a story" (529), especially, he says, if one takes into account chaotic effects (the famous butterfly effect). "The cumulative result of these considerations," he says, "is to induce a degree of epistemic humility in contemplating the place of evil in the design of the world" (530).

There are a number of serious problems with Tracy's solution to the evolutionary problem of evil. For a start, his argument from epistemic finitude to epistemic humility can easily be turned against him. Quite simply, why not go a step further and be truly humble and acknowledge that we are but animals rather than immortal gods in human flesh? In other words, why not accept, as Walter Kaufmann (1958) so strikingly put it, that "Man is the ape that wants to be a god" (354)? This humble but thoroughly evolutionary view was presaged by Darwin, when he wrote in one of his evolution notebooks that "Man in his arrogance thinks himself a great work. worthy the interposition of a deity, more humble & I believe true to consider him created from animals" (Barrett *et al*. 1987, 300).

But more to the point, Tracy's focus on particular cases of apparently gratuitous suffering is a red herring, indeed a major example of missing the forest for the trees. Like so many others who offer a theodicy, Tracy fails to fully appreciate the enormity of suffering in the animal world. The suffering recognized by Williams and many other evolutionary biologists includes not only the enormity of animal suffering observed by humans, it includes also the enormity of animal suffering contemporary with humans but that never came under their purview. But much more, it includes the hundreds of millions of years of animal suffering that preceded the evolution of humans, very little of which led to that evolution or will lead to anything like it—suffering, then, that had absolutely nothing to do with soul making.

Moreover, one could well argue that no good, no matter how "high" it is thought to be, is worth the cost, or is even really good, if it is purchased at the price of sentient beings having their flesh ripped from their living bodies during the act of predation (try to imagine the horror), or living in nearly constant fear, or suffering the agony of starvation, or of temperature extremes, or the torture caused by disease or by parasitism. For Tracy, "The 'problem' of evil is not one that we are going to 'solve'" (530). This, in a sense, is true enough, but let us be clear that the problem is a problem only for the theodicist. For Christians, he says, it boils down to "Trust in God's goodness" (530). Trust is doubtless something very important, but for the genuine lover of truth the reason for giving trust must not be the gain of psychological comfort; instead, it must be fundamentally epistemological. The question, then, is whether what we know about evolution gives us any reason for such trust. The answer is clearly *no*. As Darwin so perfectly put it in a letter to his closest friend, the botanist J. D. Hooker, "What a book a Devil's chaplain might write on the clumsy, wasteful, blundering low & horridly cruel works of Nature!" (Burkhardt and Smith 1990, 178).

In this chapter we have seen evolutionary principles applied to religion at the level of memes, then to the level of instincts, and then finally to the attempt to harmonize evolution with theology, all with devastating results. What makes this especially devastating is that for most people on this planet the meaning of life is ultimately found in religion. Does this mean, then, that evolutionary principles must spell disaster for the meaning of life too? It is time now that we turn to this, the last and arguably most important of the big questions, which may be put as whether the pages in the next chapter should all be left blank.

9

Evolution and the Meaning of Life

People often think of the meaning of life as the ultimate question of philosophy. Even though many professional philosophers avoid the question (they think it is just plain too big and therefore not worth the effort), certainly it is one of the central questions. The best definition of philosophy I've ever seen is that philosophy attempts to answer in a systematic and thorough way the question "What is x?," where x is filled by fundamental concepts. Hence, "What is science?," "What is a law of nature?," "What is a species?," etc., are central questions in philosophy of science. "What is knowledge?" is the central question of epistemology. And so on. Thus, "What is the meaning of life?" is a perfectly legitimate philosophical question, and many introduction to philosophy anthologies have a section on the meaning of life, while there are at least three anthologies devoted to the topic (Donnelly 1994; Klemke 2000; Benatar 2004). If one surveys these volumes, one finds that some believe that religion is required to give meaning to life, others find that unacceptable. Some find that death (conceived as the total annihilation of the self) takes away from the meaning of life, while others argue that death is a prerequisite. Some believe that meaning must come from outside of ourselves, others that meaning is purely something we make. And then there are some who question the very question itself, who question whether "What is the meaning of life?" is itself a meaningful question. They are concerned with the meaning of "meaning." What is remarkable is how very little evolution makes its way into these discussions, if it makes its way at all.

If we are going to bring evolution into the question (and hopefully by now the reader believes that evolution is relevant to the big questions), a number of issues arise. One is whether evolution adds any meaning to life, takes away from it, or has no relevance. Another is whether we have a basic need, an evolved instinctual need, to find meaning in our lives. These two questions shall be examined in this chapter, along with whether evolutionary biology is compatible with a major philosophical and literary movement known as *existentialism*.

With regard to the first question, one might naturally suppose that evolutionary biology in particular and modern science in general, when considered pure (that is, science as science, without any theological interpretation superimposed), entail the view that there really is no meaning to life. The universe began with a "big bang" roughly 15 billion years ago as the result of a quantum fluctuation in the primordial singularity and has been expanding ever since, destined either to wind down and cool and end in heat death, in complete and utter disorder (maximum entropy), or to oscillate between expansion and contraction endlessly throughout time. Life on Earth (negative entropy), which began almost 4 billion years ago and evolved in millions upon millions of different directions (species), is destined to end in maximum entropy long before the rest of the universe. Not only is extinction the destiny of all species—apart from physics, this is simply a fact of evolution—but the sun, which makes life on earth possible, is in midlife and will either explode or burn out in roughly 5 billion years, turning to ash and dust everything that revolves around it. This is the fate of every star. In all of this there is no rhyme or reason, no meaning. And it will do no good to try to find meaning by trying to find God or life after death in science, for science does not give us these either.[1]

Or maybe meaning is the exception. This is what Julian Huxley thought, the grandson of "Darwin's bulldog," T. H. Huxley, and a major evolutionist in his own right. Rejecting the belief in God or gods as a subjective projection onto nature, and similarly rejecting the belief in personal immortality, Huxley (1939) claims that although no purpose or meaning "inheres in the universe or in our existence," one nevertheless "can be found" (78). Indeed, Huxley argues that a purposive and worthwhile life is not only possible but that it is more possible than ever before thanks to science. It is because of science that we no longer fear epidemics and earthquakes as the product of divine wrath, it is because of science that we understand much better the subconscious roots of irrational fear, guilt, and cruelty, and it is because of science that we have gained a large measure of control over nature, a measure that is increasing all the time. Moreover, unlike the traditional Judaic-Christian view of history, which Huxley says is not only false but morally stunted, the view of history presented by science, especially evolutionary biology, is indeed grand. It is a view, for Huxley, permeated with progress, where progress is defined as "all-round specialization" (81). But most importantly, even though evolution is not teleological, with any predetermined plan, evolutionary progress passed a threshold with the

---

1   There is the field of parapsychology, which collectively has thus far shown that the putative evidence for life after death is extremely poor (Flew 1987; Blackmore 1993; Shermer 1997, chs 5 and 16; see also the *Skeptical Inquirer*). More importantly, all the hard scientific evidence points to the conclusion that mind is in some way the evolved product of the brain and cannot exist without it (Chapter 1). As Carl Sagan put it in *Cosmos*, "extraordinary claims require extraordinary evidence," and there just is no extraordinary evidence for life after death that survives close scrutiny. With regard to a logically separate issue, the existence of God, there is the intelligent design theory in physics, the argument from the apparent fine tuning of the cosmic constants (e.g., Davies 1983), but this is hardly a case of science pointing toward the existence of God, as the minority of physicists who subscribe to this theory were convinced of God's existence in the first place, but more importantly there are much simpler explanations for the apparent fine tuning, explanations that do not involve a mind at all (Weinberg 1992, chs 9 and 11, 2001, chs 20 and 21; Smolin 1997; Stenger 2007).

human brain, for here "deliberate purpose could be substituted for the blind sifting of selection; change could be speeded up ten-thousandfold. In man evolution could become conscious" (81).

While much of what Huxley says is apparently true, it flounders in the mud of evolutionary progress. His rosy belief in scientific progress, for a start, which is based on his belief in evolutionary progress, was perhaps justified in 1939, with the advances in physics, genetics, evolutionary biology, and medicine. But much has changed since 1939, such that people today more than ever are afraid of the technology that has developed out of science, of the technology and its power and of the people involved (which ultimately means human nature). These fears have been immortalized in movies such as *Dr Strangelove*, *Jurassic Park*, and *The Terminator*, and they are not without foundation.

But the problem with Huxley's view goes much deeper than that. The fundamental problem is with his belief in evolutionary progress. As the historian of biology William Provine (1988) so rightly points out in his analysis of Huxley's view, the belief in evolutionary progress is anthropocentric and hence arbitrary (see Nitecki 1988; Dawkins 1992; Ruse 1996). Whatever leads to humans, or to the key human characteristics, is evolutionary progress, on this view. But one could just as easily do the same with other criteria. In terms of the ability to adapt, for example, the family of HIV viruses not only has us beat but is second to none. In terms of ruling the world, the dinosaurs ruled for over 100 million years. In terms of importance for life on Earth, bacteria come out on top and humans are at the bottom. Indeed, from an ecological point of view, humans with their population explosion, resource exploitation, pollution, and destruction of habitats are the cause of the current mass extinction decimating biodiversity on earth, such that many ecologists view the human species not as the pinnacle of evolutionary progress but as a cancerous growth or parasite, one that is destroying its host, the biosphere (e.g., Odum 1971, 222–223; McNeill 1976, 15–23).

But even if the concept of evolutionary progress turns out not to be subjective and arbitrary, that humans really are in some way extraordinary and precious and rare, levels above anything else that evolution has ever produced on this planet, Provine (1988, 58–62) argues that the Modern Synthesis, which marks the maturity of biology as a science, was not so much a synthesis of the various fields of biology as it was a "constriction," limiting evolutionary theory to a small number of principles and mechanisms, and excluding thereby any notion of God, teleology, or purpose behind or underlying evolution. Hence he argues, implicitly against Ruse and others, that the Modern Synthesis (really "Evolutionary Constriction") created an inevitable conflict between science and religion, a conflict that religion cannot possibly win. Indeed, Provine claims that in line with this constriction "very few truly religious evolutionary biologists remain. Most are atheists, and many have been driven there by their understanding of the evolutionary process and other science" (68). As for those scientists who claim that evolution and religion are compatible, which is the official position of the United States National Academy of Sciences, Provine suspects "intellectual dishonesty" (69), just as every member of the Congress of the United

States professes to be deeply religious but surely is not. What is behind it all, says Provine, is politics motivated by the fear of a public uproar if the truth be told, votes in the case of politicians, public funding in the case of scientists. And yet in spite of it all, in spite of the fact that "nature has no detectable purposive forces of any kind" (64), that the compatibility of evolution and religion is "false" (65), and that "I can see no cosmic or ultimate meaning in human life" (70), Provine nevertheless believes that "humans can lead meaningful lives," lives "filled with meaning" (70). But not only does he mean this in the sense of living a full life, he means it in the sense of social engineering, where "behaving well" can be reinforced à la Skinner, who was "surely correct" that "we can really control the moral development of small children" (71) .

With some of this I have deep reservations. Even though an evolutionist (historian), Provine gives us no theory of what and what may not satisfy us, as humans, in terms of a meaning *in* life. Is it wide open or does our evolutionary past impose restrictions, albeit statistical? Provine's reliance on Skinner and social engineering likewise seems too wide. If there is any truth to the new kind of evolutionary ethics, then "controlling" the moral development of small children is not going to be as simple a matter as stimulus-response conditioning. Moreover, in the absence of theory, it is by no means obvious to what direction children should be conditioned. One simply cannot go around using words such as "behaving well" and "moral development," as Provine does, as if everyone knows and agrees what they mean. Even if the new evolutionary ethics provides us with some guidelines, the direction of application of those instincts is often also a matter of factual beliefs. By saying we ought to act "kindly" to our "neighbors"—indeed Provine says, "I think that what modern science tells us is: 'If you don't love your neighbor, then you're just plain stupid!'" (72)—does Provine mean to say we should not have capital punishment, that we should allow hardcore pornography and prostitution, that euthanasia ought to be made legal, and that we ought not to eat animals? And what of environmental ethics? In all of these practical matters Provine leaves us completely hanging.

At any rate, if Provine gives us a much less enthusiastic answer to the meaning of life from an evolutionary point of view than does Huxley, the Oxford zoologist Richard Dawkins gives us the most negative, though arguably the most consistent with modern evolutionary theory. I say this because of the direction that evolutionary theory has taken from the time of Darwin to the present. For Darwin, evolution by natural selection operated sometimes at the level of groups but mainly at the level of individual organisms. Since the development of classical genetics in the early part of the 20th century, however, which focused on genes, and through to the discovery of DNA and the development of molecular biology, evolutionary theory has focused more and more at the level of genes, at the level of the bits of genetic information that get passed on from generation to generation, as the ultimate level of natural selection and evolution. Of course, this smacks of reductionism and many biologists and philosophers of biology are resistant to this level of explanation as the main level. Things like organisms and ecological niches and processes like group selection count too. Of course! But it remains, for Dawkins and many others, that if you really want to understand what evolution and life are ultimately about, then you have got to look at evolution and life principally from the perspective of the gene.

This is the theme of Dawkins's modern classic *The Selfish Gene* (1976). Genes, of course, are not the sort of things that can be selfish or unselfish, but just like Darwin's metaphor of "natural selection," the equally striking metaphor of "selfish genes" directs our attention to processes with profound explanatory power. Genes behave *as if* they were selfish, selfish in the sense that their prime directive is to make more and more copies of themselves. When we look at genes in this way, for Dawkins and so many other evolutionary biologists, the lights finally and truly come on in our heads. Somehow, some way, roughly 4 billion years ago, some primitive molecules began making crude copies of themselves. Immediately natural selection kicked in, favoring some of these primitive molecules over others, as some were naturally better fitted at faithfully replicating themselves than others. Eventually some of these "naked genes" joined with other genes into "gene complexes" and started evolving simple and then increasingly complex phenotypes, since phenotypes increased their survival and reproduction. Hence the evolution of individual organisms with their adaptations and species, such that, as Darwin so eloquently put it at the very end of his *Origin of Species* (1859), "from so simple a beginning endless forms most beautiful and most wonderful have been, and are being, evolved" (490).

But what does all of this entail for the meaning of life? For many who profess to be evolutionists, the meaning of the universe is the evolution of life, and the meaning of the evolution of life on Earth is human life. But absolutely none of this follows from evolutionary biology as a science. Indeed, the matter is quite the opposite. Dawkins himself, in a later book meaningfully titled *River Out of Eden* (1995), addresses the issue squarely, uncompromisingly, and without pulling any punches (something for which Dawkins has become quite famous). For a start, Dawkins argues that part of the problem with the question of the meaning of life comes not from life or from evolution (not directly at least) but from ourselves. "We humans," he says, "have purpose on the brain" (96). Evidence of this, I would say, is ubiquitous, from mindless sayings such as "Everything happens for a reason" to Rorschach tests and coincidental phone calls. The problem is not evidence of purpose and meaning, but rather a psychological propensity to project them onto the world. When taken to an extreme, says Dawkins, we call it "paranoia," but whether moderate or extreme it remains what he calls a "nearly universal delusion."

Why the word "delusion"? An *illusion* is a false or misleading representation of reality that is caused by the environment, whereas a *delusion* is when the cause is inside our heads. Children and primitives (the latter word is accurate though not politically correct), and possibly also higher animals such as cats and dogs, share the propensity for projecting onto inanimate things and nonliving processes—such as thunder, wind, tumbling rocks, and eclipses—mind and intentionality. Hopefully the rest of us have gone beyond that. But the tendency returns when we start to ponder the existence of the universe itself, or the laws of nature, or biological evolution. And it most strongly returns, as Dawkins rightfully points out, when tragedy strikes, as in "Why, oh why, did the cancer/earthquake/hurricane have to strike *my* child?" (96).

If not mind and purpose *in* these things, the propensity is to think that there must be a mind and purpose *behind* them. But, says Dawkins, the mere fact that we can ask a question—in this case, "What is the meaning or purpose of this?"—does not mean that there has to be an answer. Like it or not, it might just be the case that there really is no meaning or purpose. In other words, the question might be the wrong question right from the start, maybe even an illegitimate question altogether.

In the case of biological organisms, on the other hand, they "seem to have purpose written all over them" (97). Hence the most famous, popular, and pervasive argument for the existence of God is the *argument from design*, which (before Darwin at any rate) focused mainly on the complex adaptations of organisms. Since these so vastly exceeded the productions of human invention, argued the archdeacon William Paley in the early 1800s in the most famous version of the argument, there must by reason of analogy be a vastly greater intelligence behind their production. All of this came to a grinding halt with Darwin, who argued not only that biological adaptation *could* be the result of blind mechanical processes (natural selection operating on random variation) but that it *is* the result of blind mechanical forces. Today there is no more doubt, at least among professional biologists. "The strong illusion of purposeful design is now well understood," says Dawkins (1995). "The true process . . . is Darwinian natural selection" (98).

To really drive home what this means for the meaning of life from an evolutionary point of view, Dawkins employs two technical concepts that at best are only implicit in Darwin, concepts that when applied collectively to biology involve the modern concept of gene (which of course was the main element lacking in Darwin's own theory of evolution). The first technical concept is *utility function*. The utility function of something is that which is being maximized. In business it is profit, in politics it is power, in utilitarian ethics it is happiness. The second technical concept is *reverse engineering*. Engineering is the design and construction of something for a purpose. Reverse engineering is the opposite; it takes something that is already made and tries to figure out what it was designed for.

When we reverse engineer biological organisms—whether the organisms as a whole, such as cheetahs and gazelles or digger wasps, or their parts and instinctive behaviors, such as eyes, wings, beaks, brains, language, or altruism—from the viewpoint of selfish gene theory the result turns out always to be the same. In one of the most striking and poignant statements in the entire history of evolutionary biology, up there with Dobzhansky's famous statement quoted at the beginning of this book, "everything makes sense," says Dawkins (1995), "once you assume that DNA survival is what is being maximized" (106). For the whole pageantry of evolution on Earth, for your life and mine no less and no more than for the sparrow flying by, or the bacteria we try to kill every time we wash our hands, or the HIV virus that is ravaging the world, the meaning is all the same. The meaning of it all boils down to genes, to their survival and reproduction.

The same applies equally for all the pain and suffering and death in the world. "Genes don't care about suffering," says Dawkins, "because they don't care about anything" (131). The meaning of pain, of course, is that of a biological adaptation. As a warning system and motivator to action it increases survival and reproduction. But there is so much pain in the world, so much suffering. It is truly, as Dawkins says, "beyond all decent contemplation" (132).[2] The *why* of pain in this very different sense, of course, cannot be that of an adaptation. Instead, its meaning must be that of a byproduct, even a necessary byproduct, but as a byproduct it no less follows from evolutionary principles. Food scarcity and the consequent hunger and starvation "must be so," say Dawkins (1995), for the simple reason that "If there is ever a time of plenty, this very fact will lead to an increase in population until the natural state of starvation and misery is restored" (132). The horror of predation, the misery of parasitism, as well as the anguish of diseases such as cancer all follow with just as much certainty, simply from the nature of competition, mutation, selfish genes, and natural selection. Moreover, the seemingly gratuitous and excessive nature of so much of the pain that we and other animals suffer follows from the fact that the processes of evolution evolved in us the mechanisms for pain sensitivity but did not evolve mechanisms to know when to stop, simply because evolution is not a mind with a purpose but a set of non-mental, non-teleological processes. The mechanism of pain sensitivity on the whole is adaptive, but it is not, it cannot, function perfectly and intelligently in every situation. Being burned to death is going to hurt like hell, even though it should not and it would not if it were truly intelligently designed.

For Dawkins (1995), then—and it does seem to be the most consistent view on the meaning of life if we take biology, the professional study of life, seriously— "The universe we observe has precisely the properties we should expect if there is, at bottom, no design, no purpose, no evil and no good, nothing but blind, pitiless indifference" (133).

Before one gets so deeply depressed by this to the point of blowing one's brains out or jumping off a tall building, one needs to read an even later book by Dawkins titled *Unweaving the Rainbow* (1998). The title might seem to suggest the unmaking of hope, but that would be a mistaken impression. Instead, the book is about "the sense of wonder in science" (x), exemplified by Sir Isaac Newton's discovery, when he used a prism to divide white light into the colors of the rainbow, that white light is not pure but mixed (a discovery that, alas, is still lost on those who favor traditional weddings). Dawkins, however, before he gets into what his book is mainly about, begins his Preface with a striking acknowledgment that his previous books have tended to have a deeply disturbing effect on the lay public, especially when he works out the implications of selfish gene theory:

> A foreign publisher of my first book [*The Selfish Gene*] confessed that he could not sleep for three nights after reading it, so troubled was he by what he saw as its cold, bleak message. Others have asked me how I can bear to get up in the mornings. A

---

2   Why anyone would disagree with what should be so obvious a truth, especially from the viewpoint of evolution (with the suffering of billions upon billions of sentient beings stretched over millions of years), is an interesting question in itself. A possible reason is theological, because the fact of so much pain, suffering, and death in the world does not rest easily with the belief in an all-knowing, all-powerful, and all-loving God. But perhaps a deeper reason is Freudian in nature. We instinctually avoid pain. Thinking about the pain of others, let alone contemplating the enormity of the pain and suffering in the world, is painful to us. Therefore, as a defense mechanism, we simply try not to think about it. We either banish the thoughts from our minds or we do our best to diminish them. Whatever the reason, the result is the same, namely, a lamentable divorce from reality.

teacher from a distant country wrote to me reproachfully that a pupil had come to him in tears after reading the same book, because it had persuaded her that life was empty and purposeless. He advised her not to show the book to any of her friends, for fear of contaminating them with the same nihilistic pessimism.

(ix)

So as to counterbalance the grim message of his earlier writings, and to show that it is "preposterously mistaken" to think that science robs ours lives of meaning and worth, Dawkins proceeds to tell us that his present book is an attempt "to try a more positive response," to show "what these complainers and naysayers are *missing*." This, he says, "is one of the things that the late Carl Sagan did so well, and for which he is sadly missed" (x).

Carl Sagan, the late Cornell University astronomer, humanist, and award-winning popularizer of science, was indeed a rare and precious gem. He definitely epitomized what Dawkins calls "science inspired by a poetic sense of wonder" (xii). But he epitomized more, much more. One has only to watch his video series *Cosmos* to see what many scientists and philosophers have been claiming for some time now, that genuine spirituality does not at all require the supernatural, that one can find meaning and deep satisfaction in the pursuit of knowledge, especially scientific knowledge, that science opens up to us the universe in such a way that through science we transcend ourselves.

This is a theme that I shall return to, the idea that science does not really deny and frustrate but actually taps into something deep in human nature, not simply an instinct for knowledge but something much deeper, namely, an instinct for the meaning of life. But before I get to that, it is interesting to notice that Dawkins in his Preface above seems to be performing the sort of move that defines a major philosophical movement known as *existentialism*. Because of its internal variation, existentialism is probably most accurately defined as a response to *nihilism* (the view that there is no purpose or meaning to life and no objective moral values). Whichever one studies first, existentialism or evolutionary biology, one will probably find it tempting to see commonalities between the two, perhaps even to the point of seeing a synthesis. But would such a synthesis be legitimate?

The evolutionary psychologist David Barash thinks so. In an article in which he coins the term *evolutionary existentialism*, Barash (2000) attempts to scratch beneath the surface and show that the two intellectual fields, "though they seem, at first, to be poles apart," have actually a lot in common and "are, in fact, a remarkably compatible pair" (1012). Existentialism certainly has a lot to do with the meaning of life. One might even say that's what it's all about. The French existentialist Jean-Paul Sartre (1947), for example, defined the common core of existentialism as the view that "existence precedes essence" (34), by which he meant that the human condition in the modern world, conceived subjectively from the inside out, "must be the starting point," that it is the most important question, not the objective nature of things. Albert Camus (1942), also a French existentialist, put the matter even more strikingly when he wrote,

There is but one truly serious philosophical problem, and that is suicide. Judging whether life is or is not worth living amounts to answering the fundamental question of philosophy. All the rest—whether or not the world has three dimensions, whether the mind has nine or twelve categories—comes afterward. These are games; one must first answer.

(3)

But is modern evolutionary biology really compatible with this philosophy of existence, often characterized by disillusionment and despair? In what follows I shall argue that any marriage between the two would be a sham marriage destined for both childlessness and divorce, indeed that "evolutionary existentialism" is an oxymoron. But I shall not argue this simply as an academic exercise. Instead, the reasons prove, I believe, to be of the profoundest importance, for they point directly to the existence in humans of a basic evolved need or instinct for finding the meaning of life, but even more importantly they point to just what the nature of that instinct is, to what will and will not satisfy it.

For a start, there are of course *some* commonalities between evolutionary biology and existentialism. Although there are some religious evolutionary biologists and some religious existentialists, both evolutionary biology *per se* and existentialism *per se* are essentially godless. Moreover, they both view the universe as ultimately absurd, not only with a contingent past and future but as having no goal or purpose. For existentialists, we are "thrust toward existence" (Sartre) or "thrown into the world" (Heidegger), all of it having no rhyme or reason, and at each and every moment we face the eternal nothingness of death. For evolutionary biologists, they view life, as Barash puts it, "as a vast existential roulette game," one where the house cannot be beat and "The only goal is to keep on playing," which is precisely the goal of selfish genes. To think of humans as the goal is pure anthropocentrism. We are merely one of roughly 30 million existing species, and life will go on evolving in a myriad of directions without us. If I may Darwinize Shakespeare in *Macbeth*, for evolutionary biologists "It [the 4 billion year pageantry of evolution on earth] is a tale told by an idiot [mutation and natural selection], full of sound and fury [biodiversity], signifying nothing [death and eternal extinction]."

Beyond these commonalities, however, and the fact that both can depress the hell out of people, I suggest that existentialism and evolutionary biology really do not have much in common. But more importantly we need to see why.

One common topic involves ethics. For existentialists, morality is manmade. Moreover, to follow and conform to any morality except our own self-created or self-chosen one is to live, as they like to call it, *inauthentically*, as a herd animal with herd mentality (Nietzsche's "human, all too human"). Because the herd constrains our individual freedom (Sartre's "hell is other people"), we need to liberate ourselves and make our own choices. But this does not mean anarchy. As Sartre (1947) put it, "moral choice is to be compared to the making of a work of art" (55). While it might be true that not only existentialists but evolutionary biologists share what Barash calls "an uncertain relationship to morality" (1014), it does not at all seem to be the case that evolutionary ethics, whether the old or the new (see Chapter 7), shares the individualism that characterizes existentialist ethics. Both the old and the new recognize humans as members of the same species, so that ethics becomes objective though not absolute. Moreover, contrary to Barash, the new evolutionary ethicists, although aware of Moore and Hume, do not typically view evolutionary ethics as running afoul, as Barash puts it, of what "Moore has labeled the 'naturalistic fallacy,' first elaborated by David Hume" (1014). (Again the false equivalence!) The new evolutionary ethics goes further than the old and views morality normatively as a supraindividual adaptation, as we have seen in the cases of Wilson and Ruse. In sum, I would say evolutionary ethics does not at all seem compatible with either the spirit or the letter of existentialist ethics.

This brings us to what Barash considers another common topic, the focus on the individual. Both existentialism and evolutionary biology focus "on the smallest possible unit of analysis," a convergence which he says is "chilling, and invigorating" (1013). For existentialists, group mentality is herd mentality (Nietzsche's phrase) and is to be avoided. The primary focus, instead, is on the individual, on his or her existence and choices. Making your own choices, overcoming what Walter Kaufmann (1973, 3) calls *decidophobia* (the fear of making fateful decisions for yourself), is being what they call *authentic*, while "going with the flow" is what they call *bad faith* (since in following the group we might think we are free but we are only pretending). Evolutionary biology also seems to be individual oriented. For Darwin, natural selection operates primarily at the level of the individual organism, rarely at the level of the group, and never for the good of the species. (Indeed, some biologists and philosophers even conceive now of species as individuals, though Darwin did not.) Today, group selection has fallen mostly into disfavor, mainly because it is less parsimonious than individual selection (Williams 1966, 1992), while, like the Wicked Witch of the East, "for the good of the species" arguments are not just dead but most sincerely dead. But Barash is misleading when he says, "Actually, it [modern biology] goes further yet, focusing when possible on genes instead" (1013). This plays down an important difference regarding the focus on individuality. First of all, in modern evolutionary biology an organism does not really reproduce itself (to the dismay of virtually every parent, one does *not* reproduce oneself by having a child). Instead, what gets passed on are genes. But genes in the physical sense of a piece of DNA are not really what get passed on. Instead, it is copies that are passed on, and hence genes in the sense of information. As I pointed out in Chapters 2 and 4, following G. C. Williams (1992), a gene is an abstract thing, a unit of information, carried by physical bits of DNA and RNA. Indeed, as Dawkins (1986) puts it, "If you want to understand life, don't think about vibrant, throbbing gels and oozes, think about information technology" (112). The upshot is that the real individual of importance in evolutionary biology is abstract, not concrete. This, however, is hardly consistent with the spirit of existentialism, which is concerned with the individual in the concrete, Sartre's "the ensemble of his acts" (see below).

A further common topic is the rejection of essences. For existentialists, not only do I have no essence, but the human species has no essence, which means that there is no human nature. As Sartre (1947) put it, at first he [man] is nothing. Only afterward will he be something, and he himself will have made what he will be. Thus, there is no human nature . . . he is only what he wills himself to be after this thrust toward existence. Man is nothing else but what he makes of himself.

(36)

Again, "Man makes himself. He isn't ready made at the start" (56). There really is a rejection of human nature here. Indeed, for Sartre we are what we do, *literally*, as in "There is no reality except in action . . . Man is nothing else than his plan . . . nothing else than the ensemble of his acts; nothing else than his life" (47). For evolutionary biologists, on the other hand, this is all poetic nonsense. Granted, ever since Darwin the idea that a species has an essence has been dead (Stamos 2003, 2007). But evolutionary biology nevertheless views humans as all of a species. Moreover, as with other species the human species has a nature. This nature, of course, is statistical at any horizontal time slice, and gradually changes vertically over time. But it is a nature nonetheless, and the subject of scientific study as with all other species. Accordingly, although we can personally resist our human nature to some extent, more so in some ways than in others, we remain human after all, so that only in a very limited and qualified sense can we will ourselves to be what we are.

A further topic is that of free will. For existentialists, human free will is a necessary postulate, Sartre's (1947) the "absoluteness of choice" (53). We need to be free to choose what we are, indeed to choose the meaning of our lives. As Sartre (1947) put it, "there is no determinism, man is free, man is freedom . . . man is condemned to be free" (41). The existentialist and Freudian psychologist Erich Fromm (1966) put the matter even more strikingly. Noting that it was God or the gods in traditional religions that defined and gave humans the meaning of their lives, the death of God à la Nietzsche, says Fromm, means that we are now free to be the meaning givers. Hence the title of his book, *You Shall Be As Gods*. But again, for evolutionary biologists this must all be nonsense. This is because evolutionary biology, along with modern physics, leaves no room for a mind that is transcendental and free of its body, and that means no room for a free will. The mind-brain has different evolved control centers, and these can sometimes conflict with one another, so that as we have seen Dawkins put it in Chapter 1, our genes exert a statistical rather than deterministic influence over us along with other influences, allowing us the possibility to rebel against the tyranny of our genes. For Barash, on the other hand, "human behavior is composed of both genetic components and a hefty dose of free will" (1015), such that genes give us only "predispositions," they only "whisper within us, they do not shout," so that "It is our job, our responsibility, to choose whether to obey" (1016). The problem with all of this is not only that our genes do a lot of shouting and not simply whispering, but that what Barash calls free will is not really *free* will. If he takes science seriously it cannot be. As John Searle (1984) so poignantly put it,

As long as we accept the bottom-up conception of physical explanation, and it is a conception on which the past three hundred years of science are based, then psychological facts about ourselves, like any other higher level facts, are entirely causally explicable in terms of and entirely realised in systems of elements at the fundamental micro-physical level. Our conception of physical reality simply does not allow for radical freedom.

(98)

To think otherwise, from the viewpoint of modern science, is both an illusion and delusion.

The belief in free will leads Barash to focus on what he believes is yet another commonality, the prescription to rebel. For existentialists, rebellion is a major theme. Camus wrote a classic of existentialism which he titled *The Rebel* (1956), the main theme of which is that, "confronted with an unjust and incomprehensible condition," we "demand order in the midst of chaos, and unity in the very heart of the ephemeral" (10). Indeed man, he says, "is the only creature who refuses to be what he is" (11). For Barash, evolutionary biology likewise involves rebellion. We have seen Dawkins advocate rebelling against the tyranny of our genes. We have seen Williams advocate rebelling against natural selection. Likewise Barash advocates rebellion. In a world with a human population explosion, taking all of human history to reach 1 billion at around 1850, reaching 6 billion around 2000, and projected to double to 12 billion by 2040 (although many claim the world will not be able to support that many people and the human population must crash before that point), Barash advocates having no children if we "want to be fully human" and that, since obedience has caused far more harm in human history than disobedience, "we need to teach more *dis*obedience. Not only disobedience to political power and social authority, but especially disobedience to some of our troublesome genetic inclinations" (1015). But this does not at all characterize the kind of rebellion one finds in evolutionary biologists *as evolutionary biologists*. When they do rebel, it is not *existential*, rebelling against chaos and an incomprehensible condition, or refusing to be what we are. Instead, when evolutionary biologists do rebel, it is typically against the destruction of ecosystems, or the loss of biodiversity, or the poverty of science education in our public schools, or the obscurantist tactics of fundamentalist religions. Of course evolutionary biologists advocate rebelling against our genes here and there (one pretty much finds that in everyone, evidence politics and religion), but there remains something fundamentally different from the *kind* of rebellion one typically finds in existentialism.

The topic of rebellion leads us to the final difference I shall focus on, the difference in mental attitude and character that distinguishes existentialists from evolutionary biologists. Needless to say, this and the statements that follow need to be taken statistically, but if I am right it would mean that an "evolutionary existentialist," if ever there really could be one, would be an oddball, having something of a split personality. I say this because not only do existentialists tend to chain smoke, but they tend to be filled with dramatics, the kind that follows from distress, forlornness, anguish, and despair (all of these words are in Sartre 1947)—*angst* is the famous term in existentialism—shaking their fist (or middle finger) at the universe and either contemplating suicide or swearing it off. But one just does not find this in evolutionary biologists, and it is not because they are philosophically naive and have never seriously stopped to think about existence (or because they make more money). Far from it. The reason for the difference goes much deeper. It is, I suggest, because in evolutionary biologists, but not in existentialists, a deep instinct for the meaning of life is more or less satisfied, and it is one that cannot be satisfied in any way one chooses. That is the fundamental mistake that existentialists make.

Let me approach this matter from a different angle. For existentialists, as Camus (1942) put it, "in a universe suddenly divested of illusions and lights, man feels an alien, a stranger. His exile is without remedy since he is deprived of the memory of a lost home or the hope of a promised land" (6). Evolutionary biologists, on the other hand, although they are men (human) and share with existentialists the view that death is the end and life is ultimately meaningless, by no means share the feeling of being a stranger in the universe, the feeling of being alien, lost, or exiled. Nor do they think that suicide is the biggest question and every other question is trivial. Why the profound difference?

The answer, I believe, was hit upon by the philosopher Robert Nozick (1981). The literature on the meaning of life is, as one might expect, enormous, but Nozick seems to have hit the nail on the head, to have captured the essence of it all, whether religious or secular, when he claimed that meaning is "relational" (599). This is true, of course, not only for words and sentences and symbols, but most importantly for human lives. "For a life to have meaning," he says, "it must connect with other things, with some things or values beyond itself. Meaning, and not merely of lives, seems to lie in such connections" (594). It is in relational meaning, not intrinsic meaning, that we go beyond ourselves, and we need to go beyond ourselves. A self-absorbed life is not a meaningful life. This is a lesson that one can find throughout history, in one biography after another. Again as Nozick puts it, "Attempts [I would add the qualifier "successful"] to find meaning in life seek to transcend the limits of an individual life. The narrower the limits of a life, the less meaningful it is" (594).

This speaks to a human instinct for the meaning of life, an instinct that cannot be satisfied in just any way. But now the question of evidence arises. One piece of evidence, I would say, is the ubiquity of the question of the meaning of life. It recurs in religion after religion and throughout the history of philosophy, existentialists make a living out of it, scientists feel the need to address it in one way or another, novels and movies torture themselves over it, and it crops up again and again at the top of public opinion polls and statistical surveys. There really does seem to be a poverty of stimulus about it, something that recurs independently in the human spirit.

But mere universality is not by itself enough to establish a genetic basis for something, in this case a particular and deep-seated need. What is required is some evidence that the satisfaction of the need increases survival and hence reproduction in one way or another. I know I am going out on a limb here (what else is one to expect from the descendant of a monkey?), but I suggest the evidence exists in spades, and a classic example comes from Auschwitz, Poland. What I have in mind are the observations of Viktor Frankl documented in the first half of his book, *Man's Search for Meaning* (1984). In his life prior to World War II, Frankl had been a successful psychiatrist in the Freudian school. But his Jewish origins led him to the Nazi death camps from 1942–1945, where he ended up at the most notorious of them all, Auschwitz. His firsthand experiences permanently changed his view on human nature. He no longer looked at sex as the basic drive and instinct of humans, but instead what he called the "will to meaning" (108). He saw that those who lost that will soon after died in the camps, not so much from the Nazis but either from suicide or from their immune systems which had become suppressed. In the horrid conditions of the camps, their lower resistance to disease quickly did them in. Those who maintained a will to meaning, however, who maintained a *why* to their existence, tended to survive. Out of this insight, Frankl went on to develop a school of psychiatry which he called *logotherapy*. Although the extreme environment of concentration camps no longer applied, Frankl found that neuroses (mental illnesses without a physical cause)— such as hysteria, obsessive-compulsive disorders, addictions to power, alcohol, or sex, as well as non-illnesses but nonetheless problems such as juvenile delinquency or boredom—were typically caused by a frustrated will to meaning, what he called an *existential frustration* (106), or worse, an *existential vacuum* (111). Hence the psychotherapy he developed was to help patients find meaning in their lives.

Frankl's school of psychiatry is explicitly existentialist, and that is part of the problem with it, as I see it. Frankl was concerned with the basic need for meaning as a *proximate* cause, and he was correct in placing this need as a basic human need, as an innate instinct, what he called "the primary motivation" (105). But he never seriously stopped to ask for its *ultimate* cause. Because of this, he pretty much appropriated the existentialist solution to the meaning of life, which is that it is open to invention and choice and is realized in the here and now. There is no answer in the abstract, he says, the meaning has to be "something very real and concrete" (85), something found in action. Moreover, it differs "from man to man, and from moment to moment" (85) and "each man is questioned by life; and he can only answer to life by *answering for his own life*" (113). The logotherapist helps the patient to see the possibilities given the patient's particular situation, to see "the whole spectrum of potential meaning" (115) in front of him, but the logotherapist absolutely does not, says Frankl, impose "value judgments on his patients." Rather he leaves it "up to the patient to decide" (114).

But there is the problem. The whole existentialist approach, in line with the SSSM, has a mistaken view of human nature. Hence its solution to the meaning of life cannot help but be mistaken. From an evolutionary point of view, humans have a nature, and as with any animal nature it is not going to be wide open but statistical, with statistical norms. Hence not just any answer to the meaning of life, no matter what the person and their situation, will do. And we can see this in our instinctual response to people such as Adolf Hitler, Charles Manson, and Osama bin Laden. They no doubt found what Frankl requires for a meaningful life, namely, "self-transcendence," in that, as Frankl puts it, each life "points, and is directed, to something, or someone, other than oneself" (115). And they might even have thought they found another of Frankl's requirements, namely, what satisfies "the depth of his being" (108). But we hardly think of them as having genuinely meaningful lives. Instead, we think of them as having wasted not only their own lives but those of many others.

Something more restricted is needed for a satisfactory answer to the meaning of life, and it is why, it seems to me, existentialists never appear to be happy but appear to be filled, instead, with angst. Once again, Nozick, I suggest, hits much closer to the mark. For Nozick (1981), there are basically two ways that the search for meaning is satisfied. One is in "personal relations" with others, where "loving another brings us most outside our own limits and narrow concerns" (595). For most, this form of meaning is found in family, friends, and romantic love, but Nozick construes personal relations broadly to include world causes and fighting injustice as well. Accordingly, finding meaning in personal relations would have to include practical ethics. And when we look to personal relations in this broad sense, we find indeed some deeply meaningful lives, such as that of Mohandas Gandhi (Gandhi 1929), who gave to the world a very effective means of fighting systemic injustice and whose life and teachings had an enormous impact on people such as Martin Luther King Jr. and the nonviolent civil rights movement that he spearheaded in the 1960s. But personal relations in Nozick's sense of meaning can be extended beyond humans to include animals as well, and even all living things (though now the concept of person would no longer apply). Indeed, this is what made Albert Schweitzer's life so meaningful along with his teaching of "reverence for life" (Schweitzer 1933). In his early thirties, Schweitzer gave up a brilliant multifaceted career as a philosopher, theologian, musician, and music scholar to become a medical doctor and devote the remainder of his life to a hospital he established and ran in equatorial Africa. He found his life deeply meaningful and his fame and Nobel Prize for Peace suggest that many others did as well. Fortunately, life is filled with examples of people who found life deeply meaningful in personal relations, though few of them are so famous as the above three.

The other way that a meaningful life is to be had, according to Nozick (1981), is in the pursuit of knowledge, the "intellectual life." "Knowledge," he says, takes us "beyond our limits," so much so that "there is nothing that cannot be thought of, theorized about, pondered" (597). In philosophy there is a tradition, going back to Plato and Aristotle, which holds that rational thought and contemplation is the highest human activity. Modern science grew out of philosophy as a series of separate specializations, and arguably, more than philosophy, science takes us to the universe. Hence we find in scientists such as Charles Darwin and Carl Sagan enormously meaningful and stimulating lives, and they each knew it.

Given the social nature of human evolution, as well as the epistemological nature of human evolution, topics we have dealt with in previous chapters (Chapters 7 and 1 respectively), Nozick's twofold answer makes perfect sense. The one is about binding relations to others, the other is about binding relations to our environment. If anywhere a satisfactory answer to the meaning of life is to be found from an evolutionary point of view, it is to be in those two kinds of relations.

Nevertheless, does not evolution still mean that our lives really are meaningless, even granting that we have an evolved instinct for meaning which can be satisfied more or less in the ways Nozick described? In a sense, the answer of course is *yes*, but in the end I want to suggest that it does not really matter. Evolution has also given us reason, from reason comes logic, and from logic comes one of the most important fallacies for our topic, namely, the fallacy of *division*, the fallacy of assuming that because a whole has a certain property some or each of its parts must also have that property. Even if physics is right and the universe as a whole has no meaning, it does not necessarily follow that one of its parts, life as a whole, has no meaning. And even if evolutionary biology is right and life as a whole has no meaning, it does not necessarily follow that one of its parts, my life or yours, has no meaning. In spite of the truths of evolution and physics, we each can still have a deeply meaningful life.

Appendix:
Common Misconceptions
About Evolution

What follows is a discussion on what seem to me the most common misconceptions about evolution. Many will not be able to get into the big questions, let alone learn to appreciate them, until they get over these hurdles. The sad thing is, they should not be hurdles. Each of them should have been taught and clarified in science classes in public schools, rather than the enormous time spent on trivia such as dissecting frogs. The body of knowledge, research, and theory known as *evolutionary biology* is one of the biggest and most important achievements of modern science. And yet scientific illiteracy continues to be high among the public. This Appendix is brief and is not meant as a replacement for what is needed, but it is a start and the present book would be lacking without it. What would be really good is to take a full-year introductory course on evolutionary biology. Autodidacts might prefer a careful study of an introductory text on evolutionary biology (e.g., Maynard Smith 1993; Ridley 1993; Freeman and Herron 1998; Futuyma 1998). In either case, one should approach the topic with an open mind, not a predisposition to disbelieve (religion is typically the problem here, in particular the politics of religion). For those who do not have the opportunity or the time for either of the above, much can still be gained by taking a quick look at the case for evolution, for which I suggest Quammen (2004), published in *National Geographic* magazine, as well as the video *Origin of Species: Beyond Genesis*, published by The Discovery Channel. There are also a number of excellent websites, of which I suggest www.pbs.org/wgbh/evolution/darwin/.

Origin of Life vs Evolution of Life

The question of the origin of life is often confused with the question of the evolution of life. But they are separate questions. How life originated in the first place is one thing, how life evolved once it began is quite another. I shall deal with processes of evolution under *Patterns* below. Of course, a common theological view, including most theistic views of evolution, is that God started life on Earth. But there is no good reason to resort to a supernatural explanation here. As the astronomer Carl Sagan put it in his video series *Cosmos*, "extraordinary claims require extraordinary evidence." One could also appeal to Ockham's Razor. There is no need to appeal to the supernatural when natural explanations, which by their very nature are simpler, will do just as well. Although there is no settled scientific theory on how life originated on Earth (there are possibly five main competing theories), there are abundantly good reasons for believing it began naturalistically and here on Earth. (Some think that life on earth began from a "seed" deposited from some other planet, perhaps via a meteor, but this only raises the question of how life began on that other planet and is best ignored.) Somehow, some way, possibly more than once, a molecule was formed that naturally made a rough and ready copy of itself, perhaps something like the way crystals form. From this humble beginning some of its descendant molecules reproduced better than others, so that the process of natural selection kicked in, resulting in ever increasing complexity. DNA could not have been the first molecule of life, since it and the genetic code it carries are much too complicated. Quite possibly the earliest molecules of life were something like RNA. Organic molecules would also have been available for the earliest genes to build *phenotypes*, outward structures called *organisms*, hence improving the survival and reproduction of the genes that made them, beginning of course with the simplest and evolving ever more complex forms as a matter of competition. (This is the concept of the *selfish gene* made famous by Richard Dawkins 1976.) All of this became much more plausible in 1953, when Stanley Miller and Harold Urey passed an electric charge (mimicking lightning or high UV) through a chamber containing chemicals thought to be abundant in the early history of our planet, namely, water, methane, and ammonia (but not free oxygen, as this is a product of photosynthetic life). The result was striking, since it included many of the components that make up DNA and RNA, including the four nucleotide letters, as well as many of the amino acids that make up proteins. There are plenty of good articles and books and chapters on theories of the origin of life, such as Cairns-Smith (1985), Dawkins (1986, ch. 6), Eigen (1992), Bada (1995), and Maynard-Smith and Szathmáry (1999).

Evolution as Fact and Evolution as Theory

It should not come as a shock to the general public let alone to college and university students, but the fact remains that most seem unaware that the view expressed by Dobzhansky at the beginning of the Introduction is uncontroversial in natural science. (Such is the poverty of science education in public schools.) The world of professional science is a worldwide community, comprised of hundreds of thousands of the most educated and intelligent people in the world, people who have studied the relevant phenomena and evidence and knowledge in their respective fields to a degree that the general public can scarcely imagine. Within that community—*and it is the only community that really counts*—evolution ceased to be debated long ago. It was primarily Darwin who turned things around. As a result of his work, and the work of legions of biologists who have added to it in innumerous ways, evolution long ago became accepted as a fact. The enormity of the evidence gathered since Darwin's time, together with the congruence of the different kinds of evidence, along with the way new evidence fits in so nicely, is simply overwhelming (see Gould 1983; Dawkins 1986; Maynard Smith 1993; Ridley 1993; Futuyma 1995, 1998; Freeman and Herron 1998).

It might help here to follow scientists in a distinction that they routinely make, namely, between hypothesis, theory, and fact. A *hypothesis* is simply a guess, a preliminary shot at an explanation. Its source does not even matter. It could be an educated guess or it could come to one in a dream. The idea that the Earth is a planet began in some such way. It was simply a hypothesis that occurred to some ancient Greeks. It did not graduate to the level of a *theory*, however, until some evidence was adduced in its favor. Eventually there comes a time when there is so much evidence that it becomes ridiculous to call it a theory anymore and necessary to use the word *fact*. (This is not to say, of course, that all theories graduate to the level of a fact; most either fade away or die a violent death.)

The same distinction can be found in detective work and courts of law. One might have a hunch about who committed the crime. If evidence is found that supports the hunch, then it becomes a theory. Eventually, and ideally, either the theory is established as a fact in a court of law or it is dismissed as insufficiently supported or even as erroneous. It is to be noticed that even in a court of law, direct eyewitness evidence of culpability is not always necessary. Sometimes the indirect evidence becomes so massive that there is no room for "reasonable doubt" anymore. In such cases the judges and jurors have both a legal and a moral obligation to pass judgment. "Proof" (a terribly misused and abused term) is really not the issue. If we want to get philosophical about it, nothing (not even Descartes' famous "I think, therefore I am") can be proved beyond the shadow of a doubt. In courts of law, as well as in everyday life, what is at issue is evidence beyond any reasonable doubt.

The same is true of science. Just as it is a fact (beyond reasonable doubt, though not absolutely provable) that the Earth is round and is a planet (the Flat-Earth Society notwithstanding), so too the idea that biological species are not fixed but evolve began as a hypothesis (again by some ancient Greeks), was raised later to the level of a theory, and finally graduated to the level of a fact (the Institute for Creation Research notwithstanding). Quite simply, the evidence is so enormous and comes from so many different fields of research that there is no room for reasonable doubt anymore. One and only one explanation stands out as the reasonable one, and Darwin was the man who first made the case and shook the world. Darwin (1859) brought together, and in a most masterful way, evidence from a wide variety of fields (some of it from his own research, most of it from the independent research of others), namely, animal and plant breeding, paleontology (fossils), biogeography, embryology, scientific classification (groups within groups within groups), comparative anatomy (including homologies and vestigial organs), and geology (the age of the Earth kind).

Each of these areas of research have been massively added to since Darwin's time, by thousands upon thousands of professional researchers, with all of it capped off and clinched by the discovery and study of DNA. In short, the evidence is individually indirect but jointly conclusive.[1]

To be sure, there are many theories and debates about evolution (mainly to do with rates, mechanisms, and paths), but these are debates *within* evolutionary biology and should not be confused with debates *about* the fact of evolution. The latter, debates about the fact of evolution, take place outside the realm of professional science. They are typically instigated by those with a religious (fundamentalist) agenda, with scientists and philosophers of science typically being the ones who provide the replies. These debates can easily mislead the public, as they no doubt have, for they have led to many public debates and an enormous literature, all of which foster the public perception that the same debates also occur within science. They do not, not even in the slightest. For those who wish to pursue this matter further, there are a number of good books that deal with creationism (e.g., Kitcher 1982; Godfrey 1983; Futuyma 1995; Pennock 1999) as well as the latest version known as *intelligent design creationism* (e.g., Miller 1999; Pennock 2001; Dembski and Ruse 2004; Young and Edis 2004; Brockman 2006).

Why Evolutionary Biology is a Genuine Science

The question of what is science, and what distinguishes it from pseudo- or non-science, is a debate in philosophy of science, and it is too large for me to deal with here. Suffice it to say that evolutionary biology is a genuine science, not merely because of the trivial fact that it employs people who are called "scientists," but because as a field it meets just about all the requirements of a genuine science that one could hope for.

For a start, evolutionary biology makes statements that are either directly or indirectly confirmable, in other words it has predictive content, for example that natural selection operates in nature, or that mutations in genes are random with respect to the environment, or that intermediate forms should be found in the fossil record, or even that the Earth is many hundreds of millions of years old.

Conversely, evolutionary biology also makes statements that are falsifiable. If, for example, a human skull were to be found in rock strata formed 100 million years ago

---

1 The comparison with detective work and law courts and the accumulation of circumstantial evidence, as good as it is, is in a way misleading. The fact is, although our life spans are short and we cannot sit back and watch the incredibly slow process of one species evolving into another (slow to us, that is, in fact so slow that many people refuse to believe what their eyes cannot see), there are nevertheless examples of evolutionary change that we can directly observe right before our eyes. One is the evolution of bacterial resistance to antibiotics such as penicillin. This example of natural selection operating on random variation has been observed over the past few decades, as with the evolution of various viruses. Another example is what is known as *ring species*, species that evolved round a barrier, for example a large lake or mountain range, such that the terminal populations are reproductively incompatible. Such species are said to be examples of vertical (temporal) evolution laid flat.

along with dinosaur fossils, the entire edifice of evolutionary biology would collapse. The same would be true if it were found that each biological species has its own unique genetic code (indeed, this would be the perfect evidence against evolution if there were a God who wished to provide it). Of course, nothing like this has ever been found to be the case.

A further feature of genuine science is that it opens up new fields of research. Darwin himself saw that this would happen if his views became accepted, and he named specifically psychology and heredity along with others. Today, of course, evolutionary psychology and molecular biology are thriving fields.

Finally, a genuine science should be consistent with what is already known. Since the discovery of DNA and the causes of mutations, it is now known beyond any doubt that evolution and the processes that cause it are part of a seamless connection with the levels of chemistry and quantum physics below it. The division of the study of nature into different sciences is merely an artificial convenience. (For more on some of the above topics, see Chapter 8.)

The only way in which biology might not measure up to other genuine sciences, such as physics, is that biology possibly does not have any laws of nature specific to it. Whether biology has genuine laws is a topic debated by biologists and philosophers of biology to no end, and I shall have no more to say about it here (I shall have more in Chapter 2), except to say that the consensus today (with which I agree) is that while laws are powerful for explanation and prediction they are not necessary for a field to count as a genuine science.

Pattern vs Processes of Evolution

A common misconception is to confuse the pattern of evolution with the processes of evolution. When one talks about evolution, one has to make clear whether one is talking about the pattern of evolution or the processes responsible for it. The *pattern* of evolution is simply the actual lines evolution has taken through its long past. Darwin conceived of the pattern of evolution as a great Tree of Life, beginning with a trunk and branching outward. It is now safe to assume that life on Earth probably began in one place, given the evidence from the standard genetic code. Many debates in evolution are about the specifics of the pattern. For example, it was long debated whether modern chimpanzees are closer in evolutionary history to humans than gorillas. Some said chimpanzees, some said gorillas. DNA analysis finally settled the issue. The branch leading to these three species split roughly nine million years, one branch leading to modern gorillas, the other leading to modern chimpanzees and modern humans. That latter branch then later split roughly five million years ago, one branch leading to modern chimpanzees, the other to modern humans. (Of course, the branching story is much more complicated than that.)

The pattern of evolution is also sometimes thought of in terms not of species but of character traits. It needs to be noted that species, once extinct, do not recur in the great Tree. (There is a minor exception to this general rule in repeated speciation by polyploidy, but I won't bother with that.) Character traits, however, often recur throughout the Tree. Wings, for example, evolved independently at least four times, while eyes evolved independently at least 40 times.

What causes the pattern of evolution, whether in terms of the phylogeny of species or in terms of character traits, is an entirely different matter. If *evolution* is defined as the change in gene frequencies within a population (this is the definition common in population genetics), then there are a number of different *processes* that cause evolution, namely, birth, death, immigration, emigration, mutation, genetic drift, and natural selection.[2]

Only natural selection produces adaptations. For this reason the concept of natural selection, as Darwin recognized, is arguably the most profound of any of the concepts found in biology. In spite of being a relatively simple concept (compare it, for example, to Einstein's concept of gravity), it is notoriously difficult to teach. Indeed, as Richard Dawkins (1986, xi) points out, it is as if our brains were wired to not get it. At any rate, it takes some work, but once one gets it the lights go on. And on and on and on (that is one of the themes of this book).

The first point to notice is that natural selection is a process, even though it is often spoken of as a force (as in "the driving force of evolution" or "selection pressure"). It is a directional process that occurs when a number of preconditions are met. Perhaps the best way to understand natural selection is to keep reading definitions and applications of the concept. The most terse definition, "nonrandom differential reproductive success," is common in population genetics, but it is not exactly learner friendly. The best definitions focus on the preconditions necessary for natural selection to occur. Although they sometimes vary slightly from author to author, they all say basically the same thing. For example, the biologist John Endler (1986, 4) focuses on three preconditions, as most do, namely, variation in individuals of a population in some particular trait, fitness differences between those traits, and heritability of those traits. One might add superfecundity and a field of competition to all of this, but one is not really adding anything new that is not already there, whether explicit or implicit.

A process that does *not* cause evolution, whether mere genetic change or adaptation, is *inheritance of acquired characteristics*, also known as *Lamarckism* (after the French biologist Jean Baptiste de Lamarck, who made much of this supposed mechanism of evolution at the beginning of the 19th century). This theory of evolution, unlike natural selection, is ridiculously easy to teach and readily sticks in the mind, so much so that even after learning about evolution by natural selection many will still retain Lamarckism and confuse it with Darwin's mechanism. Quite simply, Lamarckism is the theory that morphological or behavioral changes in an organism get passed on to the organism's progeny. Although widely held in the past even by many biologists (Darwin himself retained it as a minor mechanism in evolutionary change), Lamarckism in biology died with the discovery of DNA (although by then it had already long been dying in its death bed). To give two simple examples, no matter

2   The definition of *evolution* from population genetics is one I do not myself particularly care for, since it entails that creationists believe in evolution! Consequently, I prefer a definition that includes phylogenetic history and adaptation by natural selection.

how much you work out and build up your biceps or spend hours practicing and getting good at playing the piano, there is no way these activities change the DNA in your sperm or eggs, let alone in the very specific way needed to produce babies that will have bigger biceps or better piano skill. Known as *the central dogma of molecular biology*, the transfer of information is always from genes to proteins and never vice versa.

Evolution and Chance

One can hear ad nauseum the criticism that if you throw the pieces of a watch together you won't get a watch, or if you throw lots of different letters together you won't get a book. This criticism, varying only with its examples, betrays an enormous amount of confusion and misunderstanding and is simple-minded in the extreme. Evolution by natural selection is a *cumulative* process, not a one-shot affair. Natural selection operates on what is already there in a population, not only changing gene frequencies but also, if allowed enough time, producing in successive generations increasing complexity. It might help to think of natural selection as a sieve (a metaphor used by a number of biologists), but not a particularly good sieve. Imagine this sieve being applied repeatedly over and over again, with random variation added to the elements along the way. The sieve, again not a good one, favors some traits and variations of traits and disfavors others, so that some of the elements get through and others do not. The sieve, however, also changes over time, almost always gradually, so that what it favors and disfavors also changes. If one can imagine such a sieve, one is on one's way to understanding evolution by natural selection. It is a process that is incremental and extremely gradual, with every incremental step being useful in some way to its possessor (see Darwin 1859, ch. 6). Although the operation of natural selection is mediated through phenotypes, ultimately it works at the level of DNA letters. It is not strictly a chance process but neither is it a deterministic one. Instead, it is a statistical process (i.e., inheriting a slightly improved trait does not guarantee longer survival and greater reproduction, but it does increase the probability), one that is capable of producing complexity to such a degree that it can easily fool otherwise intelligent organisms into thinking that what it produced was produced by intelligent design.

And yet there is much chance in evolution. The mutations that feed natural selection are largely a matter of chance, as they are random with respect to the environment and even connect with quantum indeterminism in physics (see Stamos 2001). Genetic drift is also a matter of chance, though now operating at the level of populations. Other chance processes occur at the level of populations. Most species (species are composed of one or more populations) are *endemic*, meaning they are found in only one geographical area. A fire, or a virus, or the introduction of a new species of predator or parasite, can destroy a local population and consequently change forever the direction of evolution. Chance can also rule at the truly large level, well above that of populations or species. The demise of the dinosaurs is now thought to be mainly the result of chance. Dinosaurs ruled the world for over 100 million years. They had good genes. There is now strong evidence that a major impact in the Yucatan peninsula, probably a meteor, drastically changed environmental conditions across the world resulting in well over 50% of the world's species going extinct, not just the dinosaurs (some of which did actually survive—their descendants are known today as *birds*).

Missing Links and Monkey Ancestors

One often hears the implied criticism, "If we evolved from monkeys, then why are there still monkeys?" Or one hears that science has failed in finding the missing links in human evolution. These two criticisms are connected, but let us start with the first. It is a mistake to think that modern humans evolved from modern monkeys. That is like saying I was born from one of my cousins. If one cares about being biologically correct, then one will say that modern humans and modern monkeys evolved millions of years ago from an ancestral population that branched into different directions, one branch leading to modern monkeys, the other leading to modern apes (modern humans are one of the many species of apes). Modern monkeys exist because they are the ancestors of earlier monkey species *and* because they successfully exploit current places in nature. One could make much the same argument about bacteria. If humans evolved ultimately from bacteria, then why are there still bacteria? Again, humans did not evolve from *modern* bacteria. Moreover, bacteria still remain (and did not evolve away into something more complex) because they fill places in nature requiring these simple forms of organism. Among the simplest of bacteria are the *autotrophs*, which derive energy solely from non-organic sources. Interestingly, were it not for them and other kinds of bacteria, none of us would be here today. The chain of dependence is not only in the past but also in the present. Were all bacteria in the world destroyed right now, it would be game over for all life on Earth in a very short time.

The above misconception about ancestors is related to the misconception about missing links. To think that all missing links should be in existence today is as ridiculous as to expect all businesses to be in existence today. With businesses as with species, the vast majority that have ever existed are now extinct. As Darwin fully recognized, what goes hand in hand with evolution is extinction. Darwin also recognized that it is equally ridiculous to expect all transitional forms to have been recorded in the fossil record. The fossilization process is such that fossilization is relatively rare. The vast majority of organisms never get fossilized, largely because they get eaten, but also because the conditions for fossilization are relatively rare and sporadic. Even so, there are many excellent examples of transitional forms. *Archaeopteryx*, which existed roughly 75 million years before the demise of the dinosaurs, is a perfect example of the transition between dinosaurs and birds (indeed, for biologists today of cladistic persuasion, birds *are* dinosaurs), as it shares common features with both reptiles and birds. The same is true of *Australopithecus afarensis*, which existed around four million years ago. It has the same body size as modern chimpanzees, the same brain size (about a third of modern humans), and shares many other features, but unlike chimpanzees these hominids had evolved fully bipedal walking and used simple stone tools. They are naturally thought to be very close in phylogenetic history to the branching point that led eventually to modern chimpanzees on the one hand and modern humans on the other. The fossil record from *A. afarensis* onward is actually rather good, with much branching as one should expect, but the most noticeable feature being a gradual increase in brain size. Links such as *Archaeopteryx* and *A. afarensis*, fortunately for modern knowledge, are no longer missing, but to expect every little increment between forms, let alone distant forms, to be preserved in the fossil record is simply unreasonable.

Glossary

References to the key terms will be found in the Index

Adaptation The fit of an organism to its environment by virtue of a complex trait that looks like it was designed by intelligence. Adaptations can be either physical (e.g., eyes) or behavioral (e.g., the weaving of a spider's web). The concept of adaptation has an element of relativism to it. A trait that is adaptive in one environment can be maladaptive in another. Ultimately, a trait evolved as an adaptation because it increased reproductive success.

Allele Different versions of the same gene, occupying the same location in a chromosome pair (homologous chromosomes, e.g., the XY in human males or the XX in females). A gene is said to be *homozygous* if it has identical alleles opposite each other on the homologous chromosomes, and *heterozygous* if the alleles are different. When two different alleles are present, one is sometimes *dominant* over the other, masking the effect of the recessive allele. Traits of *recessive* alleles only show when the recessive alleles are homozygous.

Allometry The growth of different parts of an organism at different rates or at different times. For example, in human juveniles, brain size is ahead of body size.

Altruism Self-sacrificing behavior.

Biogeography The geographical distribution of species or their component populations or their character traits. Biogeography is a major source of evidence for evolution, such as the many endemic species found on the Galapagos Islands as studied by Charles Darwin, with similar species on the South American coast.

Biological Determinism See Determinism.

Cladism "Clade" is the Greek word for branch. Cladism is the now dominant school of professional taxonomy in biology. Based on character analysis, it classifies organisms into branching diagrams, which in theory should be roughly congruent with the actual clades of phylogenetic history.

Conspecifics Members of the same species.

Determinism The view that every event has a cause and the same causes produce the same effects. *Biological determinism* is the mistaken view that the same genes will produce the same effects. This view is mistaken because genes are not the only cause of an organism's physical and behavioral traits.

DNA Deoxyribonucleic acid, the macromolecule coiled up inside chromosomes. DNA contains *genes*, the genetic information that makes a blade of grass a blade of grass, a mosquito a mosquito, and (evolutionists would say) a human a human. The *genetic code* (see Genetic Code) is digital and contains four molecular letters (G, C, T, A), called *nucleotide bases*, strung along the DNA and which are arranged stereoscopically in organisms which have their DNA in a twisted ladder-like double helix (such that G is always coupled with C and T is always coupled with A). DNA is reproduced by the double helix splitting and the complementary letters added to make two double helices (see Mutation).

Embryology The study of embryos and a major source of evidence for evolution. The closer the common ancestry of two different species, the more their embryos resemble each other. Dogs and humans, for example, have barely indistinguishable embryos.

Epiphenomena An effect that in turn is not a cause. Some argue that mind is merely an epiphenomenon of the brain. Although our mental states seem to cause our physical states, such as when we will or choose to raise one of our arms, this is an illusion according to epiphenomenalism. It is the brain that moves the arm and the will or choice is merely an epiphenomenon.

Epistemology The study of knowledge, from the Greek words *episteme*, meaning "knowledge," and *logos*, meaning "the study of."

Ethology The field devoted to the study of animal behavior in the wild.

Exaptation The term coined by Gould and Vrba (1982) to refer to a functional shift in a trait (structural or behavioral) such that the current function of the trait was not evolved (or not entirely evolved) by natural selection. The trait could be what is commonly called in biology a *preadaptation*, a trait evolved for one purpose and function and later fortuitously co-opted (and perhaps further evolved) for another (e.g., feathers were originally evolved in small dinosaurs for thermoregulation and then later co-opted for flight)—Gould and Vrba do not like the label "preadaptation" because the "pre" smacks of teleology as in "predestined"—or it could be a trait that was originally nonfunctional but was fortuitously co-opted (and perhaps further evolved) for its current function (e.g., the panda's thumb, not a true thumb since it was formed from an extended wrist bone when Panda's adapted to a diet based on bamboo) (see Gould 1980b, ch. 1, 1991, 144 n).

Evolution Change in a species, population, or lineage of organisms that has a genetic basis. (Hence not all change in a species is evolutionary change.) "Evolution" is often defined as a change in gene frequencies. Gene frequencies have many causes of change, namely, birth, death, immigration, emigration, mutation, genetic drift, and natural selection. Only natural selection produces adaptations, and some confine the meaning of "evolution" to this more restricted view of evolutionary change.

Evolutionary Psychology The application of evolutionary biology to help explain psychological phenomena.

Fitness Fitness in biology is not to be confused with the meaning of fitness in every-day life. Rather, it refers to the ability to survive and reproduce, i.e., *reproductive success*. A trait, or a mutation, that generally increases survival and reproduction is a trait or mutation that increases fitness. Sometimes the term is applied to organisms, more usually to traits, but its most exact application is to genes. (See Dawkins 1982, ch. 10, who traces five different senses of the word in biology.) It should be noted that fitness values are often relative to the environment. What increases fitness in one environment might not increase fitness in another.

Genetic Code The *standard genetic code* (formerly called *universal* but now *standard* because there are known exceptions to it, such as in mitochondria) is the par-ticular mapping of 64 codons to 20 amino acids (chains of amino acids make up proteins and ultimately our bodies). A *codon* is a triplet of the four nucleotides in RNA, the four genetic letters (G, C, U, A). RNA is the intermediary between DNA and its phenotype. Given a four-letter code and 20 amino acids, there are going to be 64 codons available to code for those amino acids. Since a codon cannot code for more than one amino acid, it follows that there is going to be a certain amount of overlap or *synonymy* in the code. In the standard code, it so happens that 61 of the codons each code for a particular amino acid, while three codons are stop codons, coding for the termination of the amino acid chain. The result is *supervenience* (see Supervenience), such that different strings of RNA, or ultimately DNA, can code for the very same product. For example, the amino acid lysine is coded for by either AAA or AAG, the amino acid glutamine is coded for by either GAA or GAG, and the amino acid arginine is coded for by either CGU, CGC, CGA, CGG, AGA, or AGG. Hence, given an RNA molecule with the codons AAA-GAG-CGA, one can infer the amino acid chain lysine-glutamine-arginine. But given the amino acid chain lysine-glutamine-arginine, one cannot infer the codons AAA-GAG-CGA, because that amino acid chain could have been coded for by many different triplets of codons, such as, say, AAG-GAA-CGU. From all of this it follows that phenotypes supervene on geno-types and, accordingly, that biology is, in a sense, not reducible to physics/chem-istry (see Reductionism). What adds to this conclusion is that the genetic code is contingent; it is not the product of chemical laws but rather of natural selection and the contingencies of history. If life evolved elsewhere in the universe, or even independently on Earth more than once, we should expect to find different genetic codes.

Genetic Drift Random sampling error. For example, a *founder population* (a part of a population isolated in some way from its main population), or a population go-ing through a *bottleneck* in its number, might, just by chance (random sampling), carry a higher frequency of a particular allele than the population it came from. The higher frequency of this allele could be perpetuated in the resulting popula-tion, possibly increasing to *fixation* (100%), even though it confers no adaptive advantage.

Genotype The genetic program of an organism, its DNA or RNA.

Homology Similarity due to common evolutionary descent, whether or not there is any difference in structure or function. Homologies are a major source of evidence for evolution. A similar structure (or behavior, or DNA sequence, etc.) in different species that does not make sense in terms of function makes perfect sense in terms of common evolutionary descent. A standard example is the bones in the human hand and the bones in the wing of a bat. The similarity between the bones is not due to function but to the fact that humans and bats evolved from the same ancestral mammalian species many millions of years ago, the same original structure being modified for different functions. Since the function of the human hand is quite different than the function of the wing of a bat, were they designed by intelligence they would not be similar but very different. Here, as in so many other instances, nothing in biology makes sense except in the light of evolution.

Instinct Genetically determined or influenced goal-directed behavior or thoughts.

Kin Selection A form of natural selection where a gene or gene complex is favored by natural selection even though it lowers individual fitness, the reason being that the gene or gene complex raises the fitness of close kin to a degree that over-compensates the loss of fitness to the individual.

Lamarckism The theory of the *inheritance of acquired characteristics*, now dead. Build your biceps all you want or spend hours every day practicing the piano, none of this changes the DNA in your sperm or eggs to produce children with a greater propensity for developing big biceps or talent at the piano.

Meme The unit of cultural transmission, replication, and evolution.

Modern Synthesis At bottom, the synthesis of Darwinian evolution by natural selection with Mendelian genetics, which serves as the foundation for all other subdisciplines in biology, and which was forged between the 1930s and 1950s, the principal names of which are Ronald Fisher, J. B. S. Haldane, Sewell Wright, Theodosius Dobzhansky, Ernst Mayr, and G. G. Simpson.

Mutation A change in genetic information (see DNA and RNA). *Point mutations* are a change in a single letter in DNA or RNA (for example, a G might be substituted where there should be an A). Mutations can also involve larger pieces of chromosome, sometimes called *macromutations*. Many mutations are silent, meaning they do not make any difference in the product coded for by the DNA or RNA, many mutations are harmful or even lethal, while some mutations actually result in improvement, i.e., increased fitness in terms of survival and reproduction. Mutation is often said to *feed* natural selection. Even though this is not strictly true (ignoring the metaphor)—the variation that natural selection feeds upon can also come from *recombination*, a process of mixing genes up that occurs in the production of gametes (sperm and eggs)—it is nevertheless ultimately true.

Natural Selection The process that produces evolutionary change and adaptations in organisms. It is a remarkably simple concept, with enormous power and depth, and yet it is difficult to teach. One way to get it is to keep reading different definitions and explanations of it. There are a number of formulations but they all amount to pretty much the same thing. Given *heritable variation* and *differential fitness* in a *field of competition*, evolution by natural selection is bound to occur. (See Appendix for more details.)

Neoteny The evolution of juvenile features into adulthood.

Ockham's Razor The principle of explanatory parsimony (simplicity), a guiding principle widespread in both philosophy and science, attributed to the 14th-century philosopher William of Ockham. According to Ockham's Razor, if one has two or more competing explanations for the same phenomenon, and if they are roughly of equal explanatory power, then one should choose the simplest among them. This principle is sometimes stated briefly as "One should not multiply explanatory entities beyond necessity."

Ontogeny The development of an individual organism, based on its genetic program.

Phenotype The outward physical expression of the genotype (see Genotype). This can be either physical or behavioral. The phenotype is not to be confused with the morphology or an organism, although there is much overlap between the two. For example, I have five fingers on each of my hands. In this case, my morphology and my phenotype are the same. However, if one of my fingers is cut off, my morphology has changed but not my phenotype.

Phylogeny The particular lines of evolutionary history; not to be confused with *ontogeny* (see Ontogeny). The concept of phylogeny applies to species but not to individual organisms. Species evolution is contingent and open-ended and it is improper to call it *development*. Organisms, on the other hand, develop from a single cell to an adult based on their genetic program. They have ontogeny but not phylogeny. It has sometimes been said that ontogeny recapitulates phylogeny, but this is an old view of evolution with only some vestiges of truth to it.

Pleiotropy It is typically the case that a gene or gene complex (or a mutation) has two or more phenotypic traits. Natural selection can select for one of the traits while the other tags along, but strictly speaking the tag-along trait is not a product of natural selection.

Reductionism The reduction of something to something more simple. There are basically three kinds of reductionism, although they often overlap. One is *ontological* reduction, explaining something as being nothing more than something more basic (e.g., the reduction of material objects to atoms). Another is *explanatory* reduction, explaining one level of phenomenon in terms of a lower level (e.g., explaining life in terms of DNA and cellular processes). Finally, there is *theory* reduction, explaining one theory in terms of another (e.g., the reduction of Newton's theory of gravity to Einstein's theory of gravity).

Ring Species A species that evolved around a barrier, for example a large lake or mountain range, such that the terminal populations are reproductively incompatible. Ring species are a source of evidence for evolution. It is debated whether the product is one species or not, but either way it is agreed that the phenomenon is an example of vertical evolution laid flat.

RNA Ribonucleic acid, the single-stranded intermediary between DNA and the proteins (made out of amino acids) that make up our bodies. Many viruses are made up of RNA rather than DNA and are usually the simpler form of viruses.

Sociobiology The application of evolutionary biology to help explain social behavior in humans and animals.

Speciation The formation of a species.

Species A notoriously difficult concept to define, as the concept is still debated and in flux in professional biology. What can be said, however, is that the concept serves a dual purpose in biology. For one, species are the basic units of classification, the foundation of a taxonomic hierarchy (traditionally and most basically, species make up a genus, genera make up a family, families make up an order, orders make up a class, classes make up a phylum, and phyla make up a kingdom). The so-called *biological species concept*, based on reproductive barriers with a genetic basis, is fairly common in zoology, the *morphological species concept*, based on overall similarity, is fairly common in botany and bacteriology, *cladistic* or *phylogenetic species concepts*, based on evolutionary branching, are common in taxonomy. The list goes on. The other purpose of the species concept is purely evolutionary. Species have ranges, species bud off founder populations, species split, species evolve (some say more properly the populations of which species are composed evolve), species speciate, species become rare, species become endangered, and finally, species become extinct (the fate of all species). (See Stamos 2003 for a detailed discussion on the many competing species concepts, and Stamos 2007 for Darwin's species concept.)

Standard Social Science Model (SSSM) The explanatory model—common in sociology, behaviorism, cultural anthropology, Marxism, women's studies, and gay studies—that in effect denies human nature by focusing on the environment (culture and conditioning) for a complete explanation of human behavior such as homosexuality/heterosexuality, rape, gender differences, racism, and morality. In other words, if the SSSM views human nature as anything it views it not as something innate but as something quite *plastic* (moldable).

Superfecundity Reproduction that is greater than what is minimally needed to maintain the same numbers within a population or species. Superfecundity is an important part of Darwin's argument. As he noted in the *Origin*, every species exhibits superfecundity, even elephants. Superfecundity entails competition for limited resources.

Supervenience A property is said to *supervene* on a disjunctive base of subvenient properties if (i) one can infer from one of the subvenient properties the supervening property and (ii) one cannot infer any one of the subvenient properties from the supervenient property. For example, information such as $2 + 2 = 4$ can be found on a piece of paper, or on a magnetic tape, or on a CD, or in a brain, and so on. Although physically very different, the mediums constitute the disjunctive base. Because the subvenient properties are disjunctive, information cannot be reduced to them but instead supervenes on them. Other examples of supervenient properties commonly given are liquidity, goodness, consciousness, fitness, and the genetic code.

Universal Grammar (UG) "The basic design underlying the grammars of all human languages; also refers to the circuitry in children's brains that allows them to learn the grammar of their parents' language" (Pinker 1994, 483). Associated primarily with the revolutionary work of Noam Chomsky, the UG can be conceived as the innate common denominator of, and basic constraint upon, all human languages (the numerator).

Vestigial Organs Organs or traits of no service to the organism but which bear the mark of previous functionality. Vestigial organs are a major source of evidence for evolution. In fact, every organism carries vestiges from its evolutionary past, such as the pelvic bones in snakes and whales, hollow bones in the flightless wings of dodos and penguins (suggesting that these animals were originally adapted to flight), and flightless beetles with wings sealed beneath covers that never open. Humans are no different. Our wisdom teeth are vestiges of an evolutionary past with larger jaw bones, such that our jaws are now too small for these teeth. Likewise, the fine little hairs covering our bodies are vestiges of our ape ancestry, as well as screaming and getting goose bumps when startled or frightened (in gorillas and chimpanzees and many other hairy animals, hair standing on end adds an intimidation factor, increasing fitness in fight-or-flight situations dealing with members of one's own or other species). Speaking of intimidation, in some humans, particularly males, one can still see today a surprising amount of body hair, as with the former wrestler George "The Animal" Steele. Why humans and not other apes lost most of their body hair is another matter, which I shall not take up here, but there is no doubt about the meaning of the fine hairs covering our bodies. It remains to be added that sometimes vestigial structures are expressed only rarely in individuals (called *atavisms*), such as teeth in hens and toes on horses. Perhaps George should have been called George "The Atavist" Steele, but "The Animal" does sound scarier after all.

References

All references are to reprints where indicated

Aiello, Leslie C. (1998). "The Foundations of Human Language." In Jablonski and Aiello (1998, 21–34).

Alcock, John (2001). *The Triumph of Sociobiology*. Oxford: Oxford University Press.

Alland, Alexander, Jr. (2002). *Race in Mind: Race, IQ, and Other Racisms*. New York: Palgrave Macmillan.

Altman, Andrew (1996). "Making Sense of Sexual Harassment Law." *Philosophy and Public Affairs* 25, 36–64.

Anderson, Stephen R. (2004). *Doctor Dolittle's Delusion: Animal Communication and the Nature of Human Language*. New Haven, CT: Yale University Press.

Anderson, Stephen R., and Lightfoot, David W. (2002). *The Language Organ: Linguistics as Cognitive Physiology*. Cambridge: Cambridge University Press.

Andersson, Malte (1994). *Sexual Selection*. Princeton: Princeton University Press.

Andreasen, Robin O. (2004). "The Cladistic Race Concept: A Defense." *Biology & Philosophy* 19, 425–442.

Antony, Louise M., and Hornstein, Norbert, eds. (2003). *Chomsky and His Critics*. Oxford: Blackwell Publishing.

Archer, John (2006). "Testosterone and Human Aggression: An Evaluation of the Challenge Hypothesis." *Neuroscience and Biobehavioral Reviews* 30, 319–345.

Asch, Solomon E. (1955). "Opinion and Social Pressure." *Scientific American* 193, 31–35.

Atran, Scott (2002). *In Gods We Trust: The Evolutionary Landscape of Religion*. Oxford: Oxford University Press.

Aunger, Robert, ed. (2000). *Darwinizing Culture: The Status of Memetics as a Science*. Oxford: Oxford University Press.

——— (2002). *The Electric Meme: A New Theory of How We Think*. New York: Free Press.

Bada, Jeffrey L. (1995). "Cold Start." *The Sciences* 35, 21–25.

Badcock, Christopher (2000). *Evolutionary Psychology: A Critical Introduction*. Cambridge: Polity Press.

Baldwin, Thomas (1993). "Editor's Introduction." In Thomas Baldwin, ed. (1993, ix–xxxvii). *Principia Ethica: G. E. Moore*. Rev. ed. Cambridge: Cambridge University Press.

Barash, David P. (2000). "Evolutionary Existentialism, Sociobiology, and the Meaning of Life." *BioScience* 50, 1012–1017.

Barbour, Ian G. (1997). *Religion and Science*. San Francisco: Harper.

Barkow, Jerome H., Cosmides, Leda, and Tooby, John, eds. (1992). *The Adapted Mind: Evolutionary Psychology and the Generation of Culture*. Oxford: Oxford University Press.

Barlow, Horace (1987). "The Biological Role of Consciousness." In Blakemore and Greenfield (1987, 361–374).

Barrett, Paul H., *et al.* (1987). *Charles Darwin's Notebooks, 1836–1844*. Ithaca, NY: Cornell University Press.

Beatty, John (1995). "The Evolutionary Contingency Thesis." In Gereon Wolters and James G. Lennox, eds. (1995, 45–81). *Concepts, Theories, and Rationality in the Biological Sciences*. Pittsburgh: University of Pittsburgh Press.

Beauvoir, Simone de (1949). *The Second Sex*. H. M. Parshley, trans. (1953). New York: Knopf.

Beldecos, Athena, *et al.* (1988). "The Importance of Feminist Critique for Contemporary Cell Biology." *Hypatia* 3, 61–76. Reprinted in Kourany (2002, 192–203).

Belenky, Mary Field, *et al.* (1986). *Women's Ways of Knowing: The Development of Self, Voice, and Mind*. New York: Basic Books.

Benatar, David, ed. (2004). *Life, Death, & Meaning: Key Philosophical Readings on the Big Questions*. Lanham, MD: Rowman & Littlefield.

Benedict, Ruth (1934). "Anthropology and the Abnormal." *Journal of General Psychology* 10, 59–82.

Berlocher, H. Gene (2001). "The Challenges of Postmodernism." In David Stewart and H. Gene Berlocher, eds. (2001, 205–214). Upper Saddle River, NJ: Prentice-Hall.

Berman, Dr Edgar (1989). *In Africa With Schweitzer*. New York: Harper & Row.

Bhasin, S., *et al.* (2001). "Proof of the Effect of Testosterone on Skeletal Muscle." *Journal of Endocrinology* 170, 27–38.

Bickerton, Derek (1990). *Language & Species*. Chicago: University of Chicago Press.

——— (1998). "Catastrophic Evolution: The Case for a Single Step from Protolanguage to Full Human Language." In James R. Hurford, Michael Studdert-Kennedy, and Chris Knight, eds. (1998, 341–358). *Approaches to the Evolution of Language: Social and Cognitive Bases*. Cambridge: Cambridge University Press.

Blackless, M., *et al.* (2000). "How Sexually Dimorphic Are We? Review and Synthesis." *American Journal of Human Biology* 12, 151–166.

Blackmore, Susan (1993). *Dying to Live: Science and the Near-Death Experience*. London: HarperCollins.

——— (1999). *The Meme Machine*. Oxford: Oxford University Press.

Blakemore, Colin, and Greenfield, Susan, eds. (1987). *Mindwaves: Thoughts on Intelligence, Identity and Consciousness*. Oxford: Blackwell Publishing.

Bogdanoff, M. D., *et al.* (1961). "The Modifying Effect of Conforming Behavior Upon Lipid Responses Accompanying CNS Arousal." *Clinical Research* 9, 135.

Boyer, Pascal (2001). *Religion Explained: The Evolutionary Origins of Religious Thought*. New York: Basic Books.

Bradie, Michael (1994). "Epistemology from an Evolutionary Point of View." In Sober (1994, 453–475).

Bradley, F. H. (1893). *Appearance and Reality*. Oxford: Oxford University Press.

Brizendine, Louann (2006). *The Female Brain*. New York: Morgan Road Books.

Brockman, John, ed. (2006). *Intelligent Thought: Science Versus the Intelligent Design Movement*. New York: Vintage Books.

Brodie, Richard (1996). *Virus of the Mind: The New Science of the Meme*. Seattle: Integral Press.

Broom, Donald M. (2003). *The Evolution of Morality and Religion*. Cambridge: Cambridge University Press.

Brown, Donald E. (1991). *Human Universals*. New York: McGraw-Hill.

Brownmiller, Susan (1975). *Against Our Will: Men, Women, and Rape*. New York: Simon and Schuster.

Bulbulia, Joseph (2004). "The Cognitive and Evolutionary Psychology of Religion." *Biology & Philosophy* 19, 655–686.

Burkhardt, Frederick, and Smith, Sydney (1990). *The Correspondence of Charles Darwin: Volume 6, 1856–1857*. Cambridge: Cambridge University Press.

Burkhardt, Frederick, *et al.*, eds. (1993). *The Correspondence of Charles Darwin: Volume 8, 1860*. Cambridge: Cambridge University Press.

Buss, David M. (1994). "The Strategies of Human Mating." *American Scientist* 82, 238–249. Reprinted in Paul W. Sherman and John Alcock, eds. (1998, 216–227). *Exploring Animal Behavior: Readings from* American Scientist. 2nd ed. Sunderland, MA: Sinauer Associates.

——— (2003). *The Evolution of Desire: Strategies of Human Mating*. Rev. ed. New York: Basic Books.

Buss, David M., Larsen, Randy J., and Westen, Drew (1996). "Sex Differences in Jealousy: Not Gone, Not Forgotten, and Not Explained by Alternative Hypotheses." *Psychological Science* 7, 373–375.

Buunk, Bram P., Angleitner, Alois, Oubaid, Viktor, and Buss, David M. (1996). "Sex Differences in Evolutionary and Cultural Perspective: Tests from the Netherlands, Germany, and the United States." *Psychological Science* 7, 359–363.

Cabezon, Jose Ignacio, ed. (1988). *The Bodhgaya Interviews*. Ithaca, NY: Snow Lion Publications. Reprinted in part in Louis P. Pojman, ed. (2003, 528–533). *Philosophy of Religion*. Belmont, CA: Wadsworth.

Cairns-Smith, A. G. (1985). *Seven Clues to the Origin of Life: A Scientific Detective Story*. Cambridge: Cambridge University Press.

Campbell, Donald T. (1974). "Evolutionary Epistemology." In Paul A. Schilpp, ed. (1974, 413–463). *The Philosophy of Karl Popper*. La Salle, IL: Open Court. Reprinted in Radnitzky and Bartley (1987, 47–89).

Campbell, Richmond (1996). "Can Biology Make Ethics Objective?" *Biology & Philosophy* 11, 21–31.

Camus, Albert (1942). *The Myth of Sisyphus and Other Essays*. Justin O'Brien, trans. (1955). New York: Knopf.

——— (1951). *The Rebel: An Essay on Man in Revolt*. Anthony Bower, trans. (1956). New York: Knopf.

Caplan, Arthur L. (1982). "Say It Just Ain't So: Adaptational Stories and Sociobiological Explanations of Social Behavior." *The Philosophical Forum* 13, 144–160. Reprinted in Ruse (1989, 264–270).

Cavalli-Sforza, Luigi L. (1991). "Genes, Peoples and Languages." *Scientific American* 265 (5), 104–110.

Chalmers, A. F. (1999). *What Is This Thing Called Science?* Indianapolis: Hackett.

Cherrett, J. M., ed. (1989). *Ecological Concepts: The Contribution of Ecology to an Understanding of the Natural World*. Oxford: Blackwell Publishing.

Chomsky, Noam (1957). *Syntactic Structures*. The Hague: Mouton.

——— (1959). "Review of B. F. Skinner's *Verbal Behavior*." *Language* 35, 26–58.

——— (1965). *Aspects of the Theory of Syntax*. Cambridge: MIT Press.

——— (1968). *Language and Mind*. New York: Harcourt, Brace and World.

——— (1980). *Rules and Representations*. Columbia University Press.

——— (1986). *Knowledge of Language*. New York: Praeger.

——— (1988). *Language and Problems of Knowledge: The Managua Lectures*. Cambridge: MIT Press.

———— (1997). "Language from an Internalist Perspective." In Johnson and Erneling (1997, 118–135).

———— (2000). *New Horizons in the Study of Language*. Cambridge: Cambridge University Press.

———— (2002). *On Nature and Language*. Cambridge: Cambridge University Press.

Churchland, Paul M. (1988). *Matter and Consciousness*. Cambridge: MIT Press.

———— (1995). *The Engine of Reason, the Seat of the Soul*. Cambridge: MIT Press.

Churchland, Paul M., and Churchland, Patricia Smith (1990). "Stalking the Wild Epistemic Engine." In William G. Lycan, ed. (1990, 300–311). *Mind and Cognition: A Reader*. Oxford: Blackwell Publishing.

Clifford, W. K. (1879). "The Ethics of Belief." In F. Pollock, ed. (1879). *Lectures and Essays*. Vol. 2. London: Macmillan.

Code, Lorraine (1991). *What Can She Know? Feminist Theory and the Construction of Knowledge*. Ithaca, NY: Cornell University Press.

Cohen, David (1997). "Law, Society and Homosexuality in Classical Athens." *Past and Present* 117, 3–21. Reprinted in Mark Golden and Peter Toohey, eds. (2003, 151–166). *Sex and Difference in Ancient Greece and Rome*. Edinburgh: Edinburgh University Press.

Cohen, J. M., trans. (1957). *The Life of Saint Teresa of Avila by Herself*. New York: Penguin Books.

Coon, Carleton S. (1962). *The Origin of Races*. New York: Knopf.

Cottingham, John, *et al.*, eds. (1985). *The Philosophical Writings of Descartes*. Vol. I. Cambridge: Cambridge University Press.

Crawford, Charles, and Krebs, Dennis L., eds. (1998). *Handbook of Evolutionary Psychology: Ideas, Issues, and Applications*. Mahwah, NJ: Erlbaum Associates.

Crick, Francis (1994). *The Astonishing Hypothesis: The Scientific Search for the Soul*. London: Simon & Schuster.

Cromer, Alan (1993). *Uncommon Sense: The Heretical Nature of Science*. Oxford: Oxford University Press.

Cronin, Helena (1991). *The Ant and the Peacock*. Cambridge: Cambridge University Press.

———— (1992). "Sexual Selection: Historical Perspectives." In Keller and Lloyd (1992, 286–293).

Csikszentmihalyi, Mihaly (1993). *The Evolving Self: A Psychology for the Third Millenium*. New York: HarperCollins.

Damasio, Antonio R. (1994). *Descartes' Error: Emotion, Reason, and the Human Brain*. New York: G. P. Putnam's Sons.

Darwin, Charles (1859). *On the Origin of Species by Means of Natural Selection*. London: John Murray.

———— (1871). *The Descent of Man, and Selection in Relation to Sex*. London: John Murray.

———— (1872). *The Expression of the Emotions in Man and Animals*. London: John Murray.

Davies, Paul (1983). *God and the New Physics*. New York: Simon & Schuster.

———— (1992). *The Mind of God*. New York: Simon & Schuster.

Dawkins, Richard (1976). *The Selfish Gene*. Oxford: Oxford University Press.

———— (1982). *The Extended Phenotype: The Long Reach of the Gene*. Oxford: Oxford University Press.

———— (1983). "Universal Darwinism." In D. S. Bendall, ed. (1983, 403–425). *Evolution from Molecules to Men*. Cambridge: Cambridge University Press.

———— (1986). *The Blind Watchmaker*. London: Longman Scientific & Technical.

———— (1989). *The Selfish Gene*. Rev. ed. Oxford: Oxford University Press.

———— (1992). "Progress." In Keller and Lloyd (1992, 263–272).

———— (1993). "Viruses of the Mind." In Bo Dahlbom, ed. (1993, 13–27). *Dennett and His Critics: Demystifying Mind*. Oxford: Blackwell Publishing. Reprinted in Dawkins (2003, 128–145).

—— (1995). *River Out of Eden: A Darwinian View of Life*. New York: Basic Books.

—— (1997). "Obscurantism to the Rescue." *Quarterly Review of Biology* 72, 397–399.

—— (1998). *Unweaving the Rainbow: Science, Delusion, and the Appetite for Wonder*. Boston: Houghton Mifflin.

—— (2003). *A Devil's Chaplain: Reflections on Hope, Lies, Science, and Love*. Boston: Houghton Mifflin.

—— (2006). *The God Delusion*. Boston: Houghton Mifflin.

Dembski, William A, and Ruse, Michael, eds. (2004). *Debating Design: From Darwin to DNA*. Cambridge: Cambridge University Press.

Dennett, Daniel C. (1991). *Consciousness Explained*. Boston: Little, Brown and Company.

—— (1995). *Darwin's Dangerous Idea: Evolution and the Meanings of Life*. New York: Simon & Schuster.

—— (2006). *Breaking the Spell: Religion as a Natural Phenomenon*. New York: Penguin Viking.

DeSteno, David A., and Salovey, Peter (1996a). "Evolutionary Origins of Sex Differences in Jealousy? Questioning the 'Fitness' of the Model." *Psychological Science* 7, 367–372.

—— (1996b). "Genes, Jealousy, and Replication of Misspecified Models." *Psychological Science* 7, 376–377.

Diamond, M., Scheibel, A., Murphy, G., and Harvey, T. (1985). "On the Brain of a Scientist: Albert Einstein." *Experimental Neurology* 88, 198–204.

Distin, Kate (2005). *The Selfish Meme: A Critical Reassessment*. Cambridge: Cambridge University Press.

Dobzhansky, Theodosius (1968a). "Discussion." In Mead *et al.* (1968b, 165–166).

—— (1968b). "Introduction." In Mead *et al.* (1968b, 77–79).

—— (1973). "Nothing in Biology Makes Sense Except in the Light of Evolution." *American Biology Teacher* 35, 125–129.

Donnelly, John, ed. (1994). *Language, Metaphysics, and Death*. 2nd ed. New York: Fordham University Press.

Downes, Stephen M. (2000). "Truth, Selection and Scientific Inquiry." *Biology & Philosophy* 15, 425–442.

Dror, Itiel E., and Thomas, Robin D. (2005). "The Cognitive Neuroscience Laboratory: A Framework for the Science of Mind." In Christina E. Erneling and David Martel Johnson, eds. (2005, 283–292). *The Mind as a Scientific Object: Between Brain and Culture*. Oxford: Oxford University Press.

Dunson, David. B., Colombo, Bernardo, and Baird, Donna D. (2002). "Changes with Age in the Level and Duration of Fertility in the Menstrual Cycle." *Human Reproduction* 17, 1399–1403.

Dupré, John (1993). *The Disorder of Things*. Cambridge: Harvard University Press.

Eccles, John C. (1987). "Brain and Mind, Two or One?" In Blakemore and Greenfield (1987, 293–304).

—— (1989). *Evolution of the Brain: Creation of the Self*. New York: Routledge.

Edelman, Gerald M. (1992). *Bright Air, Brilliant Fire: On the Matter of the Mind*. New York: Basic Books.

Ehrlich, Paul R. (2000). *Human Natures: Genes, Cultures, and the Human Prospect*. Washington, DC: Island Press.

Eigen, Manfred (1992). *Steps Towards Life: A Perspective on Evolution*. Oxford: Oxford University Press.

Einon, Dorothy (2002). "More an Ideologically Driven Sermon than Science." *Biology & Philosophy* 17, 445–456.

Einstein, Albert (1954). *Ideas and Opinions*. New York: Crown Publishers.

Ellis, Bruce J. (1992). "The Evolution of Sexual Attraction: Evaluative Mechanisms in Women." In Barkow *et al.* (1992, 267–288).

Endler, John A. (1986). *Natural Selection in the Wild*. Princeton: Princeton University Press.

Ettlinger, G., and Blakemore, C. B. (1969). "Cross-modal Transfer Set in the Monkey." *Neuropsychology* 7, 41–47.

Fausto-Sterling, Anne (1992). *Myths of Gender: Biological Theories About Women and Men*. New York: Basic Books.

——— (1993). "The Five Sexes: Why Male and Female Are Not Enough." *The Sciences* 33, 20–25.

——— (2000). *Sexing the Body: Gender Politics and the Construction of Sexuality*. New York: Basic Books.

Feder, Kenneth L., and Park, Michael Alan (1993). *Human Antiquity: An Introduction to Physical Anthropology and Archaeology*. 2nd ed. Mountain View, CA: Mayfield Publishing Company.

Flanagan, Owen (2003). "Ethical Expressions: Why Moralists Scowl, Frown and Smile." In Jonathan Hodge and Gregory Radick, eds. (2003, 377–398). *The Cambridge Companion to Darwin*. Cambridge: Cambridge University Press.

Flew, Antony (1984). *God, Freedom, and Immortality: A Critical Analysis*. Buffalo: Prometheus Books.

———, ed. (1987). *Readings in the Philosophical Problems of Parapsychology*. Buffalo: Prometheus Books.

Frankfurt, Harry G. (1971). "Freedom of the Will and the Concept of a Person." *Journal of Philosophy* 68, 5–20.

Frankl, Viktor E. (1984). *Man's Search for Meaning: An Introduction to Logotherapy*. 3rd ed. New York: Simon & Schuster.

Freeland, Stephen J., and Hurst, Laurence D. (1998). "The Genetic Code is One in a Million." *Journal of Molecular Evolution* 47, 238–248.

Freeman, Derek (1983). *Margaret Mead and Samoa: The Making and Unmaking of an Anthropological Myth*. Cambridge: Harvard University Press.

——— (1989). "Fa'apua'a Fa'am and Margaret Mead." *American Anthropologist* 91, 1017–1022.

Freeman, Scott, and Herron, Jon C. (1998). *Evolutionary Analysis*. Upper Saddle River, NJ: Prentice Hall.

Fromm, Erich (1966). *You Shall Be As Gods: A Radical Interpretation of the Old Testament and Its Tradition*. New York: Henry Holt.

Futuyma, Douglas J. (1995). *Science on Trial: The Case for Evolution*. Sunderland, MA: Sinauer.

——— (1998). *Evolutionary Biology*. 3rd ed. Sunderland, MA: Sinauer.

Gaffney, Eugene S., *et al.* (1995). "Why Cladistics?" *Natural History* 104 (6), 33–35.

Galison, Peter (1995). "Theory Bound and Unbound: Superstrings and Experiments." In Friedel Weinert, ed. (1995, 369–408). *Laws of Nature: Essays on the Philosophical, Scientific, and Historical Dimensions*. New York: Walter de Gruyter.

Gandhi, Mohandas K. (1929). *An Autobiography: The Story of My Experiments with Truth*. Mahadev Desai, trans. (1957). Boston: Beacon Press.

Gaskin, J. C. A. (1984). *The Quest for Eternity: An Outline of the Philosophy of Religion*. Harmondsworth: Penguin.

Gazzaniga, Michael S. (1992). *Nature's Mind: The Biological Roots of Thinking, Emotions, Sexuality, Language, and Intelligence*. New York: Basic Books.

Ghiglieri, Michael P. (2000). *The Dark Side of Man: Tracing the Origins of Male Violence*. Cambridge: Perseus Books.

Glen, William, ed. (1994). *The Mass Extinction Debates*. Stanford: Stanford University Press.

Glickman, Barry W. (1987). "'The Gene Seemed as Inaccessible as the Materials of the Galaxies'." In Robson (1987, 33–57).

Godfrey, Laurie, ed. (1983). *Scientists Confront Creationism*. New York: W. W. Norton.

Gould, Stephen Jay (1974a). "The Nonscience of Human Nature." *Natural History* 83 (4), 21–25. Reprinted in Gould (1977, 237–242).

—— (1974b). "The Race Problem." *Natural History* 83 (3), 8–14. Reprinted as "Why We Should Not Name Human Races—A Biological View" in Gould (1977, 231–236)

—— (1974c). "Racist Arguments and IQ." *Natural History* 83 (5), 24–29. Reprinted in Gould (1977, 243–247).

—— (1977). *Ever Since Darwin*. New York: W. W. Norton.

—— (1980a). "Sociobiology and the Theory of Natural Selection." In G. W. Barlow and J. Silverberg, eds. (1980, 257–269). *Sociobiology: Beyond Nature/Nurture?* Boulder, CO: Westview Press. Reprinted in Ruse (1989, 253–263).

—— (1980b). *The Panda's Thumb*. New York: W. W. Norton.

—— (1981). *The Mismeasure of Man*. New York: W. W. Norton.

—— (1983). *Hen's Teeth and Horse's Toes*. New York: W. W. Norton.

—— (1989a). "Tires to Sandals." *Natural History* 98 (4), 8–15. Reprinted in Gould (1993, 313–324).

—— (1989b). *Wonderful Life*. London: Hutchinson Radius.

—— (1991). *Bully for Brontosaurus*. New York: W. W. Norton.

—— (1993). *Eight Little Piggies*. New York: W. W. Norton.

—— (1997). "Nonoverlapping Magisteria." *Natural History* 106 (2), 16–22, 60–62. Reprinted in Gould (1998, 269–283). *Leonardo's Mountain of Clams and the Diet of Worms*. New York: Three Rivers Press.

—— (1999). *Rocks of Ages: Science and Religion in the Fullness of Life*. New York: Ballantine.

Gould, Stephen Jay, and Lewontin, Richard C. (1978). "The Spandrels of San Marco and the Panglossian Paradigm: A Critique of the Adaptationist Programme." *Proceedings of the Royal Society of London* 205, 581–598. Reprinted in Sober (1994, 73–90).

Gould, Stephen Jay, and Vrba, Elisabeth S. (1982). "Exaptation—A Missing Term in the Science of Form." *Paleobiology* 8, 4–15.

Graur, Dan, and Li, Wen-Hsiung (2000). *Fundamentals of Molecular Evolution*. 2nd ed. Sunderland: Sinauer.

Graves, Joseph L., Jr. (2004). *The Race Myth: Why We Pretend Race Exists in America*. New York: Dutton.

Gray, Asa (1860). "Review of Darwin's Theory on the Origin of Species by Means of Natural Selection." *American Journal of Science and Arts* 29, 153–184.

Gray, John (1992). *Men Are from Mars, Women Are from Venus*. New York: HarperCollins.

Green, Christopher D., and Vervaeke, John (1997). "But What Have You Done for Us Lately? Some Recent Perspectives on Linguistic Nativism." In Johnson and Erneling (1997, 149–163).

Greene, John C., and Ruse, Michael (1996). "On the Nature of the Evolutionary Process: The Correspondence Between Theodosius Dobzhansky and John C. Greene." *Biology & Philosophy* 11, 445–491.

Gregory, Richard (1987). "In Defence of Artificial Intelligence—A Reply to John Searle." In Blakemore and Greenfield (1987, 235–244).

Griffin, Donald R. (1992). *Animal Minds*. Chicago: University of Chicago Press.

Gross, Paul R. (1998). "Bashful Eggs, Macho Sperm, and Tonypandy." In Koertge (1998, 59–70).

Gross, Paul R., and Levitt, Norman (1998). *Higher Superstition: The Academic Left and Its Quarrels with Science*. 2nd ed. Baltimore: Johns Hopkins University Press.

Grube, G. M. A., and C. D. C. Reeve, trans. (1992). *Plato: Republic*. Indianapolis: Hackett.

Haack, Susan (1993). "Epistemological Reflections of an Old Feminist." *Reason Papers* 18, 31–43. Reprinted in Louis P. Pojman (2003, 580–588). *Theory of Knowledge: Classical and Contemporary Readings*. Belmont, CA: Wadsworth.

Hacking, Ian (1983). *Representing and Intervening*. Cambridge: Cambridge University Press.

Hamer, Dean (2004). *The God Gene: How Faith Is Hardwired into Our Genes*. New York: Doubleday.

———, *et al*. (1993). "A Linkage Between DNA Markers on the X Chromosome and Male Sexual Orientation." *Science* 261, 321–327.

Hamer, Dean, and Copeland, Peter (1998). *Living With Our Genes: Why They Matter More Than You Think*. New York: Doubleday.

Hamilton, W. D. (1996). *Narrow Roads of Gene Land, The Collected Papers of W. D. Hamilton, Volume 1: Evolution of Social Behavior*. Oxford: Oxford University Press.

Harman, Gilbert (1977). *The Nature of Morality*. New York: Oxford University Press.

Harris, Christine R., and Christenfeld, Nicholas (1996a). "Gender, Jealousy, and Reason." *Psychological Science* 7, 364–366.

——— (1996b). "Jealousy and Rational Responses to Infidelity Across Gender and Culture." *Psychological Science* 7, 378–379.

Harris, Marvin (1989). *Our Kind: Who We Are, Where We Came From, & Where We Are Going*. New York: Harper & Row.

——— (1999). *Theories of Culture in Postmodern Times*. Walnut Creek, CA: AltaMira Press.

Hartnack, Justus (1972). "On Thinking." *Mind* 81, 543–552.

Hattiangadi, J. N. (1987). *How is Language Possible?: Philosophical Reflections on the Evolution of Language and Knowledge*. La Salle, IL: Open Court.

Haught, James A. (1990). *Holy Horrors: An Illustrated History of Religious Murder and Madness*. Buffalo: Prometheus Books.

Haynes, Robert H. (1993). "My Road to Repair in Yeast: The Importance of Being Ignorant." In M. N. Hall and P. Linder, eds. (1993, 145–171). *The Early Days of Yeast Genetics*. Cold Spring Harbor: Cold Spring Harbor Laboratory Press.

Heil, John, and Mele, Alfred, eds. (1993). *Mental Causation*. Oxford: Oxford University Press.

Hempel, Carl G. (1965). *Aspects of Scientific Explanation*. New York: Free Press.

——— (1966). *Philosophy of Natural Science*. Englewood Cliffs, NJ: Prentice-Hall.

Herrnstein, Richard J., and Murray, Charles (1994). *The Bell Curve: Intelligence and Class Structure in American Society*. New York: Free Press.

Hick, John H. (1985). *Problems of Religious Pluralism*. New York: St Martin's Press.

——— (1990). *Philosophy of Religion*. 4th ed. Englewood Cliffs: Prentice-Hall.

Hitchens, Christopher (2007). *God is Not Great: How Religion Poisons Everything*. Toronto: McLelland & Stewart.

Hoffer, Eric (1951). *The True Believer*. New York: Harper & Row.

Hoffman, Antoni (1989). "Twenty Years Later: Punctuated Equilibrium in Retrospect." In Albert Somit and Steven A. Peterson, eds. (1989, 121–138). *The Dynamics of Evolution: The Punctuated Equilibrium Debate in the Natural and Social Sciences*. Ithaca, NY: Cornell University Press.

Hofstadter, Douglas R. (1981). "Reflections." In Douglas R. Hofstadter and Daniel C. Dennett, eds. (1981, 373–382). *The Mind's I*. New York: Bantam Books.

Hrdy, Sarah Blaffer (1981). *The Woman That Never Evolved*. Cambridge: Cambridge University Press.

Hubbard, Ruth (1979). "Have Only Men Evolved?" In Ruth Hubbard, *et al*., eds. (1979, 7–36). *Women Look at Biology Looking at Women: A Collection of Feminist Critiques*. Cambridge: Schenkman Publishing. Reprinted in Kourany (2002, 153–170).

Hull, David L. (1988). *Science as a Process: An Evolutionary Account of the Social and Conceptual Development of Science*. Chicago: University of Chicago Press.

Hume, David (1739). *A Treatise of Human Nature*. Books I and II. David Fate Norton and Mary J. Norton, eds. (2007). Oxford: Oxford University Press.

—— (1740). *A Treatise of Human Nature*. Book III. David Fate Norton and Mary J. Norton, eds. (2007). Oxford: Oxford University Press.

—— (1748). *An Enquiry Concerning Human Understanding*. Tom L. Beauchamp, ed. (2000). Oxford: Oxford University Press.

—— (1779). *Dialogues Concerning Natural Religion*. Norman Kemp Smith, ed. (1947). Indianapolis: Bobbs-Merrill.

Huxley, Julian S. (1939). "The Creed of a Scientific Humanist." In Clifton Fadiman, ed. (1939, 127–136). *I Believe: The Personal Philosophies of Eminent Men and Women of Our Time*. New York: Simon & Schuster. Reprinted in Klemke (2000, 78–83).

Huxley, Thomas Henry (1859). "Mr. Darwin's 'Origin of Species'." *Macmillan's Magazine, 1859–1860* 1, 142–148.

Ingle, Dwight J. (1968). "The Need to Investigate Average Biological Differences among Racial Groups." In Mead *et al.* (1968b, 113–121).

Jablonski, Nina G., and Aiello, Leslie C., eds. (1998). *The Origin and Diversification of Language*. San Francisco: California Academy of Sciences.

Jackson, John P., Jr., and Weidman, Nadine M. (2004). *Race, Racism, and Science: Social Impact and Interaction*. Santa Barbara, CA: ABC-CLIO.

Jaynes, Julian (1976). *The Origin of Consciousness in the Breakdown of the Bicameral Mind*. Afterward (1990). Boston: Houghton Mifflin.

Johnson, David Martel (1988). "Brutes Believe Not." *Philosophical Psychology* 1, 279–294.

—— (2003). *How History Made the Mind: The Cultural Origins of Objective Thinking*. Chicago: Open Court.

Johnson, David Martel, and Erneling, Christina A., eds. (1997). *The Future of the Cognitive Revolution*. Oxford: Oxford University Press.

Jumonville, Neil (2003). "The Cultural Politics of the Sociobiology Debate." *Journal of the History of Biology* 35, 569–593.

Kahane, Howard, and Cavender, Nancy (2006). *Logic and Contemporary Rhetoric: The Use of Reason in Everyday Life*. Belmont, CA: Wadsworth.

Kant, Immanuel (1783). *Prolegomena to Any Future Metaphysics*. Gary Hatfield, ed. (1997). Cambridge: Cambridge University Press.

Kaufmann, Walter (1958). *Critique of Religion and Philosophy*. New York: Harper & Row.

—— (1973). *Without Guilt and Justice: From Decidophobia to Autonomy*. New York: Delta.

Keller, Evelyn Fox, and Lloyd, Elisabeth A., eds. (1992). *Keywords in Evolutionary Biology*. Cambridge: Harvard University Press.

Kim, Jaegwon (1993). *Supervenience and Mind: Selected Philosophical Essays*. Cambridge: Cambridge University Press.

—— (1996). *Philosophy of Mind*. Boulder: Westview Press.

Kimura, Doreen (2002). "Sex Differences in the Brain." *Scientific American* 12 (1), 32–37.

Kitcher, Philip (1982). *Abusing Science: The Case Against Creationism*. Cambridge: MIT Press.

Klein, Richard G. (2002). *The Dawn of Human Culture*. New York: John Wiley & Sons.

Klemke, E. D., ed. (2000). *The Meaning of Life*. 2nd ed. Oxford: Oxford University Press.

Knowles, Dudley, ed. (1990). *Explanation and its Limits*. Cambridge: Cambridge University Press.

Koertge, Noretta, ed. (1998). *A House Built on Sand: Exposing Postmodernist Myths About Science*. Oxford: Oxford University Press.

Kornblith, Hilary, ed. (1994). *Naturalizing Epistemology*. Cambridge: MIT Press.

Kourany, Janet A., ed. (2002). *The Gender of Science*. Upper Saddle River, NJ: Prentice Hall.

Kripke, Saul A. (1980). *Naming and Necessity*. Cambridge: Harvard University Press.

Krob, G., *et al*. (1994). "True Hermaphroditism: Geographical Distribution, Clinical Findings, Chromosomes and Gonadal Histology." *European Journal of Pediatrics* 153, 2–10.

Kuhn, Thomas S. (1970). *The Structure of Scientific Explanations*. 2nd ed. Chicago: University of Chicago Press.

——— (1973). "Objectivity, Value Judgment, and Theory Choice." In Kuhn (1977, 320–339). *The Essential Tension*. Chicago: University of Chicago Press.

——— (1983). "Rationality and Theory Choice." *Journal of Philosophy* 80, 563–570.

Labov, William (1966). *The Social Stratification of English in New York City*. Washington DC: Center for Applied Linguistics.

Lahti, David C. (2003). "Parting with Illusions in Evolutionary Ethics." *Biology & Philosophy* 18, 639–651.

Lakatos, Imre, and Musgrave, Alan, eds. (1970). *Criticism and the Growth of Knowledge*. Cambridge: Cambridge University Press.

Lakoff, George, and Johnson, Mark (1999). *Philosophy in the Flesh: The Embodied Mind and its Challenge to Western Thought*. New York: Basic Books.

Larson, Edward J., and Witham, Larry (1998). "Leading Scientists Still Reject God." *Nature* 394 (July 23), 313.

——— (1999). "Scientists and Religion in America." *Scientific American* 281, 88–93.

Lass, Roger (1997). *Historical Linguistics and Language Change*. Cambridge: Cambridge University Press.

Lefkowitz, Mary (1997). *Not Out of Africa: How Afrocentrism Became an Excuse to Teach Myth as History*. Rev. ed. New York: Basic Books.

Leopold, Aldo (1949). *A Sand County Almanac: And Sketches Here and There*. Oxford: Oxford University Press.

LeVay, Simon (1992). "A Difference in the Hypothalamic Structure between Heterosexual and Homosexual Men." *Science* 253, 1034–1037.

Lewontin, Richard C. (1972). "The Apportionment of Human Diversity." In M. K. Hecht and W. S. Steere, eds. (1972, 381–398). *Evolutionary Biology*. Vol. 6. New York: Plenum.

——— (1982). *Human Diversity*. New York: Scientific American Library.

——— (1991). *Biology as Ideology: The Doctrine of DNA*. Concord, ON: Anansi.

——— (1999). "The Problem With An Evolutionary Answer." *Nature* 400 (Aug. 19), 728–729.

Lewontin, R. C., Rose, Steven, and Kamin, Leon J. (1984). *Not in Our Genes: Biology, Ideology, and Human Nature*. New York: Pantheon Books.

Libet, Benjamin (1993). *Neurophysiology of Consciousness: Selected Papers and New Essays*. Boston: Birkhauser.

Lieberman, Philip (1989). "Some Biological Constraints on Universal Grammar and Learnability." In M. Rice and R. L. Schiefelbusch, eds. (1989, 199–225). *The Teachability of Language*. Baltimore: Paul H. Brookes.

Light, Robert E. (1968). "Forward." In Mead *et al*. (1968b, vii–vii).

Lingua Franca, eds. (2000). *The Sokal Hoax: The Sham That Shook the Academy*. Lincoln, NE: University of Nebraska Press.

Lipton, Peter (1990). "Contrastive Explanation." In Dudley Knowles, ed. (1990, 247–266). *Explanation and its Limits*. Cambridge: Cambridge University Press.

——— (1991). *Inference to the Best Explanation*. London: Routledge.

Lynch, Aaron (1996). *Thought Contagion: How Belief Spreads Through Society*. New York: Basic Books.

Magee, Bryan (1978). *Talking Philosophy*. Oxford: Oxford University Press.

Maienschein, Jane, and Creath, Richard, eds. (1999). *Biology and Epistemology*. Cambridge: Cambridge University Press.

Maienschein, Jane, and Ruse, Michael, eds. (1999). *Biology and the Foundation of Ethics*. Cambridge: Cambridge University Press.

Martin, Emily (1991). "The Egg and the Sperm: How Science has Constructed a Romance Based on Stereotypical Male-Female Roles." *Signs: Journal of Women in Culture and Society* 16, 485–501. Reprinted in Evelyn Fox Keller and Helen E. Longino, eds. (1996, 103–117). *Feminism & Science*. Oxford: Oxford University Press.

Martin, Michael (1990). *Atheism: A Philosophical Justification*. Philadelphia: Temple University Press.

Maynard Smith, John (1993). *The Theory of Evolution*. Cambridge: Cambridge University Press.

Maynard Smith, John, and Szathmáry, Eörs (1999). *The Origins of Life: From the Birth of Life to the Origin of Language*. Oxford: Oxford University Press.

Mayr, Ernst (1968). "Discussion." In Mead *et al.* (1968b, 103–105).

—— (1970). *Populations, Species, and Evolution*. Harvard: Harvard University Press.

—— (1976). *Evolution and the Diversity of Life: Selected Essays*. Cambridge: Harvard University Press.

—— (1982). *The Growth of Biological Thought: Diversity, Evolution, and Inheritance*. Cambridge: Harvard University Press.

—— (1988). *Toward A New Philosophy of Biology: Observations of an Evolutionist*. Cambridge: Harvard University Press.

McMullin, Ernan (1983). "Values in Science." In Peter D. Asquith and Thomas Nickles, eds. (1983, 3–28). *PSA 1982*. East Lansing, MI: Philosophy of Science Association.

McMullin, Ernan, and Cushing, James T., eds. (1989). *Philosophical Consequences of Quantum Theory*. Notre Dame, IN: University of Notre Dame Press.

McNeill, William H. (1976). *Plagues and Peoples*. New York: Doubleday.

Mead, Margaret (1928). *Coming of Age in Samoa*. New York: Morrow.

—— (1968a). "Introductory Remarks." In Mead *et al.* (1968b, 3–9).

——, *et al.*, eds. (1968b). *Science and the Concept of Race*. New York: Columbia University Press.

Medawar, P. B., and Medawar, J. S. (1983). *Aristotle to Zoos: A Philosophical Dictionary of Biology*. Cambridge: Harvard University Press.

Mellars, Paul A. (1998). "Neanderthals, Modern Humans and the Archaeological Evidence for Language." In Jablonski and Aiello (1998, 89–116).

Milgram, Stanley (1974). *Obedience to Authority: An Experimental View*. New York: Harper & Row.

Miller, Kenneth R. (1999). *Finding Darwin's God: A Scientist's Search for Common Ground Between God and Evolution*. New York: HarperCollins.

Monod, Jacques (1971). *Chance and Necessity*. New York: Knopf.

Montagu, Ashley, ed. (1964). *The Concept of Race*. New York: Free Press.

Moore, G. E. (1903). *Principia Ethica*. Cambridge: Cambridge University Press.

Murphy, Nancey (1997). "Divine Action in the Natural Order: Buridan's Ass and Schrödinger's Cat." In Russell *et al.* (1997, 325–357).

Nagel, Thomas (1979). *Mortal Questions*. Cambridge: Cambridge University Press.

Newberg, Andrew, d'Aquili, Eugene, and Rause, Vince (2002). *Why God Won't Go Away: Brain Science and the Biology of Belief*. New York: Ballantine Books.

Newton-Smith, W. H. (1995). "Popper, Science and Rationality." In Anthony O'Hear, ed. (1995, 13–30). *Karl Popper: Philosophy and Problems*. Cambridge: Cambridge University Press.

Nichols, Johanna (1992). *Linguistic Diversity in Space and Time*. Chicago: University of Chicago Press.

Nietzsche, Friedrich (1878). *Human, All Too Human*. Marion Faber, trans. (1984). Lincoln, NE: Bison Books.

——— (1882). *The Gay Science*. Walter Kaufmann, trans. (1974). *Friedrich Nietzsche: The Gay Science*. New York: Viking Penguin.

Nitecki, Matthew H., ed. (1988). *Evolutionary Progress*. Chicago: University of Chicago Press.

Nozick, Robert (1981). *Philosophical Explanations*. Cambridge: Belknap Press.

Odum, Eugene P. (1971). *Fundamentals of Ecology*. 3rd ed. Orlando: Saunders College Publishing.

Okin, Susan Moller (1989). *Justice, Gender, and the Family*. New York: Basic Books.

Okruhlik, Kathleen (1994). "Gender and the Biological Sciences." *Canadian Journal of Philosophy* 20 (Suppl.), 21–42. Reprinted in Martin Curd and J. A. Cover, eds. (1998, 192–208). *Philosophy of Science*. New York: W. W. Norton.

Parvin, Simon D. (1982). "Ovulation in a Cytogenetically Proved Phenotypically Male Fertile Hermaphrodite." *British Journal of Surgery* 69, 279–280.

Patai, Daphne, and Koertge, Noretta (2003). *Professing Feminism: Education and Indoctrination in Women's Studies*. Lanham, MD: Lexington Books.

Peacocke, Arthur (1985). "Biological Evolution and Christian Theology—Yesterday and Today." In John Durant, ed. (1985, 100–130). *Darwinism and Divinity*. Oxford: Basil Blackwell.

Penfield, Wilder (1975). *The Mystery of the Mind*. Princeton: Princeton University Press.

Pennock, Robert T. (1999). *Tower of Babel: The Evidence Against the New Creationism*. Cambridge: MIT Press.

———, ed. (2001). *Intelligent Design Creationism and Its Critics: Philosophical, Theological, and Scientific Perspectives*. Cambridge: MIT Press.

Penrose, Roger (1994). *Shadows of the Mind: A Search for the Missing Science of Consciousness*. Oxford University Press.

Perry, Horace M., III, *et al.* (1996). "Aging and Bone Metabolism in African American and Caucasian Women." *Journal of Clinical Endocrinology and Metabolism* 81, 1108–1117.

Persinger, Michael A. (1993). "Vectorial Cerebral Hemisphericity As Differential Sources for the Sensed Presence, Mystical Experiences and Religious Conversions." *Perceptual and Motor Skills* 76, 915–930.

——— (1997). "'I Would Kill in God's Name:' Role of Sex, Weekly Church Attendance, Report of a Religious Experience, and Limbic Lability." *Perceptual and Motor Skills* 85, 128–130.

Pigliucci, Massimo, and Kaplan, Jonathan (2003). "On the Concept of Race and Its Applicability to Humans." *PSA 2002* 70, 1161–1172.

Pinker, Steven (1994). *The Language Instinct*. New York: William Morrow and Company.

——— (1997). *How the Mind Works*. New York: W. W. Norton.

——— (1998). "The Evolution of the Human Language Faculty." In Jablonski and Aiello (1998, 117–126).

——— (2002). *The Blank Slate: The Modern Denial of Human Nature*. New York: Viking Penguin.

Pinker, Steven, and Bloom, Paul (1990). "Natural Language and Natural Selection." *Behavioral and Brain Sciences* 13, 707–784. Reprinted in Barkow *et al.* (1992, 451–493).

Pojman, Louis P. (2000). "Ethical Relativism Versus Ethical Objectivism." In Louis P. Pojman, ed. (2000, 633–646). *Introduction to Philosophy: Classical and Contemporary Readings*. 2nd ed. Belmont, CA: Wadsworth.

Pope John Paul II (1997). "Message to the Pontifical Academy of Sciences." *Quarterly Review of Biology* 72, 381–383.

Popper, Karl R. (1959). *The Logic of Scientific Discovery*. London: Hutchinson.

—— (1963). *Conjectures and Refutations*. London: Routledge & Kegan Paul.

—— (1978). "Natural Selection and the Emergence of Mind." *Dialectica* 22, 339–355. Reprinted in Radnitzky and Bartley (1987, 139–153).

—— (1979). *Objective Knowledge: An Evolutionary Approach*. Rev. ed. Oxford: Oxford University Press.

Provine, William B. (1988). "Progress in Evolution and Meaning in Life." In Nitecki (1988, 49–74).

Putnam, Hilary (1975). *Language, Mind and Knowledge*. Cambridge: Cambridge University Press.

—— (2002). *The Collapse of the Fact/Value Dichotomy and Other Essays*. Cambridge: Harvard University Press.

Quammen, David (2004). "Was Darwin Wrong?" *National Geographic* 206 (Nov.), 2–35.

Quine, W. V. O. (1969). *Ontological Relativity & Other Essays*. New York: Columbia University Press.

Quine, W. V. O, and Ullian, J. S. (1978). *The Web of Belief*. New York: Random House.

Quirk, Joe (2006). *Sperm Are from Men, Eggs Are from Women*. Philadelphia: Running Press.

Rachels, James (2003). *The Elements of Moral Philosophy*. 4th ed. New York: McGraw-Hill.

Radnitzky, Gerard, and Bartley, W. W., eds. (1987). *Evolutionary Epistemology, Rationality, and the Sociology of Knowledge*. La Salle, IL: Open Court.

Raup, David M. (1991). *Extinction: Bad Genes or Bad Luck?* New York: W. W. Norton.

Regis, Edward Jr., ed. (1985). *Extraterrestrials: Science and Alien Intelligence*. Cambridge: Cambridge University Press.

Ridley, Mark (1993). *Evolution*. Oxford: Blackwell Publishing.

Robson, John M., ed. (1987). *Origin and Evolution of the Universe: Evidence for Design?* Kingston and Montreal: McGill-Queen's University Press.

Rohrlich, Fritz (1983). "Facing Quantum Mechanical Reality." *Science* 23, 1251–1255.

Rose, Steven, Kamin, Leon J., and Lewontin, R. C. (1984). *Not in Our Genes: Biology, Ideology and Human Nature*. London: Penguin.

Rosenberg, Alexander (1994). *Instrumental Biology, or The Disunity of Science*. Chicago: University of Chicago Press.

Ross, R., *et al.*, (1986). "Serum Testosterone Levels in Healthy Young Black and White Men." *Journal of the National Cancer Institute* 76, 45–48.

Ruse, Michael (1986). *Taking Darwin Seriously: A Naturalistic Approach to Philosophy*. New York: Blackwell Publishing.

—— (1987). "Biological Species: Natural Kinds, Individuals, or What?" *British Journal for the Philosophy of Science* 38, 225–242.

——, ed. (1989). *Philosophy of Biology*. New York: Macmillan.

—— (1996). *Monad to Man: The Concept of Progress in Evolutionary Biology*. Cambridge: Harvard University Press.

—— (1997). "John Paul II and Evolution." *Quarterly Review of Biology* 72, 391–395.

—— (1998). "Is Darwinism Sexist? (And if It Is, So What?)" In Koertge (1998, 119–129).

—— (1999). *Mystery of Mysteries: Is Evolution a Social Construction?* Cambridge: Harvard University Press.

—— (2002). "Evolution and Ethics: The Sociobiological Approach." In Louis P. Pojman, ed. (2002, 647–662). *Ethical Theory: Classical and Contemporary Approaches*. 4th ed. Belmont, CA: Wadsworth.

Ruse, Michael, and Wilson, Edward O. (1985). "The Evolution of Ethics." *New Scientist* 17, 50–52. Reprinted in Ruse (1989, 313–317).

Rushton, J. Philippe (1995). *Race, Evolution, and Behavior: A Life History Perspective*. New Brunswick, NJ: Transaction Publishers.

—— (2000). *Race, Evolution, and Behavior: A Life History Perspective*. 3rd ed. Port Huron, MI: Charles Darwin Research Institute.

Russell, Bertrand (1940). *An Inquiry into Meaning and Truth*. London: George Allen & Unwin. Reprinted with an Introduction by Thomas Baldwin (1995). London: Routledge.

—— (1957). *Why I Am Not a Christian: And Other Essays on Religion and Related Subjects*. New York: Simon & Schuster.

Russell, D. S. (1963). *Between the Testaments*. London: SCM Press.

Russell, Robert John (1998). "Special Providence and Genetic Mutation: A New Defense of Theistic Evolution." In Russell *et al.* (1998, 191–223).

Russell, Robert John, Murphy, Nancey, and Peacocke, Arthur R., eds. (1997). *Chaos and Complexity: Scientific Perspectives on Divine Action*. Berkeley: Center for Theology and the Natural Sciences.

Russell, Robert John, Stoeger, William R., and Ayala, Francisco J., eds. (1998). *Evolutionary and Molecular Biology: Scientific Perspectives on Divine Action*. Berkeley: Center for Theology and the Natural Sciences.

Ryan, James A. (1997). "Taking the 'Error' Out of Ruse's Error Theory." *Biology & Philosophy* 12, 385–397.

Ryle, Gilbert (1949). *The Concept of Mind*. London: Hutchinson.

Sampson, Geoffrey (1980). *Schools of Linguistics*. Stanford: Stanford University Press.

Sapolsky, R. M. (1990). "Stress in the Wild." *Scientific American* 262 (1), 116–123.

Sartre, Jean-Paul (1947). "The Humanism of Existentialism." In Wade Baskin, ed. (1990, 31–62). *Jean-Paul Sartre: Essays in Existentialism*. New York: Citadel Press.

Sax, Leonard (2002). "How Common is Intersex? A Response to Anne Fausto-Sterling." *Journal of Sex Research* 39, 174–178.

Schweitzer, Albert (1933). *Out of My Life and Thought: An Autobiography*. Antje Bultmann Lemke, trans. (1990). New York: Henry Holt and Company.

Scott, Eugenie C. (1997). "Creationists and the Pope's Statement." *Quarterly Review of Biology* 72, 401–406.

Searle, John R. (1984). *Minds, Brains and Science*. Cambridge: Harvard University Press.

—— (1990). "Collective Intentions and Actions." In Philip R. Cohen, *et al.*, eds. (1990, 401–415). *Intentions in Communication*. Cambridge: MIT Press.

—— (1992). *The Rediscovery of the Mind*. Cambridge: MIT Press.

—— (1995). *The Constructing of Social Reality*. New York: Free Press.

Sesardic, Neven (2003). "Evolution of Human Jealousy: A Just-So Story or a Just-So Criticism?" *Philosophy of the Social Sciences* 33, 427–443.

Shermer, Michael (1997). *Why People Believe Weird Things: Pseudoscience, Superstition, and Other Confusions of Our Time*. New York: W. H. Freeman.

Skinner, B. F. (1957). *Verbal Behavior*. New York: Appleton-Century-Crofts.

—— (1972). *Beyond Freedom and Dignity*. New York: Knopf.

Simpson, George Gaylord (1949). *The Meaning of Evolution*. New Haven: Yale University Press.

Singer, Peter (1975). *Animal Liberation*. New York: Random House. 2nd ed. (1990). New York: Avalon Books.

—— (1981). *The Expanding Circle: Ethics and Sociobiology*. New York: Farrar, Straus & Giroux.

—— (1993). *Practical Ethics*. Cambridge: Cambridge University Press.

—— (1998). "Darwin for the Left." *Prospect* 31, 26–30. Reprinted in Singer (2000, 273–282). *Writings on an Ethical Life*. New York: HarperCollins.

——— (1999). *A Darwinian Left: Politics, Evolution and Cooperation*. New Haven, CT: Yale University Press.

Smith, Neil (1999). *Chomsky: Ideas and Ideals*. Cambridge: Cambridge University Press.

——— (2002). *Language, Bananas & Bonobos: Linguistic Problems, Puzzles and Polemics*. Oxford: Blackwell Publishing.

Smithurst, Michael (1995). "Popper and the Scepticisms of Evolutionary Epistemology, or, What Were Human Beings Made For?" In Anthony O'Hear, ed. (1995, 207–223). *Karl Popper: Philosophy and Problems*. Cambridge: Cambridge University Press.

Smolin, Lee (1997). *The Life of the Cosmos*. Oxford: Oxford University Press.

Snell, Bruno (1960). *The Discovery of the Mind: The Greek Origins of European Thought*. T. G. Rosenmeyer, trans. New York: Harper.

Sober, Elliott (1984). *The Nature of Selection*. Chicago: University of Chicago Press.

——— (1993). *Philosophy of Biology*. Boulder, CO: Westview Press.

———, ed. (1994). *Conceptual Issues in Evolutionary Biology*. 2nd ed. Cambridge: MIT Press.

Sober, Elliott, and Wilson, David Sloan (1998). *Unto Others: The Evolution and Psychology of Unselfish Behavior*. Harvard: Harvard University Press.

Sokal, Alan (1996). "Transgressing the Boundaries: Toward a Transformative Hermeneutics of Quantum Gravity." *Social Text* 46/47, 217–252.

Sokal, Alan, and Bricmont, Jean (1998). *Fashionable Nonsense: Postmodern Intellectuals' Abuse of Science*. New York: Picador.

Spencer, Hamish G., and Masters, Judith C. "Sexual Selection: Contemporary Debates." In Keller and Lloyd (1992, 294–301).

Springer, Sally P., and Deutsch, Georg (1993). *Left Brain, Right Brain*. 4th ed. New York: W. H. Freeman.

Stamos, David N. (1996). "Popper, Falsifiability, and Evolutionary Biology." *Biology & Philosophy* 11, 161–191.

——— (1997). "The Nature and Relation of the Three Proofs of God's Existence in Descartes' *Meditations*." *Auslegung* 22, 1–37.

——— (2001). "Quantum Indeterminism and Evolutionary Biology." *Philosophy of Science* 68, 164–184.

——— (2002). "Popper, Laws, and the Exclusion of Biology from Genuine Science." Paper delivered at the Karl Popper 2002 Centenary Conference in Vienna, Austria.

——— (2003). *The Species Problem: Biological Species, Ontology, and the Metaphysics of Biology*. Lanham, MD: Lexington Books.

——— (2007). *Darwin and the Nature of Species*. Albany, NY: State University of New York Press.

Stegmann, Ulrich E. (2004). "The Arbitrariness of the Genetic Code." *Biology & Philosophy* 19, 205–222.

Stein, Edward (1998). "Essentialism and Constructionism About Sexual Orientation." In David Hull and Michael Ruse, eds. (1998, 427–442). *The Philosophy of Biology*. Oxford: Oxford University Press.

Stenger, Victor J. (2007). *God: The Failed Hypothesis—How Science Shows That God Does Not Exist*. Amherst, NY. Prometheus Books.

Stevens, Anthony, and Price, John (1996). *Evolutionary Psychiatry: A New Beginning*. New York: Routledge.

Sulloway, Frank J. (1996). *Born to Rebel: Birth Order, Family Dynamics, and Creative Lives*. New York: Pantheon Books.

Swinburne, Richard (1987). "The Origin of Consciousness." In Robson (1987, 211–225).

Tattersall, Ian (2000). "Once We Were Not Alone." *Scientific American* 282 (1), 56–67.

Taylor, Paul W. (1981). "The Ethics of Respect for Nature." *Environmental Ethics* 3, 197–218. Reprinted in Jeffrey Olen and Vincent Barry, eds. (2002, 517–528). *Applying Ethics*. 7th ed. Belmont, CA: Wadsworth.

Terrace, Herbert S. (1987). "Thoughts Without Words." In Blakemore and Greenfield (1987, 123–137).

———, *et al.* (1979). "Can an Ape Create a Sentence?" *Science* 206, 891–902.

Thornhill, Randy, and Palmer, Craig T. (2000). *A Natural History of Rape: Biological Bases of Sexual Coercion*. Cambridge: MIT Press.

Tooby, John, and Cosmides, Leda (1992). "The Psychological Foundations of Culture." In Barkow *et al.* (1992, 19–136).

Tracy, Thomas F. (1998). "Evolution, Divine Action, and the Problem of Evil." In Russell *et al.* (1998: 511–530).

Trivers, Robert L. (1971). "The Evolution of Reciprocal Altruism." *Quarterly Review of Biology* 46, 35–57.

——— (1972). "Parental Investment and Sexual Selection." In B. Campbell, ed. (1972, 136–179). *Sexual Selection and the Descent of Man, 1871–1971*. Chicago: Aldine.

Urmson, J. O. (1990). *The Greek Philosophical Vocabulary*. London: Duckworth.

Vendler, Zeno (1972). *Res Cogitans: An Essay in Rational Psychology*. Ithaca, NY: Cornell University Press.

Waller, Niels G., *et al.* (1990). "Genetic and Environmental Influences on Religious Interests, Attitudes, and Values: A Study of Twins Reared Apart and Together." *Psychological Science* 1, 138–142.

Warnock, Mary, ed. (2003). *Utilitarianism and On Liberty*. 2nd ed. Oxford: Blackwell Publishing.

Washburn, S. L. (1964). "The Study of Race." In Montagu (1964, 242–260).

Watson, John B. (1924). *Behaviorism*. New York: W. W. Norton.

Weinberg, Steven (1992). *Dreams of a Final Theory*. New York: Pantheon Books.

——— (2001). *Facing Up: Science and Its Cultural Adversaries*. Cambridge: Harvard University Press.

Williams, George C. (1966). *Adaptation and Natural Selection*. Princeton: Princeton University Press.

——— (1989). "A Sociobiological Expansion of *Evolution and Ethics*." In James Paradis and George C. Williams, eds. (1989, 179–214). *Evolution and Ethics*. Princeton: Princeton University Press.

——— (1992). *Natural Selection: Domains, Levels, and Challenges*. Oxford: Oxford University Press.

——— (1996). *Plan & Purpose in Nature*. London: Weidenfeld & Nicholson.

Wilson, David Sloan (2002). *Darwin's Cathedral: Evolution, Religion, and the Nature of Society*. Chicago: University of Chicago Press.

Wilson, Edward O. (1975). *Sociobiology: The New Synthesis*. Abridged ed. (1980). Cambridge: Harvard University Press.

——— (1978). *On Human Nature*. Cambridge: Harvard University Press.

——— (1992). *The Diversity of Life*. Cambridge: Harvard University Press.

——— (1998). *Consilience: The Unity of Knowledge*. New York: Knopf.

Wingfield, John C., *et al.* (1987). "Testosterone and Aggression in Birds." *American Scientist* 75, 602–608.

Wittgenstein, Ludwig (1922). *Tractatus Logico-Philosophicus*. D. F. Pears and B. F. McGuiness, trans. (1961). London: Routledge & Kegan Paul.

——— (1953). *Philosophical Investigations*. G. E. M. Anscombe, trans. Oxford: Basil Blackwell. 3rd ed. (1958).

Wolf, Naomi (1990). *The Beauty Myth*. London: Chatto & Windus.

Wong, Kate (2005). "The Morning of the Modern Mind." *Scientific American* 292 (6), 86–95.

Woolcock, Peter G. (1999). "The Case Against Evolutionary Ethics Today." In Maienschein and Ruse, eds. (1999, 276–306).

Young, David (1992). *The Discovery of Evolution*. Cambridge: Cambridge University Press.

Young, Matt, and Edis, Taner, eds. (2004). *Why Intelligent Design Fails: A Scientific Critique of the New Creationism*. New Jersey: Rutgers University Press.

Yule, George (1996). *The Study of Language*. 2nd ed. Cambridge: Cambridge University Press.

Index